PROGRA
LOGIC
CONT

PROGRAMMABLE LOGIC CONTROLLERS

Fourth Edition

Frank D. Petruzella

Connect
Learn
Succeed™

Connect
Learn
Succeed™

PROGRAMMABLE LOGIC CONTROLLERS

Published by McGraw-Hill, a business unit of The McGraw-Hill Companies, Inc., 1221 Avenue of the
Americas, New York, NY, 10020. Copyright © 2011 by The McGraw-Hill Companies, Inc. All rights
reserved. Previous editions © 1989, 1998, and 2005. No part of this publication may be reproduced or
distributed in any form or by any means, or stored in a database or retrieval system, without the prior written
consent of The McGraw-Hill Companies, Inc., including, but not limited to, in any network or other electronic
storage or transmission, or broadcast for distance learning.

Some ancillaries, including electronic and print components, may not be available to customers outside the
United States.

This book is printed on acid-free paper.

1 2 3 4 5 6 7 8 9 0 DOW/DOW 1 0 9 8 7 6 5 4 3 2 1 0

ISBN 978-0-07-122135-1
MHID 0-07-122135-2

The Internet addresses listed in the text were accurate at the time of publication. The inclusion of a Web site
does not indicate an endorsement by the authors or McGraw-Hill, and McGraw-Hill does not guarantee the
accuracy of the information presented at these sites.

www.mhhe.com

Contents

Preface

Programmable logic controllers (PLCs) continue to evolve as new technologies are added to their capabilities. The PLC started out as a replacement for hardwired relay control systems. Gradually, various math and logic manipulation functions were added. Today PLCs are the controller of choice for the vast majority of automated processes. PLCs now incorporate smaller cases, faster CPUs, networking, and various Internet technologies.

This Fourth Edition of *Programmable Logic Controllers* continues to provide an up-to-date introduction to all aspects of PLC programming, installation, and maintenance procedures. No previous knowledge of PLC systems or programming is assumed. As one reviewer of this edition put it: "I honestly believe that someone with little or no background to PLC systems could take this book and teach themselves PLCs."

The primary source of information for a particular PLC is always the accompanying user manuals provided by the manufacturer. This textbook is not intended to replace the vendor's reference material but rather to complement, clarify, and expand on this information. With the current number of different types of PLCs on the market it is not practical to cover the specifics of all manufacturers and models in a single text. With this in mind, the text discusses PLCs in a generic sense. Although the content is of a nature to allow the information to be applied to a variety of PLCs from different manufacturers, this book, for the most part, uses the Allen-Bradley SLC 500 and ControlLogix controller instruction sets for the programming examples. The underlying PLC principles and concepts covered in the text are common to most manufacturers and serve to maximize the knowledge gained through attending PLC training programs offered by different vendors.

The text is written at a level and format understandable to students being introduced to PLCs for the first time. Feedback from instructors indicates that the information is well organized, to the point, and easy to understand. The content of this new Fourth Edition has been updated and reflects the changes in technology since the publication of the previous edition.

Each chapter begins with a brief introduction outlining chapter coverage and learning objectives. When applicable, the relay equivalent of the virtual programmed instruction is explained first, followed by the appropriate PLC instruction. Chapters conclude with a set of review questions and problems. The review questions are closely related to the chapter objectives and require students to recall and apply information covered in the chapter. The problems range from easy to difficult, thus challenging students at various levels of competence.

This new Fourth Edition has been revised to include a number of new features:

How Programs Operate When the operation of a program is called for, a bulleted list is used to summarize its execution. The list is used in place of lengthy paragraphs and is especially helpful when explaining the different steps in the execution of a program.

Representation of I/O Field Devices Recognition of the input and output field devices associated with the program helps in understanding the overall operation of the program. With this in mind, in addition to their symbols, we provide drawings and photos of field input and output devices.

New ControlLogix Chapter Some instructors have felt that students tend to get confused when switching back and forth from SLC 500 Logic to Logix 5000–based programming within the same chapter. For this reason, a *new Chapter 15* has been added that is devoted entirely to the Allen-Bradley ControlLogix family of controllers and the RSLogix 5000 software. Each part of the new Chapter 15 is treated as a separate unit of study and includes ControlLogix:

- Memory and Project Organization
- Bit-Level Programming
- Programming Timers
- Programming Counters
- Math, Comparison, and Move Instructions
- Function Block Programming

Chapter changes in this edition include:

Chapter 1

- Drawings and photos of real world field input and output devices have been added.
- 50% more figures have been added to this chapter to increase visual appeal and illustrate key concepts further.
- Most recent photographs from major PLC manufactures.
- Revisions to chapter review questions and problems.

Chapter 2

- Drawings and photos of real world field input and output devices have been added.
- Information on the latest selection of PLC hardware components.
- Human machine interfacing with Pico controllers added.
- Most recent photographs from major PLC manufactures.
- Revisions to chapter review questions and problems.

Chapter 3

- Improvement in sizing and placement of drawings make explanations of the different number systems easier to follow.

Chapter 4

- Improvement in sizing and placement of drawings make explanations easier to follow.
- Drawings and photos of real world field input and output devices have been added to the logic diagrams.

Chapter 5

- Information on the ControlLogix memory organization relocated to chapter 15.
- Program scan process explained in greater detail.
- Extended coverage of relay type instructions.
- Instruction addressing examined in greater detail.
- Addressing of a micro PLC illustrated.
- Revisions to chapter review questions and problems.

Chapter 6

- Drawings and photos of real world field input and output devices have been added.

- Drawings and photos of real world field input and output devices have been included in the ladder logic programs.
- Wiring of field inputs and outputs to a micro PLC illustrated.
- Additional coverage of hardwired motor control circuits and their PLC equivalent.
- Revisions to chapter review questions and problems.

Chapter 7

- Information on the ControlLogix timers relocated to chapter 15.
- Drawings and photos of real world field input and output devices have been included in the ladder logic programs.
- Bulleted lists used to summarize program execution.
- Most recent photographs from major PLC manufactures.
- Revisions to chapter review questions and problems.

Chapter 8

- Information on the ControlLogix counters relocated to chapter 15.
- Drawings and photos of real world field input and output devices have been included in the ladder logic programs.
- Bulleted lists used to summarize program execution.
- Most recent photographs from major PLC manufactures.
- Revisions to chapter review questions and problems.

Chapter 9

- Drawings and photos of real world field input and output devices have been included in the ladder logic programs.
- Forcing of inputs and outputs covered in greater detail.
- Differences between a safety PLC and a standard PLC explained.
- Bulleted lists used to summarize program execution.
- Most recent photographs from major PLC manufactures.
- Revisions to chapter review questions and problems.

Chapter 10

- Drawings and photos of real world field input and output devices have been included in the ladder logic programs.

- Analog control covered in more detail.
- PID control process explained in a simplified manner.
- Bulleted lists used to summarize program execution.
- Most recent photographs from major PLC manufactures.
- Revisions to chapter review questions and problems.

Chapter 11

- Drawings and photos of real world field input and output devices have been included in the ladder logic programs.
- Improvement in sizing and placement of drawings make explanations of the different math instructions easier to follow.
- Bulleted lists used to summarize program execution.
- Most recent photographs from major PLC manufactures.
- Revisions to chapter review questions and problems.

Chapter 12

- Drawings and photos of real world field input and output devices have been included in the sequencer programs.
- Improvements to sequencer line drawings designed to make this instruction easier to follow.
- Bulleted lists used to summarize program execution.
- Most recent photographs from major PLC manufactures.
- Revisions to chapter review questions and problems.

Chapter 13

- Drawings and photos of real world field input and output devices have been added.
- Safety issues examined in greater detail.
- Extended coverage of practical troubleshooting techniques.
- Improvement to PLC grounding diagrams makes this function easier to follow.
- Bulleted lists used to summarize program execution.
- Most recent photographs from major PLC manufactures.
- Revisions to chapter review questions and problems.

Chapter 14

- All pertinent information from Chapters 14 and 15 of the 3rd edition have been incorporated into this chapter.

- Examines communications at all levels in an industrial network in much greater detail.
- Fundamentals of PLC motion control have been added.
- Bulleted lists used to summarize program execution.
- New photographs from major PLC manufactures.
- Revisions to chapter review questions and problems.

Chapter 15

- Completely new chapter that concentrates on the fundamentals of ControlLogix technology.
- Includes Memory and Project Organization, Bit Level Programming, Timers, Counters, Math Instructions, and Function Block Programming.

Ancillaries

- *Activity Manual for Programmable Logic Controllers,* Fourth Edition.

    This manual contains:

    Tests made up of multiple choice, true/false, and completion-type questions for each of the chapters.

    Generic programming hands-on exercises designed to offer students real-world programming experience. These assignments are designed for use with any brand of PLC.

- *LogixPro PLC Lab Manual for use with Programmable Logic Controllers,* Fourth Edition

    This manual contains:

    LogixPro 500 simulator software CD. The LogixPro simulation software converts the student's computer into a PLC and allows the student to write ladder logic programs and verify their real-world operation.

    Over **250 LogixPro student lab exercises** sequenced to support material covered in the text.

- **Instructor's Resource Center** is available to instructors who adopt *Programmable Logic Controllers,* Fourth Edition. It includes:

    Textbook answers to all questions and problems.
    Activity Manual answers to all tests.
    Computer Simulation Lab Manual answers for all programming exercises.
    PowerPoint presentations for each chapter.
    EZ Test testing software with text-coordinated question banks.
    ExamView text-coordinated question banks.

Acknowledgments

I would like to thank the following reviewers for their comments and suggestions:

Wesley Allen
Jefferson State Community College

Bo Barry
University of North Carolina–Charlotte

David Barth
Edison Community College

Michael Brumbach
York Technical College

Fred Cope
Northeast State Technical Community College

Warren Dejardin
Northeast Wisconsin Technical College

Montie Fleshman
New River Community College

Steven Flinn
Illinois Central College

Brent Garner
McNeese State University

John Haney
Snead State Community College

Thomas Heraly
Milwaukee Area Technical College

John Lukowski
Michigan Technical University

John Martini
University of Arkansas–Fort Smith

Steven McPherson
Sauk Valley Community College

Max Neal
Griffin Technical College

Ralph Neidert
NECA/IBEW Local 26 JATC

Chrys Panayiotou
Indian River State College

Don Pelster
Nashville State Technical Community College

Dale Petty
Washtenaw Community College

Sal Pisciotta
Florence-Darlington Technical College

Roy E. Pruett
Bluefield State College

Melvin Roberts
Camden County College

Farris Saifkani
Northeast Wisconsin Technical College

David Setser
Johnson County Community College

Richard Skelton
Jackson State Community College

Amy Stephenson
Pitt Community College

William Sutton
ITT Technical Institute

John Wellin
Rochester Institute of Technology.

Last but not least, special thanks to Wade Wittmus of Lakeshore Technical College, not only for his extended help with reviews but also for his outstanding work on the supplements.

Frank D. Petruzella

About the Author

Frank D. Petruzella has extensive practical experience in the electrical control field, as well as many years of experience teaching and authoring textbooks. Before becoming a full time educator, he was employed as an apprentice and electrician in areas of electrical installation and maintenance. He holds a Master of Science degree from Niagara University, a Bachelor of Science degree from the State University of New York College–Buffalo, as well as diplomas in Electrical Power and Electronics from the Erie County Technical Institute.

Programmable Logic Controllers makes it easy to learn PLCs from the ground up! Up-to-the-minute revisions include all the newest developments in programming, installing, and maintaining processes. Clearly developed chapters deliver the organizing objectives, explanatory content with helpful diagrams and illustrations, and closing review problems that evaluate retention of the chapter objectives.

CHAPTER OBJECTIVES overview the chapter, letting students and instructors focus on the main points to better grasp concepts and retain information.

Chapter Objectives

After completing this chapter, you will be able to:

2.1 List and describe the function of the hardware components used in PLC systems

2.2 Describe the basic circuitry and applications for discrete and analog I/O modules, and interpret typical I/O and CPU specifications

2.3 Explain I/O addressing

2.4 Describe the general classes and types of PLC memory devices

2.5 List and describe the different types of PLC peripheral support devices available

Chapter content includes rich illustrative detail and extensive visual aids, allowing students to grasp concepts more quickly and understand practical applications

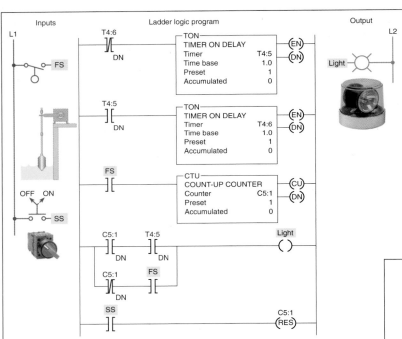

Figure 8-20 Alarm monitor program.

Here, drawings and photos of real-world input and output devices have been included

In Chapter 14, students not only read about but can also see how HMIs fit into an overall PLC system, giving them a practical introduction to the topics

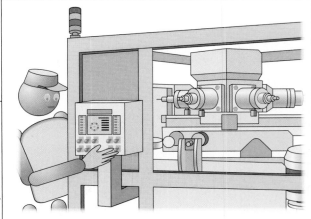

Figure 14-7 Human machine interface (HMI).

Additional coverage of communications and control networks utilizes clear graphics to demonstrate how things work

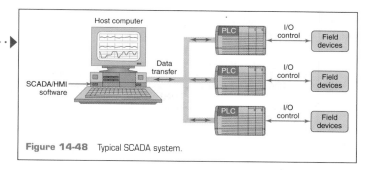

Figure 14-48 Typical SCADA system.

The scan is normally a continuous and sequential process of reading the status of inputs, evaluating the control logic, and updating the outputs. Figure 5-8 shows an overview of the data flow during the scan process. For each rung executed, the PLC processor will:

- Examine the status of the input image table bits.
- Solve the ladder logic in order to determine logical continuity.
- Update the appropriate output image table bits, if necessary.
- Copy the output image table status to all of the output terminals. Power is applied to the output device if the output image table bit has been previously set to a 1.
- Copy the status of all of the input terminals to the input image table. If an input is active (i.e., there is electrical continuity), the corresponding bit in the input image table will be set to a 1.

BULLETED LISTS break down processes to helpfully summarize execution of tasks

Diagrams, such as this one illustrating an overview of the function block programming language, help students put the pieces together

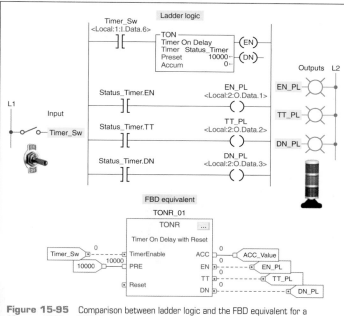

Figure 15-95 Comparison between ladder logic and the FBD equivalent for a 10 second TON and TONR timer.

Figure 15-1 Programmable automation controllers (PACs).
Source: Image Used with Permission of Rockwell Automation, Inc.

An entirely new chapter on ControlLogix has been added to familiarize students with the entire Allen-Bradley family of controllers and RSLogix 5000 software

END-OF-CHAPTER REVIEWS are structured to reinforce
chapter objectives

CHAPTER 3 REVIEW QUESTIONS

1. Convert each of the following binary numbers to decimal numbers:
   a. 10
   b. 100
   c. 111
   d. 1011
   e. 1100
   f. 10010
   g. 10101
   h. 11111
   i. 11001101
   j. 11100011

2. Convert each of the following decimal numbers to binary numbers:
   a. 7
   b. 19
   c. 28
   d. 46
   e. 57
   f. 86
   g. 94
   h. 112
   i. 148
   j. 230

3. Convert each of the following octal numbers to decimal numbers:
   a. 36
   b. 104

6. Convert each of the following hexadecimal numbers to binary numbers:
   a. 4C
   b. E8
   c. 6D2
   d. 31B

7. Convert each of the following decimal numbers to BCD:
   a. 146
   b. 389
   c. 1678
   d. 2502

8. What is the most important characteristic of the Gray code?

9. What makes the binary system so applicable to computer circuits?

10. Define the following as they apply to the binary memory locations or registers:
    a. Bit
    b. Byte
    c. Word
    d. LSB
    e. MSB

11. State the base used for each of the following number systems:

CHAPTER 3 PROBLEMS

1. The following binary PLC coded information is to be programmed using the hexadecimal code. Convert each piece of binary information to the appropriate hexadecimal code for entry into the PLC from the keyboard.
   a. 0001 1111
   b. 0010 0101
   c. 0100 1110
   d. 0011 1001

2. The encoder circuit shown in Figure 3-17 is used to convert the decimal digits on the keyboard to a binary code. State the output status (HIGH/LOW) of A-B-C-D when decimal number
   a. 2 is pressed.
   b. 5 is pressed.

   c. 7 is pressed.
   d. 8 is pressed.

3. If the bits of a 16-bit word or register are numbered according to the octal numbering system, beginning with 00, what consecutive numbers would be used to represent each of the bits?

4. Express the decimal number 18 in each of the following number codes:
   a. Binary
   b. Octal
   c. Hexadecimal
   d. BCD

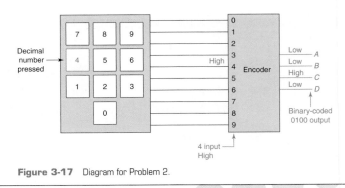

Figure 3-17   Diagram for Problem 2.

EXAMPLE PROBLEMS help bring home the applicability of chapter concepts

ANCILLARIES THAT WORK

Expanded on and updated from the previous edition, this new edition includes an outstanding instructor support package:

- ExamView and EZ Test question test banks for each chapter.
- PowerPoint lessons with animations that help visualize the actual process.
- Activity Manual contains true/false, completion, matching, and multiple-choice questions for every chapter in the text. So that students get a better understanding of programmable logic controllers, the manual also includes a wide range of programming assignments and additional practice exercises.
- On-line Instructor's Resource Center.

In addition, for students, this edition also has available:

- *LogixPro PLC Lab Manual for use with Programmable Logic Controllers* Fourth Edition, with LogixPro PLC Simulator. This manual contains:
  - LogixPro 500 simulator software CD. The LogixPro simulation software converts the student's computer into a PLC and allows the student to write ladder logic programs and verify their real-world operation.
  - Over 250 LogixPro student lab exercises sequenced to support material covered in the text.

1

Programmable Logic Controllers (PLCs)
An Overview

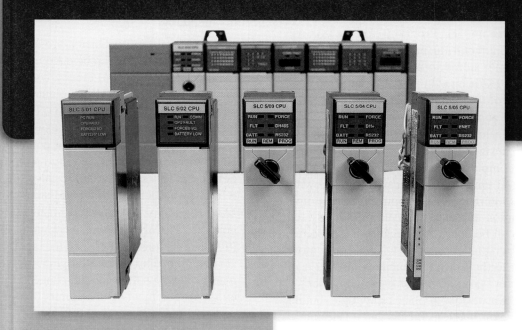

Chapter Objectives

After completing this chapter, you will be able to:

1.1 Define what a programmable logic controller (PLC) is and list its advantages over relay systems

1.2 Identify the main parts of a PLC and describe their functions

1.3 Outline the basic sequence of operation for a PLC

1.4 Identify the general classifications of PLCs

This chapter gives a brief history of the evolution of the programmable logic controller, or PLC. The reasons for changing from relay control systems to PLCs are discussed. You will learn the basic parts of a PLC, how a PLC is used to control a process, and the different kinds of PLCs and their applications. The ladder logic language, which was developed to simplify the task of programming PLCs, is introduced.

1.1 Programmable Logic Controllers

Programmable logic controllers (Figure 1-1) are now the most widely used industrial process control technology. A programmable logic controller (PLC) is an industrial grade computer that is capable of being programmed to perform control functions. The programmable controller has eliminated much of the hardwiring associated with conventional relay control circuits. Other benefits include easy programming and installation, high control speed, network compatibility, troubleshooting and testing convenience, and high reliability.

The programmable logic controller is designed for multiple input and output arrangements, extended temperature ranges, immunity to electrical noise, and resistance to vibration and impact. Programs for the control and operation of manufacturing process equipment and machinery are typically stored in battery-backed or nonvolatile memory. A PLC is an example of a real-time system since the output of the system controlled by the PLC depends on the input conditions.

The programmable logic controller is, then, basically a digital computer designed for use in machine control. Unlike a personal computer, it has been designed to operate in the industrial environment and is equipped with special input/output interfaces and a control programming language. The common abbreviation used in industry for these devices, PC, can be confusing because it is also the abbreviation for "personal computer." Therefore, most manufacturers refer to their programmable controller as a PLC, which stands for "programmable logic controller."

Initially the PLC was used to replace relay logic, but its ever-increasing range of functions means that it is found in many and more complex applications. Because the structure of a PLC is based on the same principles as those employed in computer architecture, it is capable not only of performing relay switching tasks but also of performing other applications such as timing, counting, calculating, comparing, and the processing of analog signals.

Programmable controllers offer several advantages over a conventional relay type of control. Relays have to be hardwired to perform a specific function. When the system requirements change, the relay wiring has to be changed or modified. In extreme cases, such as in the auto industry, complete control panels had to be replaced since it was not economically feasible to rewire the old panels with each model changeover. The programmable controller has eliminated much of the hardwiring associated with conventional relay control circuits (Figure 1-2). It is small and inexpensive compared to equivalent relay-based process control systems. Modern control systems still include relays, but these are rarely used for logic.

In addition to cost savings, PLCs provide many other benefits including:

- *Increased Reliability.* Once a program has been written and tested, it can be easily downloaded to other PLCs. Since all the logic is contained in the PLC's memory, there is no chance of making a logic wiring error (Figure 1-3). The program takes the place of much of the external wiring that would normally be required for control of a process. Hardwiring, though still required to connect field devices, is less intensive. PLCs also offer the reliability associated with solid-state components.

- *More Flexibility.* It is easier to create and change a program in a PLC than to wire and rewire a circuit. With a PLC the relationships between the inputs and outputs are determined by the user program instead of the manner in which they are interconnected (Figure 1-4). Original equipment manufacturers can provide system updates by simply sending out a new program. End users can modify the program in the field, or if desired, security can be provided by hardware features such as key locks and by software passwords.

- *Lower Cost.* PLCs were originally designed to replace relay control logic, and the cost savings have been so significant that relay control is becoming

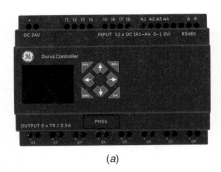
(a)

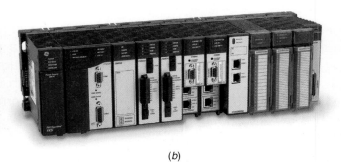
(b)

Figure 1-1 Programmable logic controller.
Source: (a–b) Courtesy GE Intelligent Platforms.

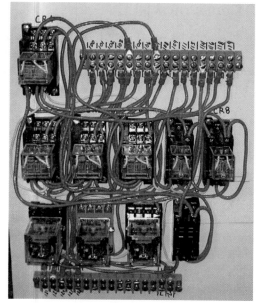

(a)

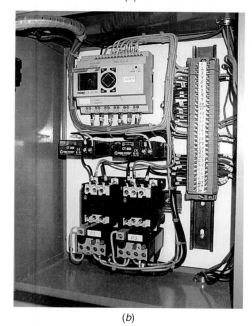

(b)

Figure 1-2 Relay- and PLC-based control panels. (a) Relay-based control panel. (b) PLC-based control panel.

Source: (a) Courtesy Mid-Illini Technical Group, Inc.; (b) Photo courtesy Ramco Electric, Ltd.

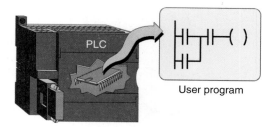

Figure 1-3 All the logic is contained in the PLC's memory.

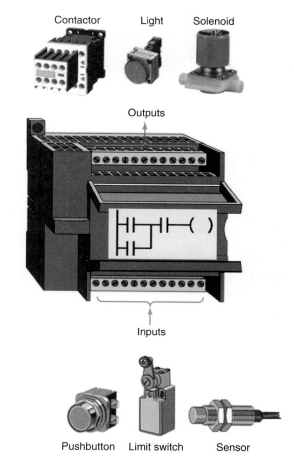

Figure 1-4 Relationships between the inputs and outputs are determined by the user program.

obsolete except for power applications. Generally, if an application has more than about a half-dozen control relays, it will probably be less expensive to install a PLC.

• *Communications Capability.* A PLC can communicate with other controllers or computer equipment to perform such functions as supervisory control, data gathering, monitoring devices and process parameters, and download and upload of programs (Figure 1-5).

• *Faster Response Time.* PLCs are designed for high-speed and real-time applications (Figure 1-6). The programmable controller operates in real time, which means that an event taking place in the field will result in the execution of an operation or output. Machines that process thousands of items per second and objects that spend only a fraction of a second in front of a sensor require the PLC's quick-response capability.

• *Easier to Troubleshoot.* PLCs have resident diagnostics and override functions that allow users to

Figure 1-5 PLC communication module.
Source: Photo courtesy Automation Direct, **www.automationdirect.com**.

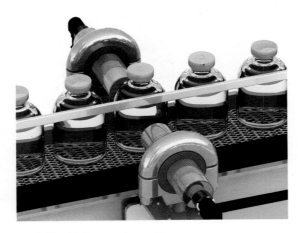

Figure 1-6 High-speed counting.
Source: Courtesy Banner Engineering Corp.

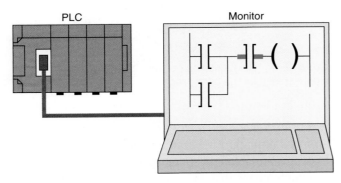

Figure 1-7 Control program can be displayed on a monitor in real time.

easily trace and correct software and hardware problems. To find and fix problems, users can display the control program on a monitor and watch it in real time as it executes (Figure 1-7).

1.2 Parts of a PLC

A typical PLC can be divided into parts, as illustrated in Figure 1-8. These are the *central processing unit (CPU)*, the *input/output (I/O)* section, the *power supply,* and the *programming device.* The term *architecture* can refer to PLC hardware, to PLC software, or to a combination of both. An *open* architecture design allows the system to be connected easily to devices and programs made by other manufacturers. Open architectures use off-the-shelf components that conform to approved standards. A system with a *closed* architecture is one whose design is proprietary, making it more difficult to connect to other systems. Most PLC systems are in fact proprietary, so you must be sure that any generic hardware or software you may use is compatible with your particular PLC. Also, although the principal concepts are the same in all methods of programming, there might be slight differences in addressing, memory allocation, retrieval, and data handling for different models. Consequently, PLC programs cannot be interchanged among different PLC manufacturers.

There are two ways in which I/Os (Inputs/Outputs) are incorporated into the PLC: fixed and modular. *Fixed I/O* (Figure 1-9) is typical of small PLCs that come in one package with no separate, removable units. The processor and I/O are packaged together, and the I/O terminals will have a fixed number of connections built in for inputs and outputs. The main advantage of this type of packaging is lower cost. The number of available I/O points varies and usually can be expanded by buying additional units of fixed I/O. One disadvantage of fixed I/O is its lack of flexibility; you are limited in what you can get in the quantities and types dictated by the packaging. Also, for some models, if any part in the unit fails, the whole unit has to be replaced.

Modular I/O (Figure 1-10) is divided by compartments into which separate modules can be plugged. This feature greatly increases your options and the unit's flexibility. You can choose from the modules available from the manufacturer and mix them any way you desire. The basic modular controller consists of a rack, power supply, processor module (CPU), input/output (I/O modules), and an operator interface for programming and monitoring. The modules plug into a rack. When a module is slid into the rack, it makes an electrical connection with a series of contacts called the backplane, located at the rear of the rack. The PLC processor is also connected to the backplane and can communicate with all the modules in the rack.

The *power supply* supplies DC power to other modules that plug into the rack (Figure 1-11). For large PLC systems, this power supply does not normally supply power to the field devices. With larger systems, power to field devices is

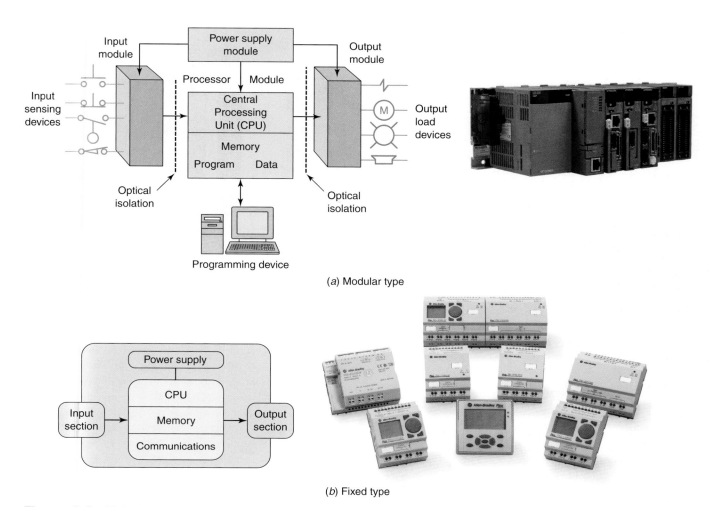

(a) Modular type

(b) Fixed type

Figure 1-8 Typical parts of a programmable logic controller.

Source: (a) Courtesy Mitsubishi Automation; (b) Image Used with Permission of Rockwell Automation, Inc.

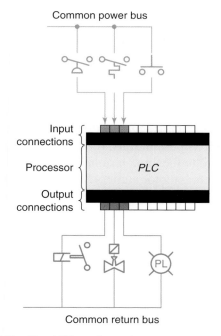

Figure 1-9 Fixed I/O configuration.

provided by external alternating current (AC) or direct current (DC) supplies. For some small micro PLC systems, the power supply may be used to power field devices.

The *processor* (CPU) is the "brain" of the PLC. A typical processor (Figure 1-12) usually consists of a microprocessor for implementing the logic and controlling the communications among the modules. The processor requires memory for storing the results of the logical operations performed by the microprocessor. Memory is also required for the program EPROM or EEPROM plus RAM.

The CPU controls all PLC activity and is designed so that the user can enter the desired program in relay ladder logic. The PLC program is executed as part of a repetitive process referred to as a scan (Figure 1-13). A typical PLC scan starts with the CPU reading the status of inputs. Then, the application program is executed. Once the program execution is completed, the CPU performs internal diagnostic and communication tasks. Next, the status of all outputs is updated. This process is repeated continuously as long as the PLC is in the run mode.

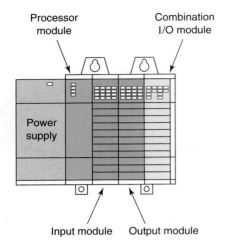

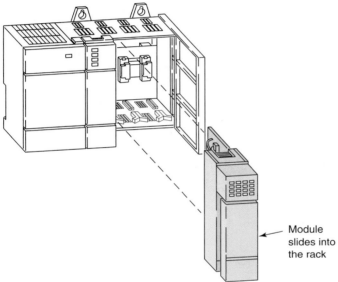

Figure 1-10 Modular I/O configuration.

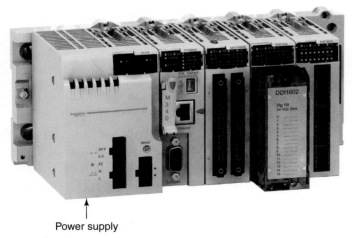

Figure 1-11 The power supply supplies DC power to other modules that plug into the rack.

Figure 1-12 Typical PLC processor modules.

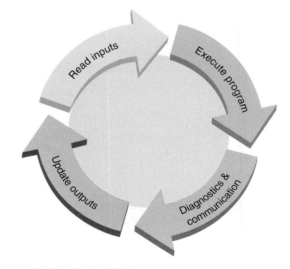

Figure 1-13 Typical PLC scan cycle.

The *I/O system* forms the interface by which field devices are connected to the controller (Figure 1-14). The purpose of this interface is to condition the various signals received from or sent to external field devices. Input devices such as pushbuttons, limit switches, and sensors are hardwired to the input terminals. Output devices such as small motors, motor starters, solenoid valves, and indicator lights are hardwired to the output terminals. To electrically isolate the internal components from the input and output terminals, PLCs commonly employ an optical isolator, which uses light to couple the circuits together. The external devices are also referred to as "field" or "real-world" inputs and outputs. The terms *field* or *real world* are used to distinguish actual external devices that exist and must be physically wired from the internal user program that duplicates the function of relays, timers, and counters.

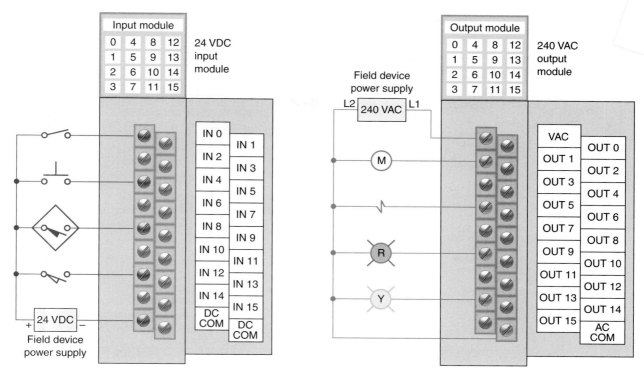

Figure 1-14 Typical PLC input/output (I/O) system connections.

A *programming device* is used to enter the desired program into the memory of the processor. The program can be entered using relay ladder logic, which is one of the most popular programming languages. Instead of words, ladder logic programming language uses graphic symbols that show their intended outcome. A program in ladder logic is similar to a schematic for a relay control circuit. It is a special language written to make it easy for people familiar with relay logic control to program the PLC. Hand-held programming devices (Figure 1-15) are sometimes used to program small PLCs because they are inexpensive and easy to use. Once plugged into the PLC, they can be used to enter and monitor programs. Both compact hand-held units and laptop computers are frequently used on the factory floor for troubleshooting equipment, modifying programs, and transferring programs to multiple machines.

A personal computer (PC) is the most commonly used programming device. Most brands of PLCs have software available so that a PC can be used as the programming device. This software allows users to create, edit, document, store, and troubleshoot ladder logic programs (Figure 1-16). The computer monitor is able to display more logic on the screen than can hand-held types, thus simplifying the interpretation of the program. The personal computer communicates with the PLC processor via a serial or parallel data communications link, or Ethernet. If

Figure 1-15 Typical hand-held programming device.
Source: Photo courtesy Automation Direct, **www.automationdirect.com**.

the programming unit is not in use, it may be unplugged and removed. Removing the programming unit will not affect the operation of the user program.

A *program* is a user-developed series of instructions that directs the PLC to execute actions. A *programming language* provides rules for combining the instructions so that they produce the desired actions. *Relay ladder logic (RLL)* is the standard programming language used

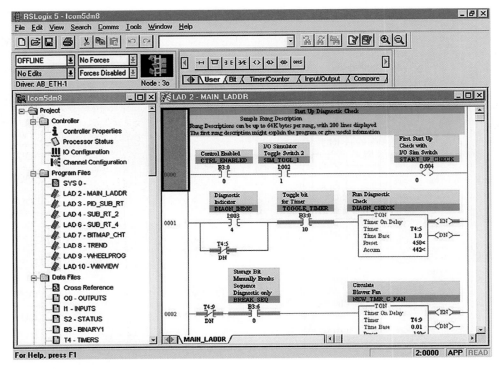

Figure 1-16 Typical PC software used to create a ladder logic program.
Source: Image Used with Permission of Rockwell Automation, Inc.

with PLCs. Its origin is based on electromechanical relay control. The relay ladder logic program graphically represents rungs of contacts, coils, and special instruction blocks. RLL was originally designed for easy use and understanding for its users and has been modified to keep up with the increasing demands of industry's control needs.

1.3 Principles of Operation

To get an idea of how a PLC operates, consider the simple process control problem illustrated in Figure 1-17. Here a mixer motor is to be used to automatically stir the liquid in a vat when the temperature and pressure reach preset values. In addition, direct manual operation of the motor is provided by means of a separate pushbutton station. The process is monitored with temperature and pressure sensor switches that close their respective contacts when conditions reach their preset values.

This control problem can be solved using the relay method for motor control shown in the relay ladder diagram of Figure 1-18. The motor starter coil (M) is energized when both the pressure and temperature switches are closed or when the manual pushbutton is pressed.

Now let's look at how a programmable logic controller might be used for this application. The same input field devices (pressure switch, temperature switch, and pushbutton) are used. These devices would be hardwired to an

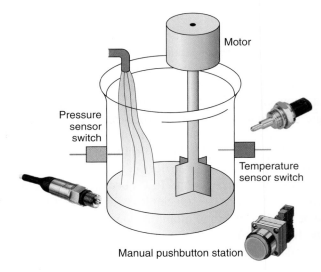

Figure 1-17 Mixer process control problem.

appropriate input module according to the manufacturer's addressing location scheme. Typical wiring connections for a 120 VAC modular configured input module is shown in Figure 1-19.

The same output field device (motor starter coil) would also be used. This device would be hardwired to an appropriate output module according to the manufacturer's addressing location scheme. Typical wiring connections for a 120 VAC modular configured output module is shown in Figure 1-20.

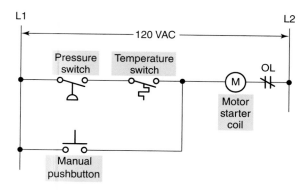

Figure 1-18 Process control relay ladder diagram.

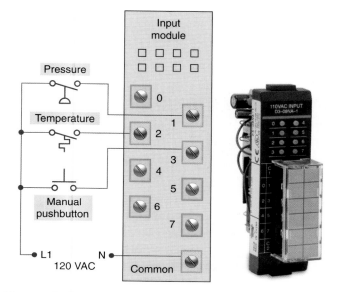

Figure 1-19 Typical wiring connections for a 120 VAC modular configured input module.
Source: Photo courtesy Automation Direct, **www.automationdirect.com**.

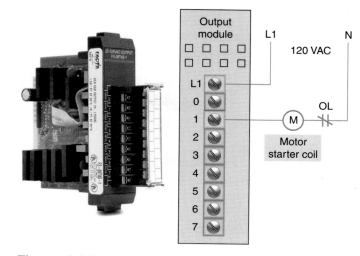

Figure 1-20 Typical wiring connections for a 120 VAC modular configured output module.
Source: Photo courtesy Automation Direct, **www.automationdirect.com**.

Next, the PLC ladder logic program would be constructed and entered into the memory of the CPU. A typical ladder logic program for this process is shown in Figure 1-21. The format used is similar to the layout of the hardwired relay ladder circuit. The individual symbols represent instructions, whereas the numbers represent the instruction location addresses. To program the controller, you enter these instructions one by one into the processor memory from the programming device. Each input and output device is given an address, which lets the PLC know where it is physically connected. Note that the I/O address format will differ, depending on the PLC model and manufacturer. Instructions are stored in the user program portion of the processor memory. During the program scan the controller monitors the inputs, executes the control program, and changes the output accordingly.

For the program to operate, the controller is placed in the RUN mode, or operating cycle. During each operating cycle, the controller examines the status of input devices, executes the user program, and changes outputs accordingly. Each ⊩ symbol can be thought of as a set of normally open contacts. The ( ) symbol is considered to represent a coil that, when energized, will close a set of contacts. In the ladder logic program of Figure 1-21, the coil O/1 is energized when contacts I/1 and I/2 are closed or when contact I/3 is closed. Either of these conditions provides a continuous logic path from left to right across the rung that includes the coil.

A programmable logic controller operates in real time in that an event taking place in the field will result in an operation or output taking place. The RUN operation for the process control scheme can be described by the following sequence of events:

- First, the pressure switch, temperature switch, and pushbutton inputs are examined and their status is recorded in the controller's memory.
- A closed contact is recorded in memory as logic 1 and an open contact as logic 0.
- Next the ladder diagram is evaluated, with each internal contact given an OPEN or CLOSED status according to its recorded 1 or 0 state.
- When the states of the input contacts provide logic continuity from left to right across the rung, the output coil memory location is given a logic 1 value and the output module interface contacts will close.
- When there is no logic continuity of the program rung, the output coil memory location is set to logic 0

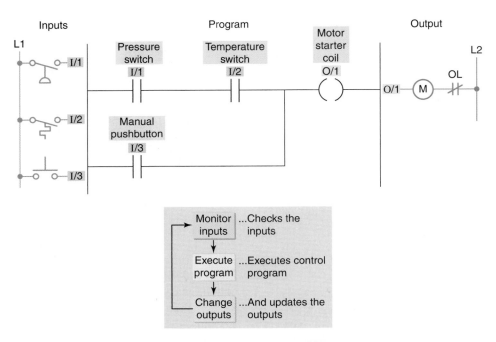

Figure 1-21 Process control PLC ladder logic program with typical addressing scheme.

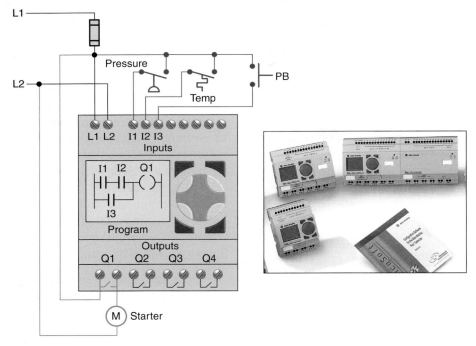

Figure 1-22 Typical wiring required to implement the process control scheme using a fixed PLC controller.

Source: Image Used with Permission of Rockwell Automation, Inc.

and the output module interface contacts will be open.

- The completion of one cycle of this sequence by the controller is called a *scan*. The scan time, the time required for one full cycle, provides a measure of the speed of response of the PLC.

- Generally, the output memory location is updated during the scan but the actual output is not updated until the end of the program scan during the I/O scan.

Figure 1-22 shows the typical wiring required to implement the process control scheme using a fixed PLC controller. In this example the Allen-Bradley Pico controller equipped with 8 inputs and 4 outputs is used to control and monitor the process. Installation can be summarized as follows:

- Fused power lines, of the specified voltage type and level, are connected to the controller's L1 and L2 terminals.
- The pressure switch, temperature switch, and pushbutton field input devices are hardwired between L1 and controller input terminals I1, I2, and I3, respectively.
- The motor starter coil connects directly to L2 and in series with Q1 relay output contacts to L1.
- The ladder logic program is entered using the front keypad and LCD display.
- Pico programming software is also available that allows you to create as well as test your program using a personal computer.

1.4 Modifying the Operation

One of the important features of a PLC is the ease with which the program can be changed. For example, assume that the original process control circuit for the mixing operation must be modified as shown in the relay ladder diagram of Figure 1-23. The change requires that the manual pushbutton control be permitted to operate at any pressure, but not unless the specified temperature setting has been reached.

If a relay system were used, it would require some rewiring of the circuit shown in Figure 1-23 to achieve the desired change. However, if a PLC system were used, no rewiring would be necessary. The inputs and outputs are still the same. All that is required is to change the PLC ladder logic program as shown in Figure 1-24.

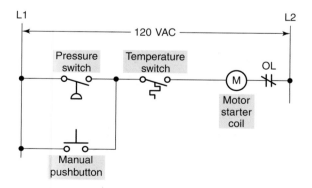

Figure 1-23 Relay ladder diagram for the modified process.

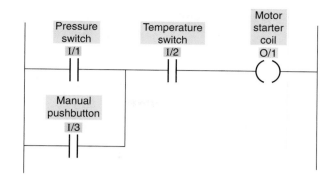

Figure 1-24 PLC ladder logic program for the modified process.

1.5 PLCs versus Computers

The architecture of a PLC is basically the same as that of a personal computer. A personal computer (PC) can be made to operate as a programmable logic controller if you provide some way for the computer to receive information from devices such as pushbuttons or switches. You also need a program to process the inputs and decide the means of turning load devices off and on.

However, some important characteristics distinguish PLCs from personal computers. First, unlike PCs, the PLC is designed to operate in the industrial environment with wide ranges of ambient temperature and humidity. A well-designed industrial PLC installation, such as that shown in Figure 1-25, is not usually affected by the electrical noise inherent in most industrial locations.

Unlike the personal computer, the PLC is programmed in relay ladder logic or other easily learned languages. The PLC comes with its program language built into its memory and has no permanently attached keyboard, CD drive, or monitor. Instead, PLCs come equipped with terminals for input and output field devices as well as communication ports.

Computers are complex computing machines capable of executing several programs or tasks simultaneously and in any order. Most PLCs, on the other hand, execute a single program in an orderly and sequential fashion from first to last instruction.

PLC control systems have been designed to be easily installed and maintained. Troubleshooting is simplified by the use of fault indicators and messaging displayed on the programmer screen. Input/output modules for connecting the field devices are easily connected and replaced.

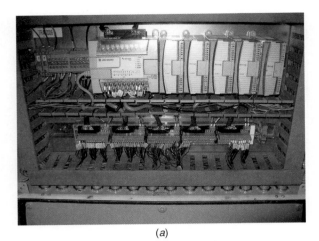

(a)

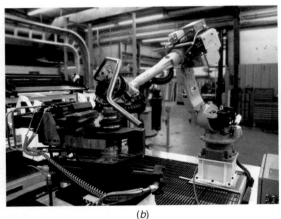

(b)

Figure 1-25 PLC installed in an industrial environment.
Source: (a–b) Courtesy Automation IG.

Figure 1-26 PLC operator interface and monitor.
Source: Courtesy Rogers Machinery Company, Inc.

Figure 1-27 Programmable automation controller (PAC).
Source: Photo courtesy Omron Industrial Automation, **www.ia.omron.com**.

Software associated with a PLC but written and run on a personal computer falls into the following two broad categories:

- PLC software that allows the user to program and document gives the user the tools to write a PLC program—using ladder logic or another programming language—and document or explain the program in as much detail as is necessary.
- PLC software that allows the user to monitor and control the process is also called a *human machine interface (HMI)*. It enables the user to view a process—or a graphical representation of a process—on a monitor, determine how the system is running, trend values, and receive alarm conditions (Figure 1-26). Many operator interfaces do not use PLC software. PLCs can be integrated with HMIs but the same software does not program both devices.

Most recently automation manufacturers have responded to the increased requirements of industrial control systems by blending the advantages of PLC-style control with that of PC-based systems. Such a device has been termed a programmable automation controller, or PAC (Figure 1-27). Programmable automation controllers combine PLC ruggedness with PC functionality. Using PACs, you can build advanced systems incorporating software capabilities such as advanced control, communication, data logging, and signal processing with rugged hardware performing logic, motion, process control, and vision.

1.6 PLC Size and Application

The criteria used in categorizing PLCs include functionality, number of inputs and outputs, cost, and physical size (Figure 1-28). Of these, the *I/O count* is the most important factor. In general, the nano is the smallest size with less than 15 I/O points. This is followed by micro types (15 to 128 I/O points), medium types (128 to 512 I/O points), and large types (over 512 I/O points).

Matching the PLC with the application is a key factor in the selection process. In general it is not advisable to

Figure 1-28 Typical range of sizes of programmable controllers.
Source: Courtesy Siemens.

Figure 1-29 Single-ended PLC application.
Source: Courtesy Rogers Machinery Company, Inc.

buy a PLC system that is larger than current needs dictate. However, future conditions should be anticipated to ensure that the system is the proper size to fill the current and possibly future requirements of an application.

There are three major types of PLC application: single-ended, multitask, and control management. A *single-ended* or stand-alone PLC application involves one PLC controlling one process (Figure 1-29). This would be a stand-alone unit and would not be used for communicating with other computers or PLCs. The size and sophistication of the process being controlled are obvious factors in determining which PLC to select. The applications could dictate a large processor, but usually this category requires a small PLC.

A *multitask* PLC application involves one PLC controlling several processes. Adequate I/O capacity is a significant factor in this type of installation. In addition, if the PLC would be a subsystem of a larger process and would have to communicate with a central PLC or computer, provisions for a data communications network are also required.

A *control management* PLC application involves one PLC controlling several others (Figure 1-30). This kind

of application requires a large PLC processor designed to communicate with other PLCs and possibly with a computer. The control management PLC supervises several PLCs by downloading programs that tell the other PLCs what has to be done. It must be capable of connection to all the PLCs so that by proper addressing it can communicate with any one it wishes to.

Memory is the part of a PLC that stores data, instructions, and the control program. Memory size is usually expressed in K values: 1 K, 6 K, 12 K, and so on. The measurement kilo, abbreviated K, normally refers to 1000 units. When dealing with computer or PLC memory, however, 1 K means 1024, because this measurement is based on the binary number system ($2^{10} = 1024$). Depending on memory type, 1 K can mean 1024 bits, 1024 bytes, or 1024 words.

Although it is common for us to measure the memory capacity of PLCs in words, we need to know the number of bits in each word before memory size can be accurately compared. Modern computers usually have a word size of 16, 32, or 64 bits. For example, a PLC that uses 8-bit words has 49,152 bits of storage with a 6 K word capacity ($8 \times 6 \times 1024 = 49{,}152$), whereas a PLC using 32-bit words has 196,608 bits of storage with the same 6 K memory ($32 \times 6 \times 1024 = 196{,}608$). The amount

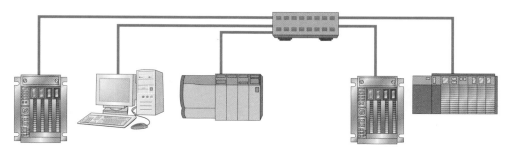

Figure 1-30 Control management PLC application.

Table 1-1 Typical PLC Instructions

Instruction	Operation
XIC (Examine ON)	Examine a bit for an ON condition
XIO (Examine OFF)	Examine a bit for an OFF condition
OTE (Output Energize)	Turn ON a bit (nonretentive)
OTL (Output Latch)	Latch a bit (retentive)
OTU (Output Unlatch)	Unlatch a bit (retentive)
TOF (Timer Off-Delay)	Turn an output ON or OFF after its rung has been OFF for a preset time interval
TON (Timer On-Delay)	Turn an output ON or OFF after its rung has been ON for a preset time interval
CTD (Count Down)	Use a software counter to count down from a specified value
CTU (Count Up)	Use a software counter to count up to a specified value

of memory required depends on the application. Factors affecting the memory size needed for a particular PLC installation include:

- Number of I/O points used
- Size of control program
- Data-collecting requirements
- Supervisory functions required
- Future expansion

The *instruction set* for a particular PLC lists the different types of instructions supported. Typically, this ranges from 15 instructions on smaller units up to 100 instructions on larger, more powerful units (see Table 1-1).

1. What is a programmable logic controller (PLC)?

2. Identify four tasks in addition to relay switching operations that PLCs are capable of performing.

3. List six distinct advantages that PLCs offer over conventional relay-based control systems.

4. Explain the differences between open and proprietary PLC architecture.

5. State two ways in which I/O is incorporated into the PLC.

6. Describe how the I/O modules connect to the processor in a modular-type PLC configuration.

7. Explain the main function of each of the following major components of a PLC:
   a. Processor module (CPU)
   b. I/O modules
   c. Programming device
   d. Power supply module

8. What are the two most common types of PLC programming devices?

9. Explain the terms *program* and *programming language* as they apply to a PLC.

10. What is the standard programming language used with PLCs?

11. Answer the following with reference to the process control relay ladder diagram of Figure 1-18 of this chapter:
    a. When do the pressure switch contacts close?
    b. When do the temperature switch contacts close?
    c. How are the pressure and temperature switches connected with respect to each other?
    d. Describe the two conditions under which the motor starter coil will become energized.
    e. What is the approximate value of the voltage drop across each of the following when their contacts are open?
    (1) Pressure switch
    (2) Temperature switch
    (3) Manual pushbutton

12. The programmable controller operates in real time. What does this mean?

13. Answer the following with reference to the process control PLC ladder logic diagram of Figure 1-21 of this chapter:
    a. What do the individual symbols represent?
    b. What do the numbers represent?
    c. What field device is the number I/2 identified with?
    d. What field device is the number O/1 identified with?
    e. What two conditions will provide a continuous path from left to right across the rung?
    f. Describe the sequence of operation of the controller for one scan of the program.

14. Compare the method by which the process control operation is changed in a relay-based system to the method used for a PLC-based system.

15. Compare the PLC and PC with regard to:
    a. Physical hardware differences
    b. Operating environment
    c. Method of programming
    d. Execution of program

16. What two categories of software written and run on PCs are used in conjunction with PLCs?

17. What is a programmable automation controller (PAC)?

18. List four criteria by which PLCs are categorized.

19. Compare the single-ended, multitask, and control management types of PLC applications.

20. What is the memory capacity, expressed in bits, for a PLC that uses 16-bit words and has an 8 K word capacity?

21. List five factors affecting the memory size needed for a particular PLC installation.

22. What does the instruction set for a particular PLC refer to?

1. Given two single-pole switches, write a program that will turn on an output when both switch A and switch B are closed.

2. Given two single-pole switches, write a program that will turn on an output when either switch A or switch B is closed.

3. Given four NO (Normally Open) pushbuttons (*A-B-C-D*), write a program that will turn a lamp on if pushbuttons *A* and *B* or *C* and *D* are closed.

4. Write a program for the relay ladder diagram shown in Figure 1-31.

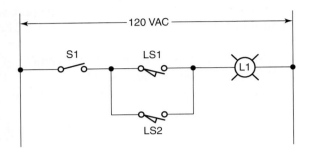

Figure 1-31 Circuit for Problem 4.

5. Write a program for the relay ladder diagram shown in Figure 1-32.

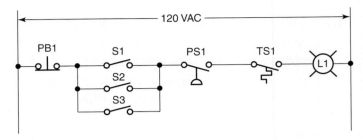

Figure 1-32 Circuit for Problem 5.

2

PLC Hardware Components

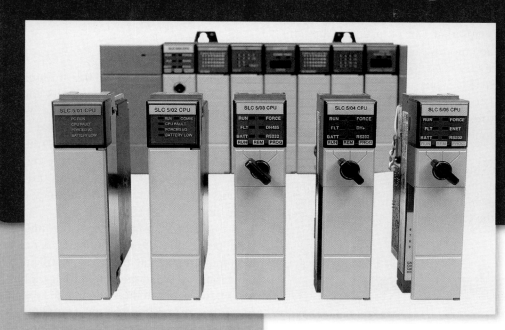

Image Used with Permission of Rockwell Automation, Inc.

Chapter Objectives

After completing this chapter, you will be able to:

2.1 List and describe the function of the hardware components used in PLC systems

2.2 Describe the basic circuitry and applications for discrete and analog I/O modules, and interpret typical I/O and CPU specifications

2.3 Explain I/O addressing

2.4 Describe the general classes and types of PLC memory devices

2.5 List and describe the different types of PLC peripheral support devices available

This chapter exposes you to the details of PLC hardware and modules that make up a PLC control system. The chapter's illustrations show the various subparts of a PLC as well as general connection paths. In this chapter we discuss the CPU and memory hardware components, including the various types of memory that are available, and we describe the hardware of the input/output section, including the difference between the discrete and analog types of modules.

2.1 The I/O Section

The input/output (I/O) section of a PLC is the section to which all field devices are connected and provides the interface between them and the CPU. Input/output arrangements are built into a fixed PLC while modular types use external I/O modules that plug into the PLC.

Figure 2-1 illustrates a rack-based I/O section made up of individual I/O modules. Input interface modules accept signals from the machine or process devices and convert them into signals that can be used by the controller. Output interface modules convert controller signals into external signals used to control the machine or process. A typical PLC has room for several I/O modules, allowing it to be customized for a particular application by selecting the appropriate modules. Each slot in the rack is capable of accommodating any type of I/O module.

The I/O system provides an interface between the hardwired components in the field and the CPU. The input interface allows *status information* regarding processes to be communicated to the CPU, and thus allows the CPU to

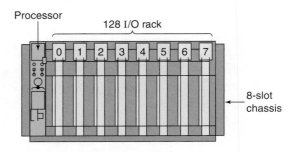

Figure 2-2   Allen-Bradley PLC chassis and rack.

communicate *operating signals* through the output interface to the process devices under its control.

Allen-Bradley controllers make a distinction between a PLC chassis and rack as illustrated in Figure 2-2. The hardware assembly that houses I/O modules, processor modules, and power supplies is referred to as the chassis. Chassis come in different sizes according to the number of slots they contain. In general, they can have 4, 8, 12, or 16 slots.

A *logical rack* is an addressable unit consisting of 128 input points and 128 output points. A rack uses 8 words in the input image table file and 8 words in the output image table file. A word in the output image table file and its corresponding word in the input image table file are called an *I/O group*. A rack can contain a maximum of 8 I/O groups (numbered from 0 through 7) for up to 128 discrete I/O. There can be more than one rack in a chassis and more than one chassis in a rack.

One benefit of a PLC system is the ability to locate the I/O modules near the field devices, as illustrated in Figure 2-3, in order to minimize the amount of wiring

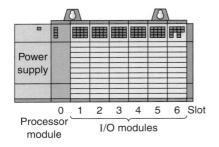

Figure 2-1   Rack-based I/O section.

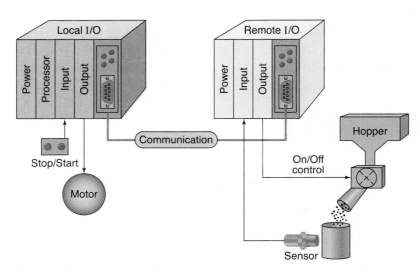

Figure 2-3   Remote I/O rack.

required. The processor receives signals from the remote input modules and sends signals back to their output modules via the communication module.

A rack is referred to as a *remote* rack when it is located away from the processor module. To communicate with the processor, the remote rack uses a special communications network. Each remote rack requires a unique station number to distinguish one from another. The remote racks are linked to the local rack through a *communications module*. Cables connect the modules with each other. If fiber optic cable is used between the CPU and I/O rack, it is possible to operate I/O points from distances greater than 20 miles with no voltage drop. Coaxial cable will allow remote I/O to be installed at distances greater than two miles. Fiber optic cable will not pick up noise caused by adjacent high power lines or equipment normally found in an industrial environment. Coaxial cable is more susceptible to this type of noise.

The PLC's memory system stores information about the status of all the inputs and outputs. To keep track of all this information, it uses a system called *addressing*. An address is a label or number that indicates where a certain piece of information is located in a PLC's memory. Just as your home address tells where you live in your city, a device's or a piece of data's address tells where information about it resides in the PLC's memory. That way, if a PLC wants to find out information about a field device, it knows to look in its corresponding address location. Examples of addressing schemes include *rack/slot-based*, versions of which are used in Allen-Bradley PLC-5 and SLC 500 controllers, *tag-based* used in Allen-Bradley ControlLogix controllers, and PC-based control used in soft PLCs.

In general, rack/slot-based addressing elements include:

Type—The type determines if an input or output is being addressed.

Slot—The slot number is the physical location of the I/O module. This may be a combination of the rack number and the slot number when using expansion racks.

Word and Bit—The word and bit are used to identify the actual terminal connection in a particular I/O module. A discrete module usually uses only one word, and each connection corresponds to a different bit that makes up the word.

With a rack/slot address system the location of a module within a rack and the terminal number of a module to which an input or output device is connected will determine the device's address. Figure 2-4 illustrates the Allen-Bradley PLC-5 controller addressing format. The following are typical examples of input and output addresses:

I1:27/17	Input, file 1, rack 2, group 7, bit 17
O0:34/07	Output, file 0, rack 3, group 4, bit 7
I1:0/0	Input, file 1, rack 0, group 0, bit 0 (Short form blank = 0)
O0:1/1	Output, file 0, rack 0, group 1, bit 1 (Short form blank = 0)

Figure 2-5 illustrates the Allen-Bradley SLC 500 controller addressing format. The address is used by the processor to identify where the device is located to monitor or control it. In addition, there is some means of connecting field wiring on the I/O module housing. Connecting the field wiring to the I/O housing allows easier disconnection and reconnection of the wiring to change modules. Lights are also added to each module to indicate the ON or OFF status of each I/O circuit. Most output modules also have blown fuse indicators. The following are typical

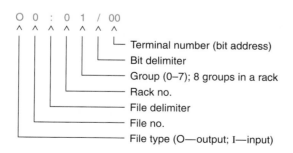

```
O  0  :  0  1  /  00
^  ^  ^  ^  ^  ^  ^
                  └── Terminal number (bit address)
               ───── Bit delimiter
            ──────── Group (0–7); 8 groups in a rack
         ─────────── Rack no.
      ────────────── File delimiter
   ───────────────── File no.
────────────────────File type (O—output; I—input)
```

Figure 2-4   Allen-Bradley PLC-5 rack/slot-based addressing format.
Source: Image Used with Permission of Rockwell Automation, Inc.

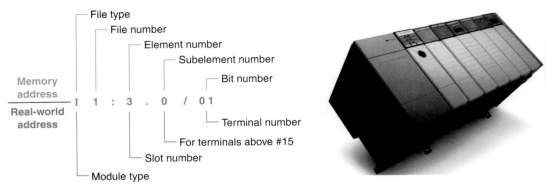

Figure 2-5 Allen-Bradley SLC 500 rack/slot-based addressing format.
Source: Image Used with Permission of Rockwell Automation, Inc.

examples of SLC 500 real-world general input and output addresses:

O:4/15	**Output module in slot 4, terminal 15**
I:3/8	**Input module in slot 3, terminal 8**
O:6.0	**Output module, slot 6**
I:5.0	**Input module, slot 5**

Every input and output device connected to a discrete I/O module is addressed to a specific *bit* in the PLC's memory. A bit is a binary digit that can be either 1 or 0. Analog I/O modules use a *word* addressing format, which allows the entire words to be addressed. The bit part of the address is usually not used; however, bits of the digital representation of the analog value can be addressed by the programmer if necessary. Figure 2-6 illustrates bit level and word level addressing as it applies to an SLC 500 controller.

Figure 2-7 illustrates the Allen-Bradley ControlLogix tag-based addressing format. With Logix5000 controllers, you use a tag (alphanumeric name) to address data (variables). Instead of a fixed numeric format the tag name itself identifies the data. The field devices are assigned tag names that are referenced when the PLC ladder logic program is developed.

PC-based control runs on personal or industrial hardened computers. Also known as soft PLCs, they simulate the functions of a PLC on a PC, allowing open architecture systems to replace proprietary PLCs. This implementation uses an input/output card (Figure 2-8) in conjunction with the PC as an interface for the field devices.

Combination I/O modules can have both input and output connections in the same physical module as illustrated in Figure 2-9. A module is made up of a printed circuit board and a terminal assembly. The printed circuit board contains the electronic circuitry used to interface the circuit of the processor with that of the input or output device. Modules are designed to plug into a slot or connector in the I/O rack or directly into the processor. The terminal assembly, which is attached to the front edge of the printed circuit board, is used for making field-wiring connections. Modules contain terminals for each input and output connection, status lights for each of the inputs and outputs, and connections to the power supply used to power the inputs and outputs. Terminal and status light arrangements vary with different manufacturers.

Most PLC modules have plug-in wiring terminal strips. The terminal block is plugged into the actual module as illustrated in Figure 2-10. If there is a problem with a module, the entire strip is removed, a new module is inserted, and the terminal strip is plugged into the new module. Unless otherwise specified, never install or remove I/O modules or terminal blocks while the PLC is powered. A module inserted into the wrong slot could be damaged by improper voltages connected through the wiring arm. Most faceplates and I/O modules are keyed to prevent putting the wrong faceplate on the wrong module. In other words, an output module cannot be placed in the slot where an input module was originally located.

Input and output modules can be placed anywhere in a rack, but they are normally grouped together for ease of wiring. I/O modules can be 8, 16, 32, or 64 point cards (Figure 2-11). The number refers to the number of inputs or outputs available. The standard I/O module has eight inputs or outputs. A *high-density* module may have up to 64 inputs or outputs. The advantage with the high-density module is that it is possible to install up to 64 inputs or outputs in one slot for greater space savings. The only disadvantage is that the high-density output modules cannot handle as much current per output.

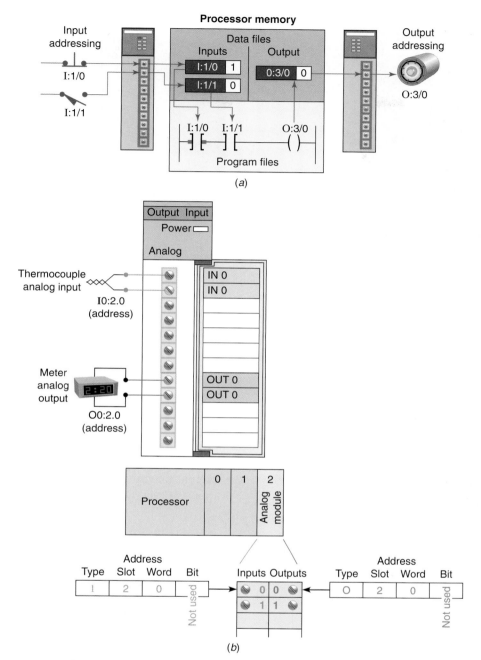

Figure 2-6 SLC 500 bit level and word level addressing. (a) Bit level addressing. (b) Word level addressing.

Figure 2-7 Allen-Bradley ControlLogix tag-based addressing format.
Source: Image Used with Permission of Rockwell Automation, Inc.

Figure 2-8 Typical PC interface card.
Source: Photo © Beckhoff Automation GmbH.

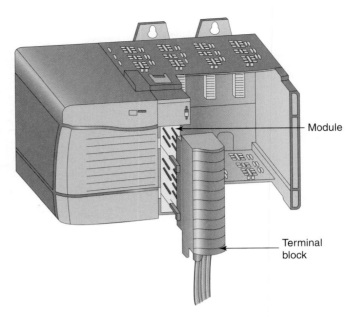

Figure 2-10 Plug-in terminal block.

2.2   Discrete I/O Modules

The most common type of I/O interface module is the *discrete* type (Figure 2-12). This type of interface connects field input devices of the ON/OFF nature such as selector switches, pushbuttons, and limit switches. Likewise, output control is limited to devices such as lights, relays, solenoids, and motor starters that require simple ON/OFF

switching. The classification of discrete I/O covers *bit-oriented* inputs and outputs. In this type of input or output, each bit represents a complete information element in itself and provides the status of some external contact or advises of the presence or absence of power in a process circuit.

Each discrete I/O module is powered by some *field-supplied* voltage source. Since these voltages can be of different magnitude or type, I/O modules are available at various AC and DC voltage ratings, as listed in Table 2-1.

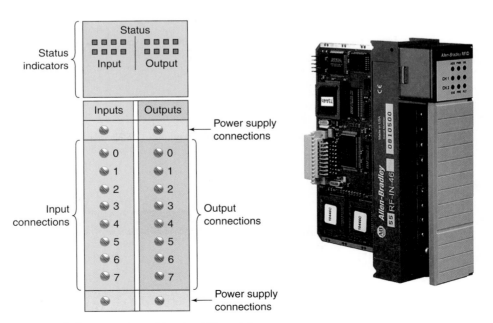

Figure 2-9 Typical combination I/O module.
Source: Image Used with Permission of Rockwell Automation, Inc.

Figure 2-11 16, 32, and 64 point I/O modules.

Source: (all) Photos courtesy Omron Industrial Automation, **www.ia.omron.com**.

The modules themselves receive their voltage and current for proper operation from the backplane of the rack enclosure into which they are inserted, as illustrated in Figure 2-13. Backplane power is provided by the PLC module power supply and is used to power the electronics that reside on the I/O module circuit board. The relatively higher

Table 2-1 Common Ratings for Discrete I/O Interface Modules

Input Interfaces	Output Interfaces
12 V **AC/DC** /24 V **AC/DC**	12–48 V **AC**
48 V **AC/DC**	120 V **AC**
120 V **AC/DC**	230 V **AC**
230 V **AC/DC**	120 V **DC**
5 V **DC** (TTL level)	230 V **DC**
	5 V **DC** (TTL level)
	24 V **DC**

Figure 2-12 Discrete input and output devices.

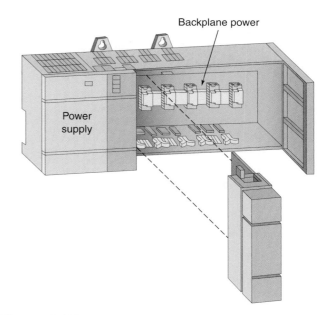

Figure 2-13 Modules receive their voltage and current from the backplane.

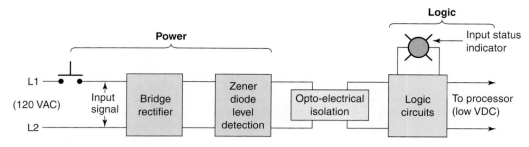

Figure 2-14 Discrete AC input module block diagram.

currents required by the loads of an output module are normally provided by user-supplied power. Module power supplies typically may be rated for 3 A, 4 A, 12 A, or 16 A depending on the type and number of modules used.

Figure 2-14 shows the block diagrams for one input of a typical alternating current (AC) *discrete input module.* The input circuit is composed of two basic sections: the *power* section and the *logic* section. An optical isolator is used to provide electrical isolation between the field wiring and the PLC backplane internal circuitry. The input LED turns on or off, indicating the status of the input device. Logic circuits process the digital signal to the processor. Internal PLC control circuitry typically operates at 5 VDC or less volts.

A simplified diagram for a single input of a discrete AC input module is shown in Figure 2-15. The operation of the circuit can be summarized as follows:

- The input noise filter consisting of the capacitor and resistors R1 and R2 removes false signals that are due to contact bounce or electrical interference.
- When the pushbutton is closed, 120 VAC is applied to the bridge rectifier input.
- This results in a low-level DC output voltage that is applied across the LED of the optical isolator.

- The zener diode (Z_D) voltage rating sets the minimum threshold level of voltage that can be detected.
- When light from the LED strikes the phototransistor, it switches into conduction and the status of the pushbutton is communicated in logic to the processor.
- The optical isolator not only separates the higher AC input voltage from the logic circuits but also prevents damage to the processor due to line voltage transients. In addition, this isolation also helps reduce the effects of electrical noise, common in the industrial environment, which can cause erratic operation of the processor.
- For fault diagnosis, an input state LED indicator is on when the input pushbutton is closed. This indicator may be wired on either side of the optical isolator.
- An AC/DC type of input module is used for both AC and DC inputs as the input polarity does not matter.
- A PLC input module will have either all inputs isolated from each other with no common input connections or groups of inputs that share a common connection.

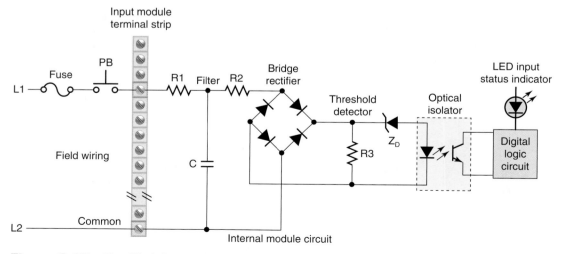

Figure 2-15 Simplified diagram for a single input of a discrete AC input module.

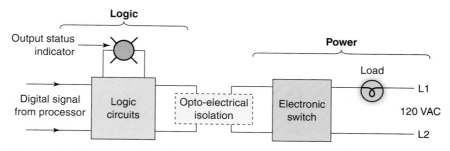

Figure 2-16 Discrete AC output module block diagram.

Discrete input modules perform four tasks in the PLC control system. They:

- Sense when a signal is received from a field device.
- Convert the input signal to the correct voltage level for the particular PLC.
- Isolate the PLC from fluctuations in the input signal's voltage or current.
- Send a signal to the processor indicating which sensor originated the signal.

Figure 2-16 shows the block diagram for one output of a typical discrete output module. Like the input module, it is composed of two basic sections: the power section and the logic section, coupled by an isolation circuit. The output interface can be thought of as an electronic switch that turns the output load device on and off. Logic circuits determine the output status. An output LED indicates the status of the output signal.

A simplified diagram for a single output of a discrete AC output module is shown in Figure 2-17. The operation of the circuit can be summarized as follows:

- As part of its normal operation, the digital logic circuits of the processor sets the output status according to the program.

- When the processor calls for an output load to be energized, a voltage is applied across the LED of the opto-isolator.
- The LED then emits light, which switches the phototransistor into conduction.
- This in turn triggers the triac AC semiconductor switch into conduction allowing current to flow to the output load.
- Since the triac conducts in either direction, the output to the load is alternating current.
- The triac, rather than having ON and OFF status, actually has LOW and HIGH resistance levels, respectively. In its OFF state (HIGH resistance), a small leakage current of a few milliamperes still flows through the triac.
- As with input circuits, the output interface is usually provided with LEDs that indicate the status of each output.
- Fuses are normally required for the output module, and they are provided on a per circuit basis, thus allowing for each circuit to be protected and operated separately. Some modules also provide visual indicators for fuse status.
- The triac cannot be used to switch a DC load.

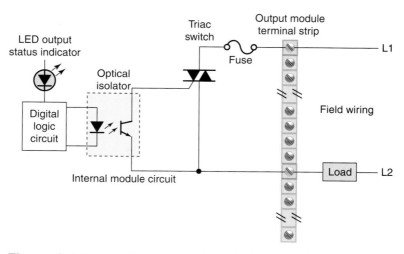

Figure 2-17 Simplified diagram for a single output of a discrete AC output module.

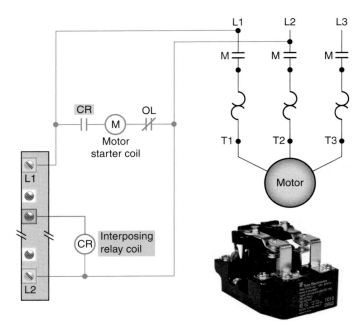

Figure 2-18 Interposing relay connection.
Source: Photo courtesy Tyco Electronics, **www.tycoelectronics.com**.

- For fault diagnosis, the LED output status indicator is on whenever the PLC is commanding that the output load be switched on.

Individual AC outputs are usually limited by the size of the triac to 1 A or 2 A. The maximum current load for any one module is also specified. To protect the output module circuits, specified current ratings should not be exceeded. For controlling larger loads, such as large motors, a standard control relay is connected to the output module. The contacts of the relay can then be used to control a larger load or motor starter, as shown in Figure 2-18. When a control relay is used in this manner, it is called an *interposing* relay.

Discrete output modules are used to turn field output devices either on or off. These modules can be used to control any two-state device, and they are available in AC and DC versions and in various voltage ranges and current ratings. Output modules can be purchased with *transistor, triac,* or *relay* output as illustrated in Figure 2-19. Triac outputs can be used only for control of AC devices,

whereas transistor outputs can be used only for control of DC devices. The discrete relay contact output module uses electromechanical as the switching element. These relay outputs can be used with AC or DC devices, but they have a much slower switching time compared to solid-state outputs. Allen-Bradley modules are color-coded for identification as follows:

Color	Type of I/O
Red	AC inputs/outputs
Blue	DC inputs/outputs
Orange	Relay outputs
Green	Specialty modules
Black	I/O wiring; terminal blocks are not removable

Certain DC I/O modules specify whether the module is designed for interfacing with current-source or current-sink devices. If the module is a current-sourcing module, then the input or output device must be a current-sinking device. Conversely, if the module is specified as current-sinking, then the connected device must be current-sourcing. Some modules allow the user to select whether the module will act as current sinking or current sourcing, thereby allowing it to be set to whatever the field devices require.

The internal circuitry of some field devices requires that they be used in sinking or sourcing circuits. In general, *sinking (NPN)* and *sourcing (PNP)* are terms used to describe a current signal flow relationship between field input and output devices in a control system and their power supply. Figure 2-20 illustrates the current flow relationship between sinking and sourcing inputs to a DC input module.

Figure 2-21 illustrates the current flow relationship between sinking and sourcing outputs to a DC output module. DC input and output circuits are commonly connected with field devices that have some form of internal

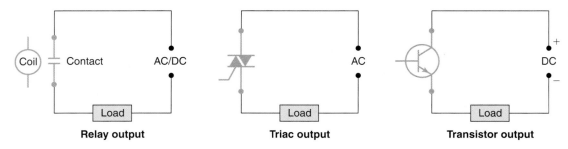

Figure 2-19 Relay, transistor, and triac switching elements.

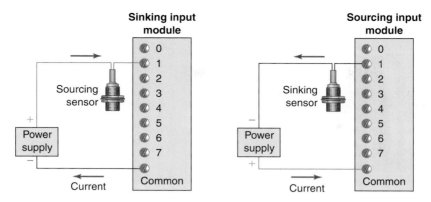

Figure 2-20 Sinking and sourcing inputs.

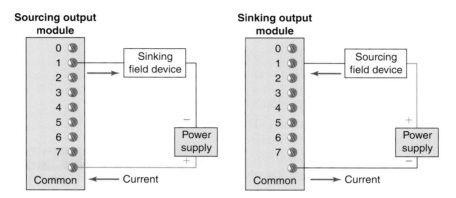

Figure 2-21 Sinking and sourcing outputs.

solid-state circuitry that needs a DC signal voltage to function. Field devices connected to the positive (+) side of the field power supply are classified as sourcing field devices. Conversely, field devices connected to the negative (−) side or DC common of the field power supply are sinking field devices.

2.3 Analog I/O Modules

Earlier PLCs were limited to discrete or digital I/O interfaces, which allowed only on/off-type devices to be connected. This limitation meant that the PLC could have only partial control of many process applications. Today, however, a complete range of both discrete and analog interfaces are available that will allow controllers to be applied to practically any type of control process.

Discrete devices are inputs and outputs that have only two states: on and off. In comparison, analog devices represent physical quantities that can have an infinite number of values. Typical analog inputs and outputs vary from 0 to 20 milliamps, 4 to 20 milliamps, or 0 to 10 volts. Figure 2-22 illustrates how PLC analog input and output modules are used in measuring and displaying the level of fluid in a tank. The analog input interface module contains the circuitry necessary to accept an analog voltage or

current signal from the level transmitter field device. This input is converted from an analog to a digital value for use by the processor. The circuitry of the analog output module accepts the digital value from the processor and converts it back to an analog signal that drives the field tank level meter.

Analog input modules normally have multiple input channels that allow 4, 8, or 16 devices to be interface to the PLC. The two basic types of analog input modules are *voltage* sensing and *current* sensing. Analog sensors measure a varying physical quantity over a specific range and generate a corresponding voltage or current signal. Common physical quantities measured by a PLC analog module include temperature, speed, level, flow, weight, pressure, and position. For example, a sensor may

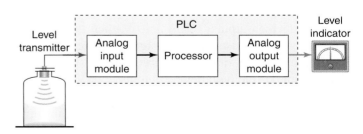

Figure 2-22 Analog input and output to a PLC.

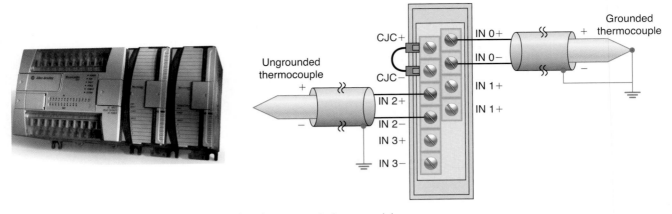

Figure 2-23 MicroLogix 4-channel analog thermocouple input module.
Source: Image Used with Permission of Rockwell Automation, Inc.

measure temperature over a range of 0 to 500°C, and output a corresponding voltage signal that varies between 0 and 50 mV.

Figure 2-23 illustrates an example of a voltage sensing input analog module used to measure temperature. The connection diagram applies to an Allen-Bradley Micro-Logic 4-channel analog thermocouple input module. A varying DC voltage in the low millivolt range, proportional to the temperature being monitored, is produced by the thermocouple. This voltage is amplified and digitized by the analog input module and then sent to the processor on command from a program instruction. Because of the low voltage level of the input signal, a twisted shielded pair cable is used in wiring the circuit to reduce unwanted electrical noise signals that can be induced in the conductors from other wiring. When using an ungrounded thermocouple, the shield must be connected to ground at the module end. To obtain accurate readings from each of the channels, the temperature between the thermocouple wire and the input channel must be compensated for. A cold junction compensating (CJC) thermistor is integrated in the terminal block for this purpose.

The transition of an analog signal to digital values is accomplished by an analog-to-digital (A/D) converter, the main element of the analog input module. Analog voltage input modules are available in two types: unipolar and bipolar. *Unipolar* modules can accept an input signal that varies in the positive direction only. For example, if the field device outputs 0 V to +10 V, then the unipolar modules would be used. Bipolar signals swing between a maximum negative value and a maximum positive value. For example, if the field device outputs −10 V to +10 V a bipolar module would be used. The *resolution* of an analog input channel refers to the smallest change in input signal value that can be sensed and is based on the number of bits used in the digital representation. Analog input

modules must produce a range of digital values between a maximum and minimum value to represent the analog signal over its entire span. Typical specifications are as follows:

Span of analog input	Bipolar	10 V	−10 to +10 V
		5 V	−5 to +5 V
	Unipolar	10 V	0 to +10 V
		5 V	0 to +5 V
Resolution			0.3 mV

When connecting voltage sensing inputs, close adherence to specified requirements regarding wire length is important to minimize signal degrading and the effects of electromagnetic noise interference induced along the connecting conductors. Current input signals, which are not as sensitive to noise as voltage signals, are typically not distance limited. Current sensing input modules typically accept analog data over the range of 4 mA to 20 mA, but can accommodate signal ranges of −20 mA to +20 mA. The loop power may be supplied by the sensor or may be provided by the analog output module as illustrated in Figure 2-24. Shielded twisted pair cable is normally recommended for connecting any type analog input signal.

The *analog output interface module* receives from the processor digital data, which are converted into a proportional voltage or current to control an analog field device. The transition of a digital signal to analog values is accomplished by a digital-to-analog (D/A) converter, the main

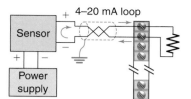

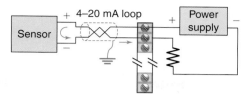

Figure 2-24 Sensor and analog module supplied power.

element of the analog output module. An analog output signal is a continuous and changing signal that is varied under the control of the PLC program. Common devices controlled by a PLC analog output module include instruments, control valves, chart recorder, electronic drives, and other types of control devices that respond to analog signals.

Figure 2-25 illustrates the use of analog I/O modules in a typical PLC control system. In this application the PLC controls the amount of fluid placed in a holding tank by adjusting the percentage of the valve opening. The analog output from the PLC is used to control the flow by controlling the amount of the valve opening. The valve is initially open 100 percent. As the fluid level in the tank approaches the preset point, the processor modifies the output, which adjusts the valve to maintain a set point.

2.4 Special I/O Modules

Many different types of I/O modules have been developed to meet special needs. These include:

HIGH-SPEED COUNTER MODULE

The high-speed counter module is used to provide an interface for applications requiring counter speeds that surpass the capability of the PLC ladder program. High-speed counter modules are used to count pulses (Figure 2-26) from sensors, encoders, and switches that operate at very high speeds. They have the electronics needed to count independently of the processor. A typical count rate

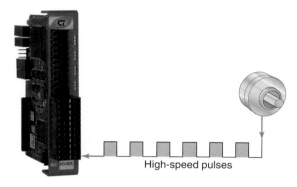

High-speed pulses

Figure 2-26 High-speed counter module.
Source: Courtesy Control Technology Corporation.

available is 0 to 100 kHz, which means the module would be able to count 100,000 pulses per second.

THUMBWHEEL MODULE

The thumbwheel module allows the use of thumbwheel switches (Figure 2-27) for feeding information to the PLC to be used in the control program.

TTL MODULE

The TTL module (Figure 2-28) allows the transmitting and receiving of TTL (Transistor-Transistor-Logic) signals. This module allows devices that produce TTL-level signals to communicate with the PLC's processor.

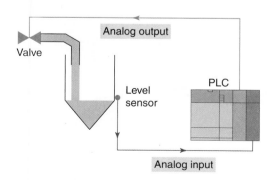

Figure 2-25 Typical analog I/O control system.

Figure 2-27 Thumbwheel switch.
Source: Photo courtesy Omron Industrial Automation, **www.ia.omron.com**.

Figure 2-28 TTL module.
Source: Courtesy Control Technology, Inc.

Figure 2-30 BASIC module.
Source: Image Used with Permission of Rockwell Automation, Inc.

ENCODER-COUNTER MODULE

An encoder-counter module allows the user to read the signal from an encoder (Figure 2-29) on a real-time basis and stores this information so it can be read later by the processor.

BASIC OR ASCII MODULE

The BASIC or ASCII module (Figure 2-30) runs user-written BASIC and C programs. These programs are independent of the PLC processor and provide an easy, fast interface between remote foreign devices and the PLC

processor. Typical applications include interfaces to bar code readers, robots, printers, and displays.

STEPPER-MOTOR MODULE

The stepper-motor module provides pulse trains to a stepper-motor translator, which enables control of a stepper motor (Figure 2-31). The commands for the module are determined by the control program in the PLC.

BCD-OUTPUT MODULE

The BCD-output module enables a PLC to operate devices that require BCD-coded signals such as seven-segment displays (Figure 2-32).

Figure 2-29 Encoder.
Source: Photo courtesy of Allied Motion Technologies, Inc.

Figure 2-31 Stepper-motor.
Source: Courtesy Sherline Products.

Figure 2-32 Seven-segment display.
Source: Courtesy Red Lion Controls.

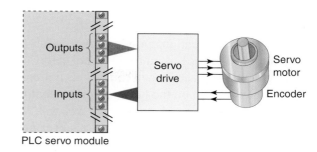

PLC servo module

Figure 2-34 PLC servo module.

Some special modules are referred to as *intelligent I/O* because they have their own microprocessors on board that can function in parallel with the PLC. These include:

PID MODULE

The proportional-integral-derivative (PID) module (Figure 2-33) is used in process control applications that incorporate PID algorithms. An algorithm is a complex program based on mathematical calculations. A PID module allows process control to take place outside the CPU. This arrangement prevents the CPU from being burdened with complex calculations. The basic function of this module is to provide the control action required to maintain a process variable such as temperature, flow, level, or speed within set limits of a specified set point.

MOTION AND POSITION CONTROL MODULE

Motion and position control modules are used in applications involving accurate high-speed machining and packaging operations. Intelligent position and motion control modules permit PLCs to control stepper and servo motors.

These systems require a drive, which contains the power electronics that translate the signals from the PLC module into signals required by the motor (Figure 2-34).

COMMUNICATION MODULES

Serial communications modules (Figure 2-35) are used to establish point-to-point connections with other intelligent devices for the exchange of data. Such connections are normally established with computers, operator stations, process control systems, and other PLCs. Communication modules allow the user to connect the PLC to high-speed local networks that may be different from the network communication provided with the PLC.

Figure 2-33 PID module.
Source: Courtesy Red Lion Controls.

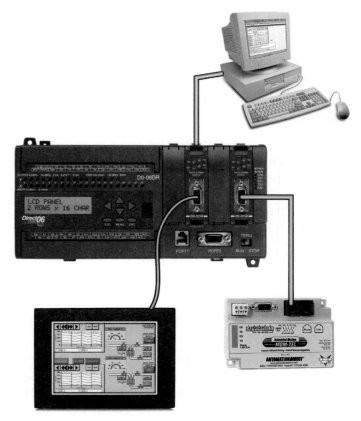

Figure 2-35 Serial communications module.
Source: Photo courtesy Automation Direct, **www.automationdirect.com**.

2.5 I/O Specifications

Manufacturers' specifications provide information about how an interface device is correctly and safely used. These specifications place certain limitations not only on the I/O module but also on the field equipment that it can operate. Some PLC systems support *hot swappable* I/O modules designed to be changed with the power on and the PLC operating. The following is a list of some typical manufacturers' I/O specifications, along with a short description of what is specified.

Typical Discrete I/O Module Specifications

NOMINAL INPUT VOLTAGE

This discrete input module voltage value specifies the magnitude (e.g., 5 V, 24 V, 230 V) and type (AC or DC) of user-supplied voltage that a module is designed to accept. Input modules are typically designed to operate correctly without damage within a range of plus or minus 10 percent of the input voltage rating. With DC input modules, the input voltage may also be expressed as an operating range (e.g., 24–60 volts DC) over which the module will operate.

INPUT THRESHOLD VOLTAGES

This discrete input module specification specifies two values: a minimum ON-state voltage that is the minimum voltage at which logic 1 is recognized as absolutely ON; and a maximum OFF-state voltage which is the voltage at which logic 0 is recognized as absolutely OFF.

NOMINAL CURRENT PER INPUT

This value specifies the minimum input current that the discrete input devices must be capable of driving to operate the input circuit. This input current value, in conjunction with the input voltage, functions as a threshold to protect against detecting noise or leakage currents as valid signals.

AMBIENT TEMPERATURE RATING

This value specifies what the maximum temperature of the air surrounding the I/O modules should be for best operating conditions.

INPUT ON/OFF DELAY

Also known as *response time,* this value specifies the maximum time duration required by an input module's circuitry to recognize that a field device has switched ON (input ON-delay) or switched OFF (input OFF-delay). This delay is a result of filtering circuitry provided to protect against contact bounce and voltage transients.

This input delay is typically in the 9 to 25 millisecond range.

OUTPUT VOLTAGE

This AC or DC value specifies the magnitude (e.g., 5 V, 115 V, 230 V) and type (AC or DC) of user-supplied voltage at which a discrete output module is designed to operate. The output field device that the module interfaces to the PLC must be matched to this specification. Output modules are typically designed to operate within a range of plus or minus 10 percent of the nominal output voltage rating.

OUTPUT CURRENT

These values specify the maximum current that a single output and the module as a whole can safely carry under load (at rated voltage). This rating is a function of the module's components and heat dissipation characteristics. A device drawing more than the rated output current results in overloading, causing the output fuse to blow. As an example, the specification may give each output a current limit of 1 A. The overall rating of the module current will normally be less than the total of the individuals. The overall rating might be 6 A because each of the eight devices would not normally draw their 1 A at the same time. Other names for the output current rating are *maximum continuous current* and *maximum load current.*

INRUSH CURRENT

An inrush current is a momentary surge of current that an AC or DC output circuit encounters when energizing inductive, capacitive, or filament loads. This value specifies the maximum inrush current and duration (e.g., 20 A for 0.1 s) for which an output circuit can exceed its maximum continuous current rating.

SHORT CIRCUIT PROTECTION

Short circuit protection is provided for AC and DC output modules by either fuses or some other current-limiting circuitry. This specification will designate whether the particular module's design has individual protection for each circuit or if fuse protection is provided for groups (e.g., 4 or 8) of outputs.

LEAKAGE CURRENT

This value specifies the amount of current still conducting through an output circuit even after the output has been turned off. Leakage current is a characteristic exhibited by solid-state switching devices such as transistors and triacs and is normally less than 5 milliamperes. Leakage current is normally not large enough to falsely trigger an output device but must be taken into consideration when switching very low current sensitive devices.

ELECTRICAL ISOLATION

Recall that I/O module circuitry is electrically isolated to protect the low-level internal circuitry of the PLC from high voltages that can be encountered from field device connections. The specification for electrical isolation, typically 1500 or 2500 volts AC, rates the module's capacity for sustaining an excessive voltage at its input or output terminals. Although this isolation protects the logic side of the module from excessive input or output voltages or current, the power circuitry of the module may be damaged.

POINTS PER MODULE

This specification defines the number of field inputs or outputs that can be connected to a single module. Most commonly, a discrete module will have 8, 16, or 32 circuits; however, low-end controllers may have only 2 or 4 circuits. Modules with 32 or 64 input or output bits are referred to as *high-density* modules. Some modules provide more than one common terminal, which allows the user to use different voltage ranges on the same card as well as to distribute the current more effectively.

BACKPLANE CURRENT DRAW

This value indicates the amount of current the module requires from the backplane. The sum of the backplane current drawn for all modules in a chassis is used to select the appropriate chassis power supply rating.

Typical Analog I/O Module Specifications

CHANNELS PER MODULE

Whereas individual circuits on discrete I/O modules are referred to as points, circuits on analog I/O modules are often referred to as channels. These modules normally have 4, 8, or 16 channels. Analog modules may allow for either single-ended or differential connections. *Single-ended* connections use a single ground terminal for all channels or for groups of channels. *Differential* connections use a separate positive and negative terminal for each channel. If the module normally allows 16 single-ended connections, it will generally allow only 8 differential connections. Single-ended connections are more susceptible to electrical noise.

INPUT CURRENT/VOLTAGE RANGE(S)

These are the voltage or current signal ranges that an analog input module is designed to accept. The input ranges must be matched accordingly to the varying current or voltage signals generated by the analog sensors.

OUTPUT CURRENT/VOLTAGE RANGE(S)

This specification defines the current or voltage signal ranges that a particular analog output module is designed to output under program control. The output ranges must be matched according to the varying voltage or current signals that will be required to drive the analog output devices.

INPUT PROTECTION

Analog input circuits are usually protected against accidentally connecting a voltage that exceeds the specified input voltage range.

RESOLUTION

The resolution of an analog I/O module specifies how accurately an analog value can be represented digitally. This specification determines the smallest measurable unit of current or voltage. The higher the resolution (typically specified in bits), the more accurately an analog value can be represented.

INPUT IMPEDANCE AND CAPACITANCE

For analog I/Os, these values must be matched to the external device connected to the module. Typical ratings are in Megohm (MΩ) and picofarads (pF).

COMMON-MODE REJECTION

Noise is generally caused by electromagnetic interference, radio frequency interference, and ground loops. Common-mode noise rejection applies only to differential inputs and refers to an analog module's ability to prevent noise from interfering with data integrity on a single channel and from channel to channel on the module. Noise that is picked up equally in parallel wires is rejected because the difference is zero. Twisted pair wires are used to ensure that this type of noise is equal on both wires. Common-mode rejection is normally expressed in decibels or as a ratio.

2.6 The Central Processing Unit (CPU)

The central processing unit (CPU) is built into single-unit fixed PLCs while modular rack types typically use a plug-in module. CPU, controller, and processor are all terms used by different manufacturers to denote the same module that performs basically the same functions. Processors vary in processing speed and memory options. A processor module can be divided into two sections: the CPU section and the memory section (Figure 2-36). The CPU section executes the program and makes the decisions needed by the PLC to operate and communicate

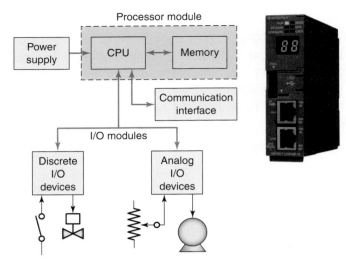

Figure 2-36 Sections of a PLC processor module.
Source: Courtesy Mitsubishi Automation.

with other modules. The memory section electronically stores the PLC program along with other retrievable digital information.

The PLC power supply provides the necessary power (typically 5 VDC) to the processor and I/O modules plugged into the backplane of the rack (Figure 2-37). Power supplies are available for most voltage sources encountered. The power supply converts 115 VAC or 230 VAC into the usable DC voltage required by the CPU, memory, and I/O electronic circuitry. PLC power supplies are normally designed to withstand momentary losses of power without affecting the operation of the PLC. *Hold-up time,* which is the length of time a PLC can tolerate a power loss, typically ranges from 10 milliseconds to 3 seconds.

The CPU contains the similar type of microprocessor found in a personal computer. The difference is that the program used with the microprocessor is designed to facilitate industrial control rather than provide general-purpose computing. The CPU executes the operating system, manages memory, monitors inputs, evaluates the

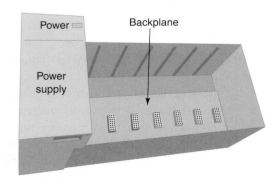

Figure 2-37 PLC power supply.

user logic (ladder program), and turns on the appropriate outputs.

The CPU of a PLC system may contain more than one processor. One advantage of using multiprocessing is that the overall operating speed is improved. Each processor has its own memory and programs, which operate simultaneously and independently. In such configurations the scan of each processor is parallel and independent thus reducing the total response time. Fault-tolerant PLC systems support dual processors for critical processes. These systems allow the user to configure the system with *redundant* (two) processors, which allows transfer of control to the second processor in the event of a processor fault.

Associated with the processor unit will be a number of status LED indicators to provide system diagnostic information to the operator (Figure 2-38). Also, a keyswitch may be provided that allows you to select one of the following three modes of operation: RUN, PROG, and REM.

RUN Position

- Places the processor in the Run mode
- Executes the ladder program and energizes output devices
- Prevents you from performing online program editing in this position
- Prevents you from using a programmer/operator interface device to change the processor mode

PROG Position

- Places the processor in the Program mode
- Prevents the processor from scanning or executing the ladder program, and the controller outputs are de-energized
- Allows you to perform program entry and editing
- Prevents you from using a programmer/operator interface device to change the processor mode

REM Position

- Places the processor in the Remote mode: either the REMote Run, REMote Program, or REMote Test mode
- Allows you to change the processor mode from a programmer/operator interface device
- Allows you to perform online program editing

The processor module also contains circuitry to communicate with the programming device. Somewhere on the module you will find a connector that allows the PLC to be connected to an external programming device. The decision-making capabilities of PLC processors go far beyond simple logic processing. The processor performs

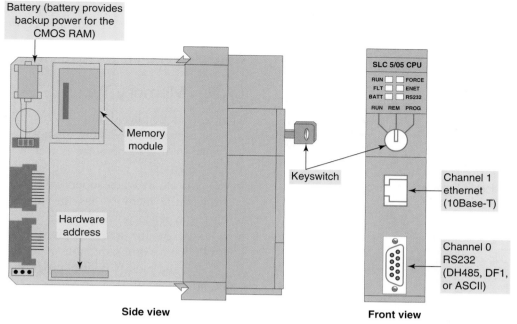

Battery (battery provides
backup power for the
CMOS RAM)

Memory
module

Hardware
address

Side view

SLC 5/05 CPU

RUN ☐☐ FORCE
FLT ☐☐ ENET
BATT ☐☐ RS232

RUN REM PROG

Keyswitch

Channel 1
ethernet
(10Base-T)

Channel 0
RS232
(DH485, DF1,
or ASCII)

Front view

Figure 2-38 Typical processor module.

other functions such as timing, counting, latching, comparing, motion control and complex math functions.

PLC processors have changed constantly due to advancements in computer technology and greater demand from applications. Today, processors are faster and have additional instructions added as new models are introduced. Because PLCs are microprocessor based, they can be made to perform tasks that a computer can do. In addition to their control functions, PLCs can be networked to do supervisory control and data acquisition (SCADA).

Many electronic components found in processors and other types of PLC modules are sensitive to *electrostatic* voltages that can degrade their performance or damage them. The following static control procedures should be followed when handling and working with static-sensitive devices and modules:

- Ground yourself by touching a conductive surface before handling static-sensitive components.
- Wear a wrist strap that provides a path to bleed off any charge that may build up during work.
- Be careful not to touch the backplane connector or connector pins of the PLC system (always handle the circuit cards by the edge if possible).
- Be careful not to touch other circuit components in a module when you configure or replace its internal components.
- When not in use, store modules in its static-shield bag.
- If available, use a static-safe work station.

2.7 Memory Design

Memory is the element that stores information, programs, and data in a PLC. The user memory of a PLC includes space for the user program as well as addressable memory locations for storage of data. Data are stored in memory locations by a process called *writing*. Data are retrieved from memory by what is referred to as *reading*.

The complexity of the program determines the amount of memory required. Memory elements store individual pieces of information called *bits* (for *binary digits*). The amount of memory capacity is specified in increments of 1000 or in "K" increments, where 1 K is 1024 bytes of memory storage (a byte is 8 bits).

The program is stored in the memory as 1s and 0s, which are typically assembled in the form of 16-bit words. Memory sizes are commonly expressed in thousands of words that can be stored in the system; thus 2 K is a memory of 2000 words, and 64 K is a memory of 64,000 words. The memory size varies from as small as 1 K for small systems to 32 MB for very large systems (Figure 2-39). Memory capacity is an important prerequisite for determining whether a particular processor will handle the requirements of the specific application.

Memory *location* refers to an address in the CPU's memory where a binary word can be stored. A word usually consists of 16 bits. Each binary piece of data is a bit and eight bits make up one byte (Figure 2-40). Memory *utilization* refers to the number of memory locations required to store each type of instruction. A rule of thumb

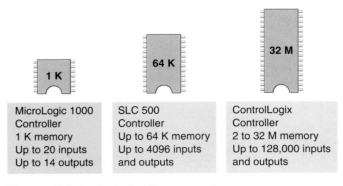

Figure 2-39   Typical PLC memory sizes.

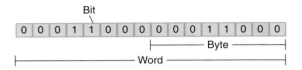

Figure 2-40   Memory bit, byte, and word.

for memory locations is one location per coil or contact. One K of memory would then allow a program containing 1000 coils and contacts to be stored in memory.

The memory of a PLC may be broken into sections that have specific functions. Sections of memory used to store the status of inputs and outputs are called input status files or tables and output status files or tables (Figure 2-41). These terms simply refer to a location where the status of an input or output device is stored. Each bit is either a 1 or 0, depending on whether the input is open or closed. A closed contact would have a binary 1 stored in its respective location in the input table, whereas an open contact would have a 0 stored. A lamp that is ON would have a 1 stored in its respective location in the output table, whereas a lamp that is OFF would have a 0 stored. Input and output image tables are constantly being revised by the CPU. Each time a memory location is examined, the table changes if the contact or coil has changed state.

PLCs execute memory-checking routines to be sure that the PLC memory has not been corrupted. This memory checking is undertaken for safety reasons. It helps ensure that the PLC will not execute if memory is corrupted.

2.8   Memory Types

Memory can be placed into two general categories: volatile and nonvolatile. Volatile memory will lose its stored information if all operating power is lost or removed. Volatile memory is easily altered and is quite suitable for most applications when supported by battery backup.

Nonvolatile memory has the ability to retain stored information when power is removed accidentally or intentionally. As the name implies, programmable logic controllers have programmable memory that allows users to develop and modify control programs. This memory is made nonvolatile so that if power is lost, the PLC holds its programming.

Read Only Memory (ROM) stores programs, and data cannot be changed after the memory chip has been manufactured. ROM is normally used to store the programs and data that define the capabilities of the PLC. ROM memory is nonvolatile, meaning that its contents will not be lost if power is lost. ROM is used by the PLC for the operating system. The operating system is burned into ROM by the PLC manufacturer and controls the system software that the user uses to program the PLC.

Random Access Memory (RAM), sometimes referred to as *read-write (R/W) memory,* is designed so that information can be written into or read from the memory. RAM is used as a temporary storage area of data that may need to be quickly changed. RAM is volatile, meaning that the data stored in RAM will be lost if power is lost. A battery backup is required to avoid losing data in the event of a power loss (Figure 2-42). Most PLCs use CMOS-RAM technology for user memory. CMOS-RAM chips have very low current draw and can maintain memory with a lithium battery for an extended time, two to five

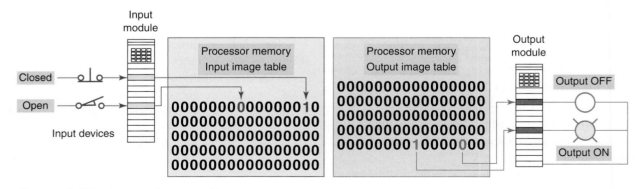

Figure 2-41   Input and output tables.

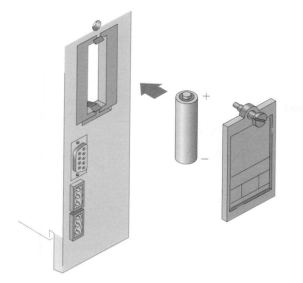

Figure 2-42 Battery used to back up processor RAM.

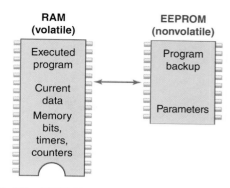

Figure 2-43 EEPROM memory module is used to store, back up, or transfer PLC programs.

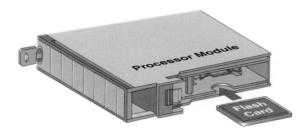

Figure 2-44 Flash memory card installed in a socket on the processor.

years in many cases. Some processors have a capacitor that provides at least 30 minutes of battery backup when the battery is disconnected and power is OFF.

Erasable Programmable Read-Only Memory (EPROM) provides some level of security against unauthorized or unwanted changes in a program. EPROMs are designed so that data stored in them can be read, but not easily altered without special equipment. For example, UV EPROMs (ultraviolet erasable programmable read-only memory) can only be erased with an ultraviolet light. EPROM memory is used to back up, store, or transfer PLC programs.

Electrically erasable programmable read-only memory (EEPROM) is a nonvolatile memory that offers the same programming flexibility as does RAM. The EEPROM can be electrically overwritten with new data instead of being erased with ultraviolet light. Because the EEPROM is nonvolatile memory, it does not require battery backup. It provides permanent storage of the program and can be changed easily using standard programming devices. Typically, an EEPROM memory module is used to store, back up, or transfer PLC programs (Figure 2-43).

Flash EEPROMs are similar to EEPROMs in that they can only be used for backup storage. The main difference comes in the flash memory: they are extremely fast at saving and retrieving files. In addition, they do not need to be physically removed from the processor for reprogramming; this can be done using the circuitry within the processor module in which they reside. Flash memory is also sometimes built into the processor module (Figure 2-44), where it automatically backs up parts of RAM. If power fails while a PLC with flash memory is running, the PLC

will resume running without having lost any working data after power is restored.

2.9 Programming Terminal Devices

A programming terminal device is needed to enter, modify, and troubleshoot the PLC program. PLC manufacturers use various types of programming devices. The simplest type is the hand-held type programmer shown in Figure 2-45. This proprietary programming device has a connecting cable so that it can be plugged into a PLC's programming port. Certain controllers use a plug-in panel rather than a hand-held device.

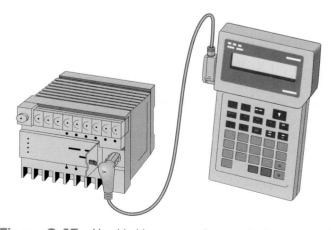

Figure 2-45 Hand-held programming terminal.

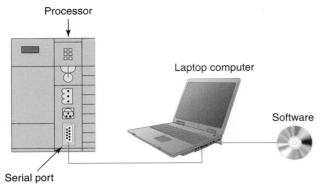

Figure 2-46 Personal computer used as the programming device.

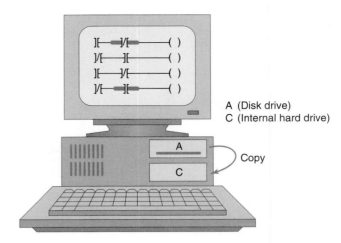

A (Disk drive)
C (Internal hard drive)

Copy

Figure 2-47 Copying programs to a computer hard drive.

Hand-held programmers are compact, inexpensive, and easy to use. These units contain multifunction keys and a liquid-crystal display (LCD) or light-emitting diode (LED) window. There are usually keys for instruction entering and editing, and navigation keys for moving around the program. Hand-held programmers have limited display capabilities. Some units will display only the last instruction that has been programmed, whereas other units will display from two to four rungs of ladder logic. So-called intelligent hand-held programmers are designed to support a certain family of PLCs from a specific manufacturer.

The most popular method of PLC programming is to use a personal computer (PC) in conjunction with the manufacturer's programming software (Figure 2-46). Typical capabilities of the programming software include online and offline program editing, online program monitoring, program documentation, diagnosing malfunctions in the PLC, and troubleshooting the controlled system. Hard-copy reports generated in the software can be printed on the computer's printer. Most software packages will not allow you to develop programs on another manufacturer's PLC. In some cases, a single manufacturer will have multiple PLC families, each requiring its own software to program.

2.10 Recording and Retrieving Data

Printers are used to provide hard-copy printouts of the processor's memory in ladder program format. Lengthy ladder programs cannot be shown completely on a screen. Typically, a screen shows a maximum of five rungs at a time. A printout can show programs of any length and analyze the complete program.

The PLC can have only one program in memory at a time. To change the program in the PLC, it is necessary either to enter a new program directly from the

keyboard or to download one from the computer hard drive (Figure 2-47). Some CPUs support the use of a memory cartridge that provides portable EEPROM storage for the user program (Figure 2-48). The cartridge can be used to copy a program from one PLC to another similar type PLC.

2.11 Human Machine Interfaces (HMIs)

A *human machine interface (HMI)* can be connected to communicate with a PLC and to replace pushbuttons, selector switches, pilot lights, thumbwheels, and other operator control panel devices (Figure 2-49). Luminescent touch-screen keypads provide an operator interface that operates like traditional hardwired control panels.

Human machine interfaces give the ability to the operator and to management to view the operation in real

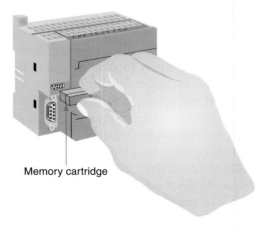

Memory cartridge

Figure 2-48 Memory cartridge provides portable storage for user program.

Figure 2-49 Human Machine Interfaces (HMIs).
Source: Photo courtesy Omron Industrial Automation, **www.ia.omron.com**.

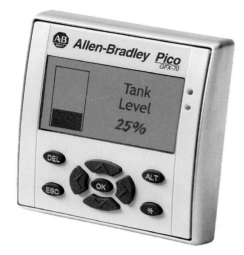

Figure 2-50 Allen-Bradley Pico GFX-70 controller.
Source: Image Used with Permission of Rockwell Automation, Inc.

time. Through personal computer–based setup software, you can configure display screens to:

- Replace hardwired pushbuttons and pilot lights with realistic-looking icons. The machine operator need only touch the display panel to activate the pushbuttons.
- Show operations in graphic format for easier viewing.
- Allow the operator to change timer and counter presets by touching the numeric keypad graphic on the touch screen.
- Show alarms, complete with time of occurrence and location.
- Display variables as they change over time.

The Allen-Bradley Pico GFX-70 controller, shown in Figure 2-50, serves as a controller with HMI capabilities.

This device consists of three modular parts: an HMI, processor/power supply, and I/O modules.

The display/keypad can be used as an operator interface or can be linked to control operations to provide real-time feedback. It has the ability to show text, date and time, as well as custom messages and bitmap graphics, allowing operators to acknowledge fault messages, enter values, and initiate actions. Users can create both the control program and HMI functionality using a personal computer with PicoSoft Pro software installed or the controller's on-board display buttons.

1. What is the function of a PLC input interface module?

2. What is the function of a PLC output interface module?

3. Define the term *logical rack*.

4. With reference to a PLC rack:
   a. What is a *remote rack?*
   b. Why are remote racks used?

5. How does the processor identify the location of a specific input or output device?

6. List the three basic elements of rack/slot-based addressing.

7. Compare bit level and word level addressing.

8. In what way does tag-based addressing differ from rack/slot-based addressing?

9. What do PC-based control systems use to interface with field devices?

10. What type of I/O modules have both inputs and outputs connected to them?

11. In addition to field devices what other connections are made to a PLC module?

12. Most PLC modules use plug-in wiring terminal strips. Why?

13. What are the advantage and the disadvantage of using high-density modules?

14. With reference to PLC discrete input modules:
    a. What types of field input devices are suitable for use with them?
    b. List three examples of discrete input devices.

15. With reference to PLC discrete output modules:
    a. What types of field output devices are suitable for use with them?
    b. List three examples of discrete output devices.

16. Explain the function of the backplane of a PLC rack.

17. What is the function of the optical isolator circuit used in discrete I/O module circuits?

18. Name the two distinct sections of an I/O module.

19. List four tasks performed by a discrete input module.

20. What electronic element can be used as the switching device for a 120 VAC discrete output interface module?

21. With reference to discrete output module current ratings:
    a. What is the maximum current rating for a typical 120 VAC output module?
    b. Explain one method of handling outputs with larger current requirements.

22. What electronic element can be used as the switching device for DC discrete output modules?

23. A discrete relay type output module can be used to switch either AC or DC load devices. Why?

24. With reference to sourcing and sinking I/O modules:
    a. What current relationship are the terms *sourcing* and *sinking* used to describe?
    b. If an I/O module is specified as a current-sinking type, then which type of field device (sinking or sourcing) it is electrically compatible with?

25. Compare discrete and analog I/O modules with respect to the type of input or output devices with which they can be used.

26. Explain the function of the analog-to-digital (A/D) converter circuit used in analog input modules.

27. Explain the function of the digital-to-analog (D/A) converter circuit used in analog output modules.

28. Name the two general sensing classifications for analog input modules.

29. List five common physical quantities measured by a PLC analog input module.

30. What type of cable is used when connecting a thermocouple to a voltage sensing analog input module? Why?

31. Explain the difference between a unipolar and bipolar analog input module.

32. The resolution of an analog input channel is specified as 0.3 mV. What does this tell you?

33. In what two ways can the loop power for current sensing input modules be supplied?

34. List three field devices that are commonly controlled by a PLC analog output module.

35. State one application for each of the following special I/O modules:
    a. High-speed counter module
    b. Thumbwheel module
    c. TTL module

d. Encoder-counter module
e. BASIC or ASCII module
f. Stepper-motor module
g. BCD-output module

36. List one application for each of the following intelligent I/O modules:
a. PID module
b. Motion and position control module
c. Communication module

37. Write a short explanation for each of the following discrete I/O module specifications:
a. Nominal input voltage
b. Input threshold voltages
c. Nominal current per input
d. Ambient temperature rating
e. Input ON/OFF delay
f. Output voltage
g. Output current
h. Inrush current
i. Short circuit protection
j. Leakage current
k. Electrical isolation
l. Points per module
m. Backplane current draw

38. Write a short explanation for each of the following analog I/O module specifications:
a. Channels per module
b. Input current/voltage range(s)
c. Output current/voltage range(s)
d. Input protection
e. Resolution
f. Input impedance and capacitance
g. Common-mode rejection

39. Compare the function of the CPU and memory sections of a PLC processor.

40. With reference to the PLC chassis power supply:
a. What conversion of power takes place within the power supply circuit?
b. Explain the term *hold-up time* as it applies to the power supply.

41. Explain the purpose of a redundant PLC processor.

42. Describe three typical modes of operation that can be selected by the keyswitch of a processor.

43. State five other functions, in addition to simple logic processing, that PLC processors are capable of performing.

44. List five important procedures to follow when handling static-sensitive PLC components.

45. Define each of the following terms as they apply to the memory element of a PLC:
a. writing
b. reading
c. bits
d. location
e. utilization

46. With reference to the I/O image tables:
a. What information is stored in PLC input and output tables?
b. What is the input status of a closed switch stored as?
c. What is the input status of an open switch stored as?
d. What is the status of an output that is ON stored as?
e. What is the status of an output that is OFF stored as?

47. Why do PLCs execute memory-checking routines?

48. Compare the memory storage characteristics of volatile and nonvolatile memory elements.

49. What information is normally stored in the ROM memory of a PLC?

50. What information is normally stored in the RAM memory of a PLC?

51. What information is normally stored in an EEPROM memory module?

52. What are the advantages of a processor that utilizes a flash memory card?

53. List three functions of a PLC programming terminal device.

54. Give one advantage and one limitation to the use of hand-held programming devices.

55. What is required for a personal computer to be used as a PLC programming terminal?

56. List four important capabilities of PLC programming software.

57. How many programs can a PLC have stored in memory at any one time?

58. Outline four functions that an HMI interface screen can be configured to perform.

1. A discrete 120 VAC output module is to be used to control a 230 VDC solenoid valve. Draw a diagram showing how this could be accomplished using an interposing relay.

2. Assume a thermocouple, which supplies the input to an analog input module, generates a linear voltage of from 20 mV to 50 mV when the temperature changes from 750°F to 1250°F. How much voltage will be generated when the temperature of the thermocouple is at 1000°F?

3. With reference to I/O module specifications:
   a. If the ON-delay time of a given discrete input module is specified as 12 milliseconds, how much is this expressed in seconds?
   b. If the output leakage current of a discrete output module is specified as 950 μA, how much is this expressed in amperes?

   c. If the ambient temperature rating for an I/O module is specified as 60°C, how much is this expressed in degrees Fahrenheit?

4. Create a five-digit code using the SLC 500 rack/slot-based addressing format for each of the following:
   a. A pushbutton connected to terminal 5 of module group 2 located on rack 1.
   b. A lamp connected to terminal 3 of module group 0 located on rack 2.

5. Assume the triac of an AC discrete output module fails in the shorted state. How would this affect the device connected to this output?

6. A personal computer is to be used to program several different PLCs from different manufacturers. What would be required?

3
Number Systems and Codes

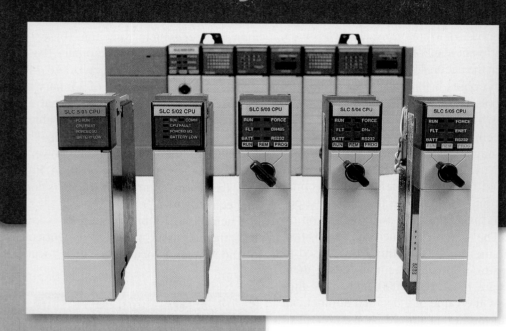

Image Used with Permission of Rockwell Automation, Inc.

Chapter Objectives

After completing this chapter, you will be able to:

3.1 Define the decimal, binary, octal, and hexadecimal numbering systems and be able to convert from one numbering or coding system to another

3.2 Explain the BCD, Gray, and ASCII code systems

3.3 Define the terms *bit, byte, word, least significant bit (LSB),* and *most significant bit (MSB)* as they apply to binary memory locations

3.4 Add, subtract, multiply, and divide binary numbers

Using PLCs requires us to become familiar with other number systems besides decimal. Some PLC models and individual PLC functions use other numbering systems. This chapter deals with some of these numbering systems, including binary, octal, hexadecimal, BCD, Gray, and ASCII. The basics of each system, as well as conversion from one system to another, are explained.

3.1 Decimal System

Knowledge of different number systems and digital codes is quite useful when working with PLCs or with most any type of digital computer. This is true because a basic requirement of these devices is to represent, store, and operate on numbers. In general, PLCs work on binary numbers in one form or another; these are used to represent various codes or quantities.

The *decimal system,* which is most common to us, has a base of 10. The radix or base of a number system determines the total number of different symbols or digits used by that system. For instance, in the decimal system, 10 unique numbers or digits—i.e., the digits 0 through 9—are used: the total number of symbols is the same as the base, and the symbol with the largest value is 1 less than the base.

The value of a decimal number depends on the digits that make up the number and the place value of each digit. A place (weight) value is assigned to each position that a digit would hold from right to left. In the decimal system the first position, starting from the rightmost position, is 0; the second is 1; the third is 2; and so on up to the last position. The weighted value of each position can be expressed as the base (10 in this case) raised to the power of the position. For the decimal system then, the position weights are 1, 10, 100, 1000, and so on. Figure 3-1 illustrates how the value of a decimal number can be calculated by multiplying each digit by the weight of its position and summing the results.

3.2 Binary System

The *binary system* uses the number 2 as the base. The only allowable digits are 0 and 1. With digital circuits it is easy to distinguish between two voltage levels (i.e., +5 V and 0 V), which can be related to the binary digits 1 and 0 (Figure 3-2). Therefore the binary system can be applied quite easily to PLCs and computer systems.

Since the binary system uses only two digits, each position of a binary number can go through only two

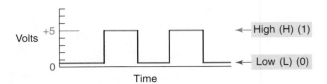

Figure 3-2 Digital signal waveform.

changes, and then a 1 is carried to the immediate left position. Table 3-1 shows a comparison among four common number systems: decimal (base 10), octal (base 8), hexadecimal (base 16), and binary (base 2). Note that all numbering systems start at *zero.*

The decimal equivalent of a binary number can be determined in a manner similar to that used for a decimal number. This time the weighted values of the positions are 1, 2, 4, 8, 16, 32, 64, and so on. The weighted value, instead of being 10 raised to the power of the position, is 2 raised to the power of the position. Figure 3-3 illustrates how the binary number 10101101 is converted to its decimal equivalent: 173.

Each digit of a binary number is known as a *bit.* In a PLC the processor-memory element consists of hundreds or thousands of locations. These locations, or registers,

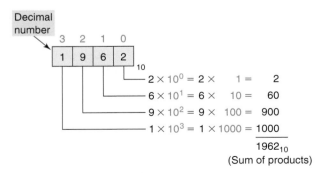

Figure 3-1 Weighted value in the decimal system.

Table 3-1 Number System Comparisons			
Decimal	Octal	Hexadecimal	Binary
0	0	0	0
1	1	1	1
2	2	2	10
3	3	3	11
4	4	4	100
5	5	5	101
6	6	6	110
7	7	7	111
8	10	8	1000
9	11	9	1001
10	12	A	1010
11	13	B	1011
12	14	C	1100
13	15	D	1101
14	16	E	1110
15	17	F	1111
16	20	10	10000
17	21	11	10001
18	22	12	10010
19	23	13	10011
20	24	14	10100

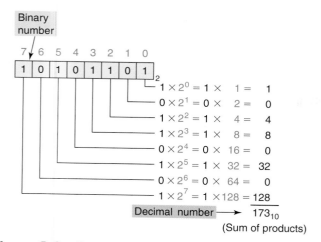

Figure 3-3 Converting a binary number to a decimal number.

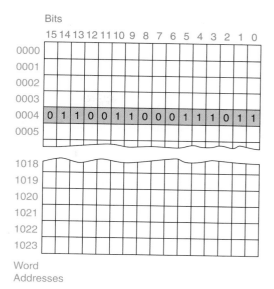

Figure 3-5 1-K word memory.

are referred to as words. Each *word* is capable of storing data in the form of binary digits, or bits. The number of bits that a word can store depends on the type of PLC system used. Sixteen-bit and 32-bit words are the most common. Bits can also be grouped within a word into bytes. A group of 8 bits is a byte, and a group of 2 or more bytes is a word. Figure 3-4 illustrates a 16-bit word made up of 2 bytes. The least significant bit (LSB) is the digit that represents the smallest value, and the most significant bit (MSB) is the digit that represents the largest value. A bit within the word can exist only in two states: a logical 1 (or ON) condition, or a logical 0 (or OFF) condition.

PLC memory is organized using bytes, single words, or double words. Older PLCs use 8-bit or 16-bit memory words while newer systems, such as the ControlLogix platform from Allen-Bradley, use 32-bit double words. The size of the programmable controller memory relates to the amount of user program that can be stored. If the memory size is 1 K word (Figure 3-5), it can store 1024 words or 16,384 (1024×16) bits of information using 16-bit words, or 32,768 (1024×32) bits using 32-bit words.

To convert a decimal number to its binary equivalent, we must perform a series of divisions by 2. Figure 3-6 illustrates the conversion of the decimal number 47 to binary. We start by dividing the decimal number by 2. If there is a remainder, it is placed in the LSB of the binary

number. If there is no remainder, a 0 is placed in the LSB. The result of the division is brought down and the process is repeated until the result of successive divisions has been reduced to 0.

Even though the binary system has only two digits, it can be used to represent any quantity that can be represented in the decimal system. All PLCs work internally in the binary system. The processor, being a digital device, understands only 0s and 1s, or binary.

Computer memory is, then, a series of binary 1s and 0s. Figure 3-7 shows the output status file for an Allen-Bradley SLC 500 modular chassis, which is made up of single bits grouped into 16-bit words. One 16-bit output file word is reserved for each slot in the chassis. Each bit represents the ON or OFF state of one output point. These points are numbered 0 through 15 across the top row from

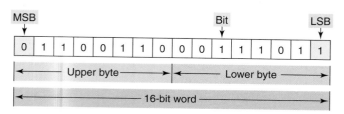

Figure 3-4 A 16-bit word.

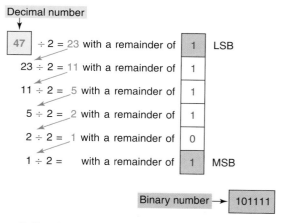

Figure 3-6 Converting a decimal number to a binary number.

15	14	13	12	11	10	9	8	7	6	5	4	3	2	1	0	Address
1	1	0	0	0	0	1	0	1	1	1	1	0	0	0	1	O:1
0	0	1	1	0	1	0	0	0	0	0	0	1	1	1	1	O:2
1	0	1	0	1	1	0	0	1	1	1	0	0	0	0	1	O:3
0	0	0	0	0	0	0	0	0	1	0	0	1	0	0	0	O:4
1	1	1	0	1	0	0	1	1	1	0	0	1	1	0	1	O:5

Figure 3-7 SLC 500 output status file.

right to left. The column on the far right lists the output module address. Although the table in Figure 3-7 illustrates sequentially addressed output status file words, in reality a word is created in the table only if the processor finds an output module residing in a particular slot. If the slot is empty, no word will be created.

3.3 Negative Numbers

If a decimal number is positive, it has a plus sign; if a number is negative, it has a minus sign. In binary number systems, such as used in a PLC, it is not possible to use positive and negative symbols to represent the polarity of a number. One method of representing a binary number as either a positive or negative value is to use an extra digit, or *sign bit,* at the MSB side of the number. In the sign bit position, a 0 indicates that the number is positive, and a 1 indicates a negative number (Table 3-2).

Another method of expressing a negative number in a digital system is by using the complement of a binary number. To complement a binary number, change all the 1s to 0s and all the 0s to 1s. This is known as the 1's complement form of a binary number. For example, the 1's complement of 1001 is 0110.

The most common way to express a negative binary number is to show it as a 2's complement number. The 2's complement is the binary number that results when 1 is added to the 1's complement. This system is shown in Table 3-3. A zero sign bit means a positive number, whereas a 1 sign bit means a negative number.

Using the 2's complement makes it easier for the PLC to perform mathematical operations. The correct sign bit is generated by forming the 2's complement. The PLC knows that a number retrieved from memory is a negative number if the MSB is 1. Whenever a negative number is entered from a keyboard, the PLC stores it as a 2's complement. What follows is the original number in true binary followed by its 1's complement, its 2's complement, and finally, its decimal equivalent.

Table 3-2 Signed Binary Numbers

Magnitude ———┐
Sign ——┐

Binary	Decimal Value
0111	+7
0110	+6
0101	+5
0100	+4
0011	+3
0010	+2
0001	+1
0000	0
1001	−1
1010	−2
1011	−3
1100	−4
1101	−5
1110	−6
1111	−7

Same as binary numbers

Table 3-3 1's and 2's Complement Representation of Positive and Negative Numbers

Signed Decimal	1's Complement	2's Complement
+7	0111	0111
+6	0110	0110
+5	0101	0101
+4	0100	0100
+3	0011	0011
+2	0010	0010
+1	0001	0001
0	0000	0000
−1	1110	1111
−2	1101	1110
−3	1100	1101
−4	1011	1100
−5	1010	1011
−6	1001	1010
−7	1000	1001

Same as binary numbers

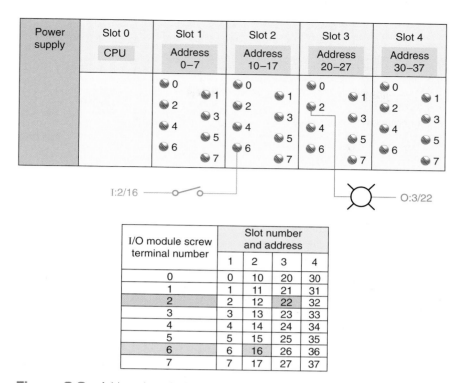

I/O module screw terminal number	Slot number and address			
	1	2	3	4
0	0	10	20	30
1	1	11	21	31
2	2	12	22	32
3	3	13	23	33
4	4	14	24	34
5	5	15	25	35
6	6	16	26	36
7	7	17	27	37

Figure 3-8   Addressing of I/O modules using the octal numbering system.

3.4   Octal System

To express the number in the binary system requires many more digits than in the decimal system. Too many binary digits can become cumbersome to read or write. To solve this problem, other related numbering systems are used.

The *octal numbering system,* a base 8 system, is used because 8 data bits make up a byte of information that can be addressed. Figure 3-8 illustrates the addressing of I/O modules using the octal numbering system. The digits range from 0 to 7; therefore, numbers 8 and 9 are not allowed. Allen-Bradley PLC-5 processors use octal-based I/O addressing while the SLC 500 and Logix controllers use decimal-base 10 addressing.

Octal is a convenient means of handling large binary numbers. As shown in Table 3-4, one octal digit can be used to express three binary digits. As in all other numbering systems, each digit in an octal number has a weighted decimal value according to its position. Figure 3-9 illustrates how the octal number 462 is converted to its decimal equivalent: 306.

Octal converts easily to binary equivalents. For example, the octal number 462 is converted to its binary equivalent by assembling the 3-bit groups, as illustrated in Figure 3-10. Notice the simplicity of the notation: the octal 462 is much easier to read and write than its binary equivalent is.

Table 3-4 Binary and Related Octal Code

Binary	Octal
000	0
001	1
010	2
011	3
100	4
101	5
110	6
111	7

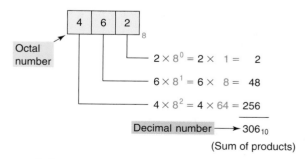

Figure 3-9   Converting an octal number to a decimal number.

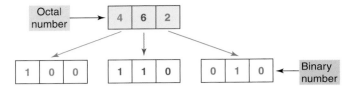

Figure 3-10 Converting an octal number to a binary number.

3.5 Hexadecimal System

The hexadecimal (hex) numbering system is used in programmable controllers because a word of data consists of 16 data bits, or two 8-bit bytes. The hexadecimal system is a base 16 system, with A to F used to represent decimal numbers 10 to 15 (Table 3-5). The hexadecimal numbering system allows the status of a large number of binary bits to be represented in a small space, such as on a computer screen or PLC programming device display.

The techniques used when converting hexadecimal to decimal and decimal to hexadecimal are the same as those used for binary and octal. To convert a hexadecimal number to its decimal equivalent, the hexadecimal digits in the columns are multiplied by the base 16 weight, depending on digit significance. Figure 3-11 illustrates how the conversion would be done for the hex number 1B7.

Hexadecimal numbers can easily be converted to binary numbers. Conversion is accomplished by writing the

Table 3-5 Hexadecimal Numbering System		
Hexadecimal	**Binary**	**Decimal**
0	0000	0
1	0001	1
2	0010	2
3	0011	3
4	0100	4
5	0101	5
6	0110	6
7	0111	7
8	1000	8
9	1001	9
A	1010	10
B	1011	11
C	1100	12
D	1101	13
E	1110	14
F	1111	15

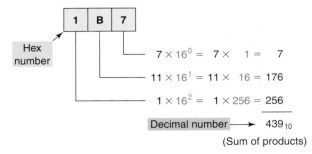

Figure 3-11 Converting a hexadecimal number to a decimal number.

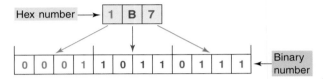

Figure 3-12 Converting a hexadecimal number to a binary number.

4-bit binary equivalent of the hex digit for each position, as illustrated in Figure 3-12.

3.6 Binary Coded Decimal (BCD) System

The binary coded decimal (BCD) system provides a convenient way of handling large numbers that need to be input to or output from a PLC. As you can see from looking at the various number systems, there is no easy way to go from binary to decimal and back. The BCD system provides a means of converting a code readily handled by humans (decimal) to a code readily handled by the equipment (binary). PLC thumbwheel switches and LED displays are examples of PLC devices that make use of the BCD number system. Table 3-6 shows examples of numeric values in decimal, binary, BCD, and hexadecimal representation.

The BCD system uses 4 bits to represent each decimal digit. The 4 bits used are the binary equivalents of the numbers from 0 to 9. In the BCD system, the largest decimal number that can be displayed by any four digits is 9.

The BCD representation of a decimal number is obtained by replacing each decimal digit by its BCD equivalent. To distinguish the BCD numbering system from a binary system, a BCD designation is placed to the right of the units digit. The BCD representation of the decimal number 7863 is shown in Figure 3-13.

A thumbwheel switch is one example of an input device that uses BCD. Figure 3-14 shows a single-digit BCD thumbwheel. The circuit board attached to the thumbwheel has one connection for each bit's weight plus

Table 3-6 Numeric Values in Decimal, Binary, BCD, and Hexadecimal Representation

Decimal	Binary	BCD	Hexadecimal
0	0	0000	0
1	1	0001	1
2	10	0010	2
3	11	0011	3
4	100	0100	4
5	101	0101	5
6	110	0110	6
7	111	0111	7
8	1000	1000	8
9	1001	1001	9
10	1010	0001 0000	A
11	1011	0001 0001	B
12	1100	0001 0010	C
13	1101	0001 0011	D
14	1110	0001 0100	E
15	1111	0001 0101	F
16	1 0000	0001 0110	10
17	1 0001	0001 0111	11
18	1 0010	0001 1000	12
19	1 0011	0001 1001	13
20	1 0100	0010 0000	14
126	111 1110	0001 0010 0110	7E
127	111 1111	0001 0010 0111	7F
128	1000 0000	0001 0010 1000	80
510	1 1111 1110	0101 0001 0000	1FE
511	1 1111 1111	0101 0001 0001	1FF
512	10 0000 0000	0101 0001 0010	200

a common connection. The operator dials in a decimal digit between 0 and 9, and the thumbwheel switch outputs the equivalent 4 bits of BCD data. In this example, the number eight is dialed to produce the input bit pattern of 1000. A four-digit thumbwheel switch, similar to the one shown, would control a total of 16 (4 × 4) PLC inputs.

Scientific calculators are available to convert numbers back and forth between decimal, binary, octal, and

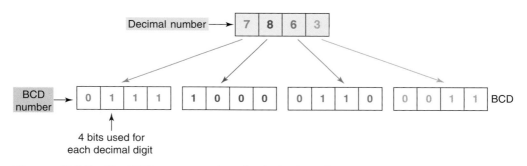

Figure 3-13 The BCD representation of a decimal number.

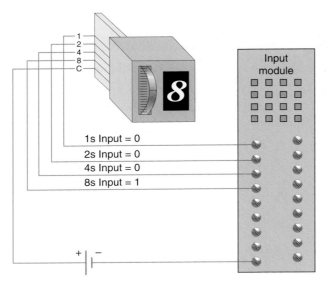

Figure 3-14 BCD thumbwheel switch interfaced to a PLC.

1s Input = 0
2s Input = 0
4s Input = 0
8s Input = 1

Table 3-7 Gray Code and Binary Equivalent

Table 3-7 Gray Code and Binary Equivalent

Gray Code	Binary
0000 0000
0001 0001
0011 0010
0010 0011
0110 0100
0111 0101
0101 0110
0100 0111
1100 1000
1101 1001
1111 1010
1110 1011
1010 1100
1011 1101
1001 1110
1000 1111

hexadecimal. In addition, PLCs contain number conversion functions such as illustrated in Figure 3-15. BCD-to-binary conversion is required for the input while binary-to-BCD conversion is required for the output. The PLC convert-to-decimal instruction will convert the binary bit pattern at the source address, N7:23, into a BCD bit pattern of the same decimal value as the destination address, O:20. The instruction executes every time it is scanned, and the instruction is true.

Many PLCs allow you to change the format of the data that the data monitor displays. For example, the change radix function found on Allen-Bradley controllers allows you to change the display format of data to binary, octal, decimal, hexadecimal, or ASCII.

3.7 Gray Code

The Gray code is a special type of binary code that does not use position weighting. In other words, each position does not have a definite weight. The Gray code is set up so that as we progress from one number to the next, only one bit changes. This can be quite confusing for counting circuits, but it is ideal for encoder circuits. For example, absolute encoders are position transducers that use the Gray code to determine angular position. The Gray code has the advantage that for each "count" (each transition from one number to the next) only one digit changes. Table 3-7 shows the Gray code and the binary equivalent for comparison.

In binary, as many as four digits could change for a single "count." For example, the transition from binary 0111 to 1000 (decimal 7 to 8) involves a change in all four digits. This kind of change increases the possibility for error in certain digital circuits. For this reason, the Gray code is considered to be an error-minimizing code. Because only one bit

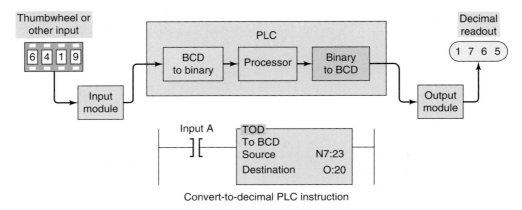

Figure 3-15 PLC number conversion.

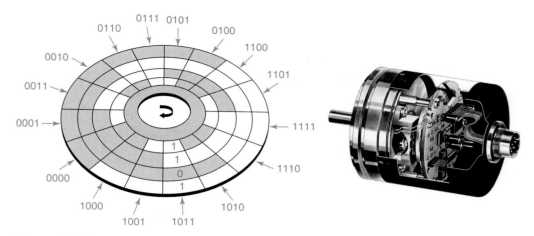

Figure 3-16 Optical encoder disk.
Source: Photo courtesy Baumer Electric.

changes at a time, the speed of transition for the Gray code is considerably faster than that for codes such as BCD.

Gray codes are used with position encoders for accurate control of the motion of robots, machine tools, and servomechanisms. Figure 3-16 shows an optical encoder disk that uses a 4-bit Gray code to detect changes in angular position. In this example, the encoder disk is attached to a rotating shaft and outputs a digital Gray code signal that is used to determine the position of the shaft. A fixed array of photo diodes senses the reflected light from each of the cells across a row of the encoder path. Depending on the amount of light reflected, each cell will output a voltage corresponding to a binary 1 or 0. Thus, a different 4-bit word is generated for each row of the disk.

3.8 ASCII Code

ASCII stands for American Standard Code for Information Interchange. It is an alphanumeric code because it includes letters as well as numbers. The characters accessed by the ASCII code include 10 numeric digits; 26 lowercase and 26 uppercase letters of the alphabet; and about 25 special characters, including those found on a standard typewriter. Table 3-8 shows a partial listing of the ASCII code. It is used to interface the PLC CPU with alphanumeric keyboards and printers.

The keystrokes on the keyboard of a computer are converted directly into ASCII for processing by the computer. Each time you press a key on a computer keyboard, a 7- or 8-bit word is stored in computer memory to represent the alphanumeric, function, or control data represented by the specific keyboard key that was depressed. ASCII input modules convert ASCII code input information from an external device to alphanumeric information that the PLC can process. The communication interfacing is done through either an RS-232 or RS-422 protocol. Modules

are available that will transmit and receive ASCII files and that can be used to create an operator interface. The user writes a program in the BASIC language that operates in conjunction with the ladder logic as the program runs.

3.9 Parity Bit

Some PLC communication systems use a binary digit to check the accuracy of data transmission. For example, when data are transferred between PLCs, one of the binary digits may be accidentally changed from a 1 to a 0. This can happen because of a transient or a noise or because of a failure in some portion of the transmission network. A parity bit is used to detect errors that may occur while a word is moved.

Parity is a system in which each character transmitted contains one additional bit. That bit is known as a parity bit. The bit may be a binary 0 or binary 1, depending on the number of 1s and 0s in the character itself. Two systems of parity are normally used: odd and even. Odd parity means that the total number of binary 1 bits in the character, including the parity bit, is odd. Even parity means that the number of binary 1 bits in the character, including the parity bit, is even. Examples of odd and even parity are shown in Table 3-9.

3.10 Binary Arithmetic

Arithmetic circuit units form a part of the CPU. Mathematical operations include addition, subtraction, multiplication, and division. Binary addition follows rules similar to decimal addition. When adding with binary numbers, there are only four conditions that can occur:

$$
\begin{array}{cccc}
0 & 1 & 0 & 1 \\
+0 & +0 & +1 & +1 \\
\hline
0 & 1 & 1 & 0 \text{ carry } 1
\end{array}
$$

Table 3-8 Partial Listing of ASCII Code

Character	7-Bit ASCII	Character	7-Bit ASCII
A	100 0001	X	101 1000
B	100 0010	Y	101 1001
C	100 0011	Z	101 1010
D	100 0100	0	011 0000
E	100 0101	1	011 0001
F	100 0110	2	011 0010
G	100 0111	3	011 0011
H	100 1000	4	011 0100
I	100 1001	5	011 0101
J	100 1010	6	011 0110
K	100 1011	7	011 0111
L	100 1100	8	011 1000
M	100 1101	9	011 1001
N	100 1110	blank	010 0000
O	100 1111	.	010 1110
P	101 0000	,	010 1100
Q	101 0001	+	010 1011
R	101 0010	−	010 1101
S	101 0011	#	010 0011
T	101 0100	(	010 1000
U	101 0101	%	010 0101
V	101 0110	=	011 1101
W	101 0111		

The first three conditions are easy because they are like adding decimals, but the last condition is slightly different. In decimal, $1 + 1 = 2$. In binary, a 2 is written 10. Therefore, in binary, $1 + 1 = 0$, with a carry of 1 to the next most significant place value. When adding larger binary numbers, the resulting 1s are carried into higher-order columns, as shown in the following examples.

Decimal	Equivalent binary
5	101
+2	+ 10
7	111

carry
1

10	10 | 10
+ 3	+ | 11
13	11 | 01

carry carry
1   1

26	1 | 1010
+12	+ | 1100
38	1 | 0 | 0110

In arithmetic functions, the initial numeric quantities that are to be combined by subtraction are the minuend

Table 3-9 Odd and Even Parity

Character	Even Parity Bit	Odd Parity Bit
0000	0	1
0001	1	0
0010	1	0
0011	0	1
0100	1	0
0101	0	1
0110	0	1
0111	1	0
1000	1	0
1001	0	1

and subtrahend. The result of the subtraction process is called the difference, represented as:

$$A \text{ (minuend)}$$
$$-B \text{ (subtrahend)}$$
$$C \text{ (difference)}$$

To subtract from larger binary numbers, subtract column by column, borrowing from the adjacent column when necessary. Remember that when borrowing from the adjacent column, there are now two digits, i.e., 0 borrow 1 gives 10.

EXAMPLE

Subtract 1001 from 1101.

$$\begin{array}{r} 1101 \\ -1001 \\ \hline 0100 \end{array}$$

Subtract 0111 from 1011.

$$\begin{array}{r} 1011 \\ -0111 \\ \hline 0100 \end{array}$$

Binary numbers can also be negative. The procedure for this calculation is identical to that of decimal numbers because the smaller value is subtracted from the larger value and a negative sign is placed in front of the result.

EXAMPLE

Subtract 111 from 100.

$$\begin{array}{r} 111 \\ -100 \\ \hline -011 \end{array}$$

Subtract 11011 from 10111.

$$\begin{array}{r} 11011 \\ -10111 \\ \hline -00100 \end{array}$$

There are other methods available for doing subtraction:

1's complement
2's complement

The procedure for subtracting numbers using the 1's complement is as follows:

Step 1 Change the subtrahend to 1's complement.

Step 2 Add the two numbers.

Step 3 Remove the last carry and add it to the number (end-around carry).

Decimal	Binary
10	1010
− 6	−0110
4	100

$$\begin{array}{r} 1010 \\ \xrightarrow{\text{1's complement}} \\ +1001 \\ \hline 10011 \\ \text{End-around carry} \quad +1 \\ \hline 100 \end{array}$$

When there is a carry at the end of the result, the result is positive. When there is no carry, then the result is negative and a minus sign has to be placed in front of it.

EXAMPLE

Subtract 11011 from 01101.

$$\begin{array}{r} 01101 \\ + \cancel{0}\,00100 \\ \hline 10001 \end{array}$$

The 1's complement

There is no carry, so we take the 1's complement and add the minus sign:

$$-01110$$

For subtraction using the 2's complement, the 2's complement is added instead of subtracting the numbers. In the result, if the carry is a 1, then the result is positive; if the carry is a 0, then the result is negative and requires a minus sign.

EXAMPLE

Subtract 101 from 111.

$$\begin{array}{r} 111 \\ + \cancel{0}\,011 \\ \hline 1010 \end{array}$$

The 2's complement

The first 1 indicates that the result is positive, so it is disregarded:

$$010$$

Subtract 101 from 111.

$$
\begin{array}{r}
111 \\
+ \ \cancel{\emptyset}\,011 \\
\hline
1010
\end{array}
$$

+ ⊘011   The 2's complement

1010   The first 1 indicates that the result is positive, so it is disregarded:

010

Binary numbers are multiplied in the same manner as decimal numbers. When multiplying binary numbers, there are only four conditions that can occur:

$$0 \times 0 = 0$$
$$0 \times 1 = 0$$
$$1 \times 0 = 0$$
$$1 \times 1 = 1$$

To multiply numbers with more than one digit, form partial products and add them together, as shown in the following example.

Decimal	Equivalent binary
5	101
×6	×110
30	000
	101
	101
	11110

The process for dividing one binary number by another is the same for both binary and decimal numbers, as shown in the following example.

Decimal	Equivalent binary
7	111
2)14	10)1110
	10
	11
	10
	10
	10
	00

The basic function of a comparator is to compare the relative magnitude of two quantities. PLC data comparison instructions are used to compare the data stored in two words (or registers). At times, devices may need to be controlled when they are less than, equal to, or greater than other data values or set points used in the application, such as timer and counter values. The basic compare instructions are as follows:

$$A = B \ (A \text{ equals } B)$$
$$A > B \ (A \text{ is greater than } B)$$
$$A < B \ (A \text{ is less than } B)$$

1. Convert each of the following binary numbers to decimal numbers:
   a. 10
   b. 100
   c. 111
   d. 1011
   e. 1100
   f. 10010
   g. 10101
   h. 11111
   i. 11001101
   j. 11100011

2. Convert each of the following decimal numbers to binary numbers:
   a. 7
   b. 19
   c. 28
   d. 46
   e. 57
   f. 86
   g. 94
   h. 112
   i. 148
   j. 230

3. Convert each of the following octal numbers to decimal numbers:
   a. 36
   b. 104
   c. 120
   d. 216
   e. 360
   f. 1516

4. Convert each of the following octal numbers to binary numbers:
   a. 74
   b. 130
   c. 250
   d. 1510
   e. 2551
   f. 2634

5. Convert each of the following hexadecimal numbers to decimal numbers:
   a. 5A
   b. C7
   c. 9B5
   d. 1A6

6. Convert each of the following hexadecimal numbers to binary numbers:
   a. 4C
   b. E8
   c. 6D2
   d. 31B

7. Convert each of the following decimal numbers to BCD:
   a. 146
   b. 389
   c. 1678
   d. 2502

8. What is the most important characteristic of the Gray code?

9. What makes the binary system so applicable to computer circuits?

10. Define the following as they apply to the binary memory locations or registers:
    a. Bit
    b. Byte
    c. Word
    d. LSB
    e. MSB

11. State the base used for each of the following number systems:
    a. Octal
    b. Decimal
    c. Binary
    d. Hexadecimal

12. Define the term *sign bit*.

13. Explain the difference between the 1's complement of a number and the 2's complement.

14. What is ASCII code?

15. Why are parity bits used?

16. Add the following binary numbers:
    a. 110 + 111
    b. 101 + 011
    c. 1100 + 1011

17. Subtract the following binary numbers:
    a. 1101 − 101
    b. 1001 − 110
    c. 10111 − 10010

18. Multiply the following binary numbers:
 a. 110 × 110
 b. 010 × 101
 c. 101 × 11

19. Divide the following unsigned binary numbers:
 a. 1010 ÷ 10
 b. 1100 ÷ 11
 c. 110110 ÷ 10

CHAPTER 3 PROBLEMS

1. The following binary PLC coded information is to be programmed using the hexadecimal code. Convert each piece of binary information to the appropriate hexadecimal code for entry into the PLC from the keyboard.
 a. 0001 1111
 b. 0010 0101
 c. 0100 1110
 d. 0011 1001

2. The encoder circuit shown in Figure 3-17 is used to convert the decimal digits on the keyboard to a binary code. State the output status (HIGH/LOW) of A-B-C-D when decimal number
 a. 2 is pressed.
 b. 5 is pressed.
 c. 7 is pressed.
 d. 8 is pressed.

3. If the bits of a 16-bit word or register are numbered according to the octal numbering system, beginning with 00, what consecutive numbers would be used to represent each of the bits?

4. Express the decimal number 18 in each of the following number codes:
 a. Binary
 b. Octal
 c. Hexadecimal
 d. BCD

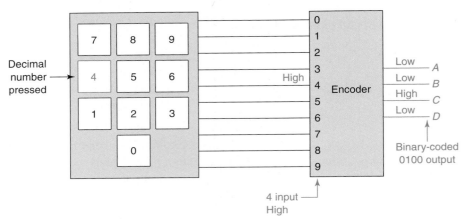

Figure 3-17 Diagram for Problem 2.

4

Fundamentals of Logic

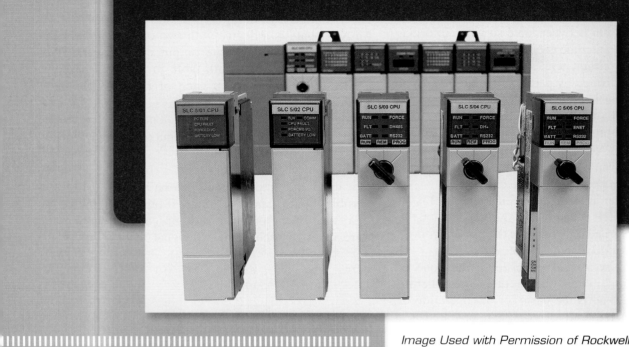

Chapter Objectives

After completing this chapter, you will be able to:

4.1 Describe the binary concept and the functions of gates

4.2 Draw the logic symbol, construct a truth table, and state the Boolean equation for the AND, OR, and NOT functions

4.3 Construct circuits from Boolean expressions and derive Boolean equations for given logic circuits

4.4 Convert relay ladder schematics to ladder logic programs

4.5 Develop elementary programs based on logic gate functions

4.6 Program instructions that perform logical operations

This chapter gives an overview of digital logic gates and illustrates how to duplicate this type of control on a PLC. Boolean algebra, which is a shorthand way of writing digital gate diagrams, is discussed briefly. Some small hand-held programmers have digital logic keys, such as AND, OR, and NOT, and are programmed using Boolean expressions.

4.1 The Binary Concept

The PLC, like all digital equipment, operates on the binary principle. The term *binary principle* refers to the idea that many things can be thought of as existing in only one of two states. These states are 1 and 0. The 1 and 0 can represent ON or OFF, open or closed, true or false, high or low, or any other two conditions. The key to the speed and accuracy with which binary information can be processed is that there are only two states, each of which is distinctly different. There is no in-between state so when information is processed the outcome is either yes or no.

A *logic gate* is a circuit with several inputs but only one output that is activated by particular combinations of input conditions. The two-state binary concept, applied to gates, can be the basis for making decisions. The high beam automobile lighting circuit of Figure 4-1 is an example of a logical AND decision. For this application, the high beam light can be turned on only when the light switch AND the high beam switch are closed.

The dome light automobile circuit of Figure 4-2 is an example of a logical OR decision. For this application, the dome light will be turned on whenever the passenger door switch OR the driver door switch is activated.

Logic is the ability to make decisions when one or more different factors must be taken into account before an action is taken. This is the basis for the operation of the PLC, where it is required for a device to operate when certain conditions have been met.

4.2 AND, OR, and NOT Functions

The operations performed by digital equipment are based on three fundamental logic functions: AND, OR, and NOT. Each function has a rule that will determine the outcome and a *symbol* that represents the operation. For the purpose of this discussion, the outcome or output is called Y and the signal inputs are called *A, B, C,* and so on. Also, binary 1 represents the presence of a signal or the occurrence of some event, and binary 0 represents the absence of the signal or nonoccurrence of the event.

The AND Function

The symbol drawn in Figure 4-3 is that of an AND gate. An AND gate is a device with two or more inputs and one output. The AND gate output is 1 only if all inputs are 1. The AND truth table in Figure 4-3 shows the resulting output from each of the possible input combinations.

Logic gate *truth tables* show each possible input to the gate or circuit and the resultant output depending upon the combination of the input(s).

Since logic gates are digital ICs (Integrated Circuits) their input and output signals can be in only one of two possible digital states, i.e., logic 0 or logic 1. Thus, the logic state of the output of a logic gate depends on the logic states of each of its individual inputs. Figure 4-4

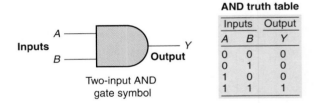

AND truth table

Inputs		Output
A	B	Y
0	0	0
0	1	0
1	0	0
1	1	1

Figure 4-3 AND gate.

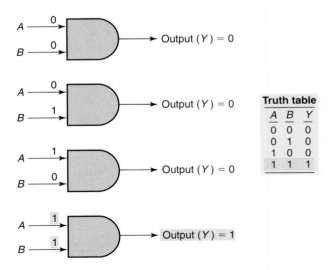

Truth table

A	B	Y
0	0	0
0	1	0
1	0	0
1	1	1

Figure 4-4 AND logic gate digital signal states.

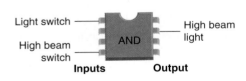

Figure 4-1 The logical AND.

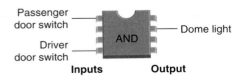

Figure 4-2 The logical OR.

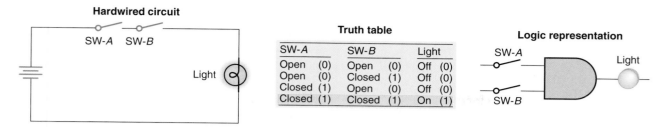

Figure 4-5   AND logic gate operates similarly to control devices connected in series.

illustrates the four possible combinations of inputs for a 2-input AND gate. The basic rules that apply to an AND gate are:

- If all inputs are 1, the output will be 1.
- If any input is 0, the output will be 0.

The AND logic gate operates similarly to control devices connected in *series,* as illustrated in Figure 4-5. The light will be on only when both switch A and switch B are closed.

The OR Function

The symbol drawn in Figure 4-6 is that of an OR gate. An OR gate can have any number of inputs but only one output. The OR gate output is 1 if one or more inputs are 1. The truth table in Figure 4-6 shows the resulting output Y from each possible input combination.

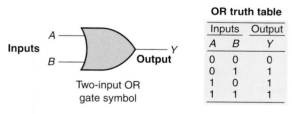

OR truth table

Inputs		Output
A	B	Y
0	0	0
0	1	1
1	0	1
1	1	1

Figure 4-6   OR gate.

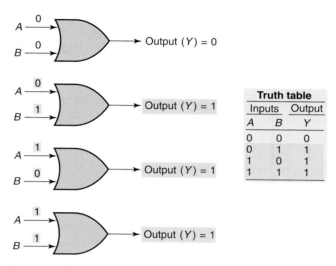

Truth table

Inputs		Output
A	B	Y
0	0	0
0	1	1
1	0	1
1	1	1

Figure 4-7   OR logic gate digital signal states.

Figure 4-7 illustrates the four possible combinations of inputs for a 2-input OR gate. The basic rules that apply to an OR gate are:

- If one or more inputs are 1, the output is 1.
- If all inputs are 0, the output will be 0.

The OR logic gate operates similarly to control devices connected in *parallel,* as illustrated in Figure 4-8. The light will be on if switch A or switch B or both are closed.

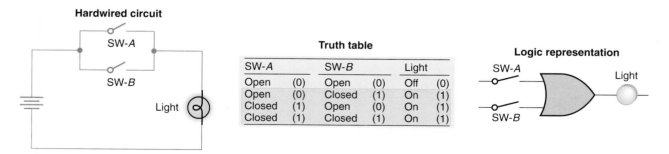

Figure 4-8   OR logic gate operates similarly to control devices connected in parallel.

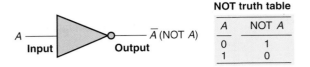

NOT truth table	
A	NOT A
0	1
1	0

Figure 4-9   NOT function.

The NOT Function

The symbol drawn in Figure 4-9 is that of a NOT function. Unlike the AND and OR functions, the NOT function can have *only one* input. The NOT output is 1 if the input is 0. The output is 0 if the input is 1. The result of the NOT operation is always the inverse of the input, and the NOT function is, therefore, called an *inverter.* The NOT function is often depicted by using a bar across the top of the letter, indicating an inverted output. The small circle at the output of the inverter is termed a state indicator and indicates that an inversion of the logical function has taken place.

The logical NOT function can be performed on a contact input simply by using a normally closed instead of a normally open contact. Figure 4-10 shows an example

of the NOT function constructed using a normally closed pushbutton in series with a lamp. When the input pushbutton is *not* actuated, the output lamp is ON. When the input pushbutton is actuated, the output lamp switches OFF.

The NOT function is most often used in conjunction with the AND or the OR gate. Figure 4-11 shows the NOT function connected to one input of an AND gate for a low-pressure indicator circuit. If the power is on (1) and the pressure switch is not closed (0), the warning light will be on (1).

The NOT symbol placed at the output of an AND gate would invert the normal output result. An AND gate with an inverted output is called a *NAND* gate. The NAND gate symbol and truth table are shown in Figure 4-12. The NAND function is often used in integrated circuit logic arrays and can be used in programmable controllers to solve complex logic.

The same rule about inverting the normal output result applies if a NOT symbol is placed at the output of the OR gate. The normal output is inverted, and the function is referred to as a *NOR* gate. The NOR gate symbol and truth table are shown in Figure 4-13.

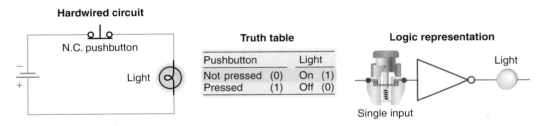

Figure 4-10   NOT function constructed using a normally closed pushbutton.

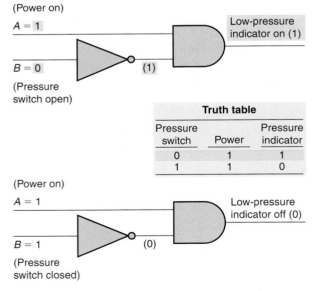

Figure 4-11   NOT function is most often used in conjunction with an AND gate.

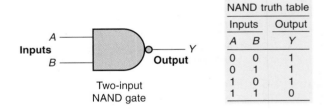

Figure 4-12   NAND gate symbol and truth table.

NAND truth table		
Inputs		Output
A	B	Y
0	0	1
0	1	1
1	0	1
1	1	0

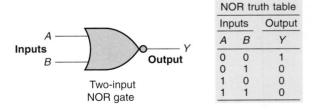

Figure 4-13   NOR gate symbol and truth table.

NOR truth table		
Inputs		Output
A	B	Y
0	0	1
0	1	0
1	0	0
1	1	0

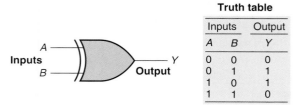

Truth table

Inputs		Output
A	B	Y
0	0	0
0	1	1
1	0	1
1	1	0

Figure 4.14 The XOR gate symbol and truth table.

The Exclusive-OR (XOR) Function

An often-used combination of gates is the exclusive-OR (XOR) function. The *XOR* gate symbol and truth table are shown in Figure 4-14. The output of this circuit is HIGH only when one input or the other is HIGH, but not both. The exclusive-OR gate is commonly used for the *comparison* of two binary numbers.

4.3 Boolean Algebra

The mathematical study of the binary number system and logic is called *Boolean algebra.* The purpose of this algebra is to provide a simple way of writing complicated combinations of logic statements. There are many applications where Boolean algebra could be applied to solving PLC programming problems.

A typical Boolean instruction list (also referred to as a statement list) is shown in Table 4-1. The instructions

Table 4-1 Typical Boolean Instruction or Statement List

Boolean Instruction and Function	Graphic Symbol
Store (STR)–Load (LD) Begins a new rung or an additional branch in a rung with a normally open contact.	—‖—
Store Not (STR NOT)–Load Not (LD NOT) Begins a new rung or an additional branch in a rung with a normally closed contact.	—⫽—
Or (OR) Logically ORs a normally open contact in parallel with another contact in a rung.	—‖⌐ —‖—
Or Not (OR NOT) Logically ORs a normally closed contact in parallel with another contact in a rung.	—‖⌐ —⫽—
And (AND) Logically ANDs a normally open contact in series with another contact in a rung.	—‖—‖—
And Not (AND NOT) Logically ANDs a normally closed contact in series with another contact in a rung.	—‖— ⫽—
And Store (AND STR)–And Load (AND LD) Logically ANDs two branches of a rung in series.	—‖⌐‖⌐ —‖—‖—
Or Store (OR STR)–Or Load (OR LOAD) Logically ORs two branches of a rung in parallel.	—‖—‖⌐ —‖—‖—
Out (OUT) Reflects the status of the rung (on/off) and outputs the discrete (ON/OFF) state to the specified image register point or memory location.	—(OUT)— —◯—
Or Out (OR OUT) Reflects the status of the rung and outputs the discrete (ON/OFF) state to the image register. Multiple OR OUT instructions referencing the same discrete point can be used in the program.	—(OROUT)—
Output Not (OUT NOT) Reflects the status of the rung and turns the output OFF for an ON execution condition; turns the output ON for an OFF execution condition.	—∅—

Logic symbol	Logic statement	Boolean equation	Boolean notations	
A, B — AND — Y	Y is 1 if A and B are 1	$Y = A \cdot B$ or $Y = AB$	Symbol	Meaning
			$\cdot$	and
A, B — OR — Y	Y is 1 if A or B is 1	$Y = A + B$	$+$	or
			$-$	not
			$\circ$	invert
A — NOT — Y	Y is 1 if A is 0 Y is 0 if A is 1	$Y = \overline{A}$	$=$	result in

Figure 4-15 Boolean algebra as related to AND, OR, and NOT functions.

are based on the basic Boolean operators of AND, OR, and NOT. Although these instructions are programmed in list format similar to BASIC and other text languages, they implement the same logic as relay ladder logic.

Figure 4-15 summarizes the basic operators of Boolean algebra as they relate to the basic AND, OR, and NOT functions. Inputs are represented by capital letters A, B, C, and so on, and the output by a capital Y. The multiplication sign ($\times$) or dot ($\cdot$) represents the AND operation, an addition sign ($+$) represents the OR operation, the circle with an addition sign $\oplus$ represents the exclusive-OR operation, and a bar over the letter $\overline{A}$ represents the NOT operation. The Boolean equations are used to express the mathematical function of the logic gate.

PLC digital systems may be designed using Boolean algebra. Circuit functions are represented by Boolean equations. Figure 4-16 illustrates how logic operators AND, NAND, OR, NOR, and NOT are used singly to form logical statements. Figure 4-17 illustrates how basic logic operators are used in combination to form Boolean equations.

An understanding of the technique of writing simplified Boolean equations for complex logical statements is a useful tool when creating PLC control programs. Some laws of Boolean algebra are different from those of ordinary algebra. These three basic laws illustrate the close comparison between Boolean algebra and

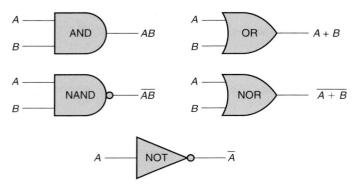

Figure 4-16 Logic operators used singly to form logical statements.

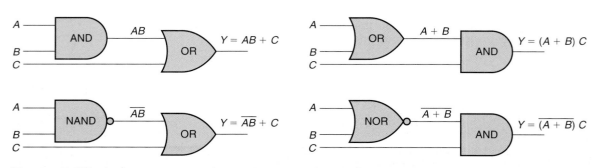

Figure 4-17 Logic operators used in combination to form Boolean equations.

ordinary algebra, as well as one major difference between the two:

COMMUTATIVE LAW

$A + B = B + A$

$A \cdot B = B \cdot A$

ASSOCIATIVE LAW

$(A + B) + C = A + (B + C)$

$(A \cdot B) \cdot C = A \cdot (B \cdot C)$

DISTRIBUTIVE LAW

$A \cdot (B + C) = (A \cdot B) + (A \cdot C)$

$A + (B \cdot C) = (A + B) \cdot (A + C)$

This law holds true only in Boolean algebra.

4.4 Developing Logic Gate Circuits from Boolean Expressions

As logic gate circuits become more complex, the need to express these circuits in Boolean form becomes greater. Figure 4-18 shows a logic gate circuit developed from the Boolean expression $Y = AB + C$. The procedure is as follows:

Boolean expression: $Y = AB + C$

Gates required: (by inspection)

   1 - AND gate with input A and B
   1 - OR gate with input C and output from previous AND gate

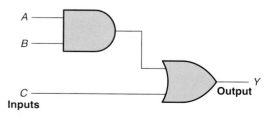

Figure 4-18 Logic gate circuit developed from the Boolean expression $Y = AB + C$.

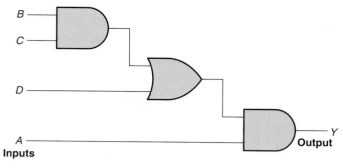

Figure 4-19 Logic gate circuit developed from the Boolean expression $Y = A(BC + D)$.

Figure 4-19 shows a logic gate circuit developed from the Boolean expression $Y = A(BC + D)$. The procedure is as follows:

Boolean expression: $Y = A(BC + D)$

Gates required: (by inspection)

   1 - AND gate with input B and C
   1 - OR gate with inputs B, C, and D
   1 - AND gate with inputs A and the output from the OR gate

4.5 Producing the Boolean Equation for a Given Logic Gate Circuit

A simple logic gate is quite straightforward in its operation. However, by grouping these gates into combinations, it becomes more difficult to determine which combinations of inputs will produce an output. The Boolean equation for the logic circuit of Figure 4-20 is determined as follows:

- The output of the OR gate is $A + B$
- The output of the inverter is $\overline{D}$
- Based on the input combination applied to the AND gate the Boolean equation for the circuit is $Y = C\,\overline{D}(A + B)$

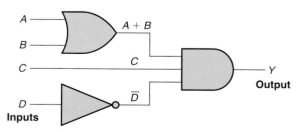

Figure 4-20 Determining the Boolean equation for a logic circuit.

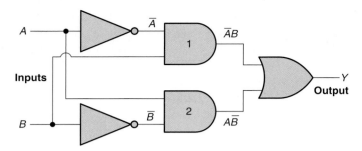

Figure 4-21 Determining the Boolean equation for a logic circuit.

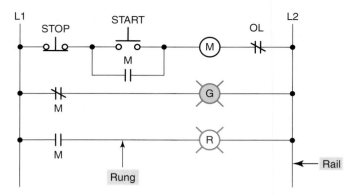

Figure 4-22 Motor stop/start relay ladder schematic.

The Boolean equation for the logic circuit of Figure 4-21 is determined as follows:

- The output of AND gate 1 is $\overline{A}B$
- The output of AND gate 2 is $A\overline{B}$
- Based on the combination of inputs applied to the OR gate the Boolean equation for the circuit is $Y = \overline{A}B + A\overline{B}$

4.6 Hardwired Logic versus Programmed Logic

The term *hardwired logic* refers to logic control functions that are determined by the way devices are electrically interconnected. Hardwired logic can be implemented using relays and relay ladder schematics. Relay ladder schematics are universally used and understood in industry. Figure 4-22 shows a typical relay ladder schematic of a motor stop/start control station with pilot lights. The control scheme is drawn between two vertical supply lines. All the components are placed between these two lines, called rails or legs, connecting the two power lines with what look like rungs of a ladder—thus the name, relay ladder schematic.

Hardwired logic is fixed; it is changeable only by altering the way devices are electrically interconnected. In contrast, programmable control is based on the basic logic functions, which are programmable and easily changed. These functions (AND, OR, NOT) are used either singly or in combinations to form instructions that will determine if a device is to be switched on or off. The form in which these instructions are implemented to convey commands to the PLC is called the *language*. The most common PLC language is *ladder logic*. Figure 4-23 shows a typical *ladder logic program* for the motor start/stop circuit. The instructions used are the relay equivalent of normally open (NO) and normally closed (NC) contacts and coils.

PLC contact symbolism is a simple way of expressing the control logic in terms of symbols. These symbols are

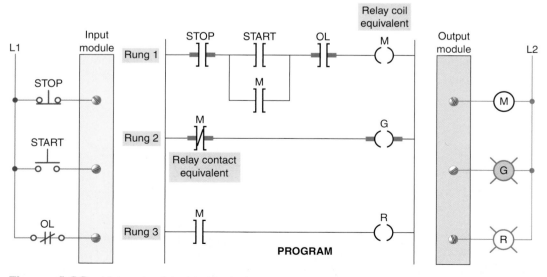

Figure 4-23 Motor stop/start ladder logic program.

basically the same as those used for representing hardwired relay control circuits. A rung is the contact symbolism required to control an output. Some PLCs allow a rung to have multiple outputs while others allow only one output per rung. A complete ladder logic program then consists of several rungs, each of which controls an output. In programmed logic all mechanical switch contacts are represented by a software contact symbol and all electromagnetic coils are represented by a software coil symbol.

Because the PLC uses ladder logic diagrams, the conversion from any existing relay logic to programmed logic is simplified. Each rung is a combination of input conditions (symbols) connected from left to right, with the symbol that represents the output at the far right. The symbols that represent the inputs are connected in series, parallel, or some combination of the two to obtain the desired logic. The following group of examples illustrates the relationship between the relay ladder schematic, the ladder logic program, and the equivalent logic gate circuit.

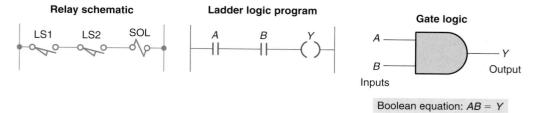

Example 4-1 Two limit switches connected in series and used to control a solenoid valve.

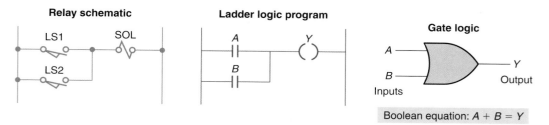

Example 4-2 Two limit switches connected in parallel and used to control a solenoid valve.

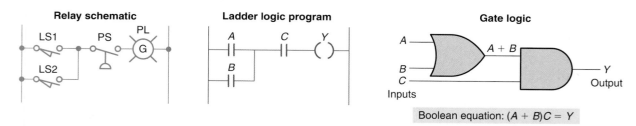

Example 4-3 Two limit switches connected in parallel with each other and in series with a pressure switch.

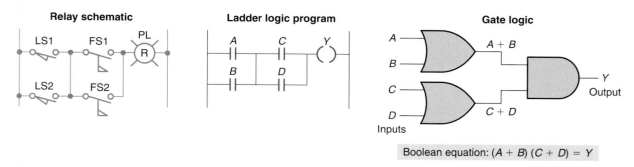

Example 4-4 Two limit switches connected in parallel with each other and in series with two sets of flow switches (that are connected in parallel with each other), and used to control a pilot light.

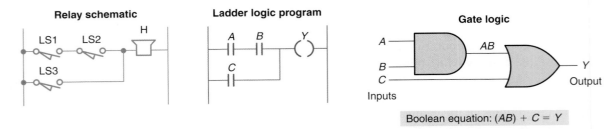

Example 4-5 Two limit switches connected in series with each other and in parallel with a third limit switch, and used to control a warning horn.

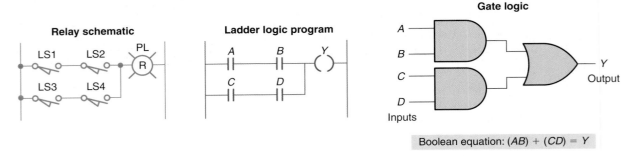

Example 4-6 Two limit switches connected in series with each other and in parallel with two other limit switches (that are connected in series with each other), and used to control a pilot light.

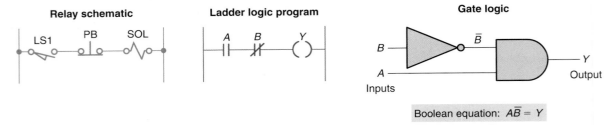

Example 4-7 One limit switch connected in series with a normally closed pushbutton and used to control a solenoid valve. This circuit is programmed so that the output solenoid will be turned on when the limit switch is closed and the pushbutton is *not pushed.*

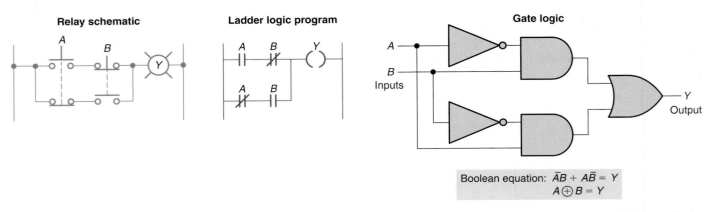

Example 4-8 Exclusive-OR circuit. The output lamp of this circuit is ON only when pushbutton A or B is pressed, but not both. This circuit has been programmed using only the normally open A and B pushbutton contacts as the inputs to the program.

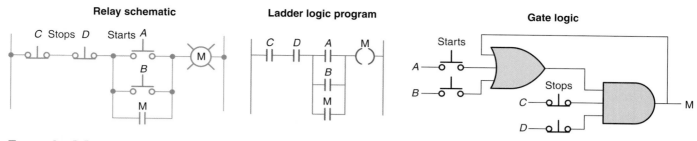

Relay schematic   Ladder logic program   Gate logic

Example 4-9 A motor control circuit with two start/stop buttons. When either start button is depressed, the motor runs. By use of a seal-in contact, it continues to run when the start button is released. Either stop button stops the motor when it is depressed.

4.7 Programming Word Level Logic Instructions

Most PLCs provide word-level logic instructions as part of their instruction set. Table 4-2 shows how to select the correct word logic instruction for different situations.

Figure 4-24 illustrates the operation of the AND instruction to perform a word-level AND operation using the bits in the two source addresses. This instruction tells the processor to perform an AND operation on B3:5 and

Table 4-2 Selecting Logic Instructions

If you want to . . .	. . . use this instruction.
Know when matching bits in two different words are both ON	AND
Know when one or both matching bits in two different words are ON	OR
Know when one or the other bit of matching bits in two different words is ON	XOR
Reverse the state of bits in a word	NOT

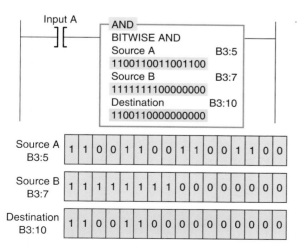

Figure 4-24 Word-level AND instruction.

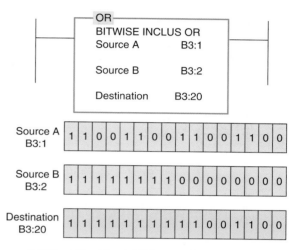

Figure 4-25 Word-level OR instruction.

B3:7 and to store the result in destination B3:10 when input device A is true. The destination bits are a result of the logical AND operation.

Figure 4-25 illustrates the operation of a word-level OR instruction, which ORs the data in Source A, bit by bit, with the data in Source B and stores the result at the destination address. The address of Source A is B3:1, the address of Source B is B3:2, and the destination address is B3:20. The instruction may be programmed conditionally, with input instruction(s) preceding it, or unconditionally, as shown, without any input instructions preceding it.

Figure 4-26 illustrates the operation of a word-level XOR instruction. In this example, data from input I:1.0 are compared, bit by bit, with data from input I:3.0. Any mismatches energize the corresponding bit in word O:4.0. As you can see, there is a 1 in every bit location in the destination corresponding to the bit locations where Source A and Source B are different, and a 0 in the destination where Source A and Source B are the same. The XOR is often used in diagnostics, where real-world inputs, such as rotary cam limit switches, are compared with their desired states.

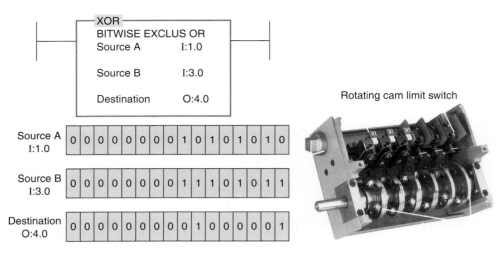

Rotating cam limit switch

```
┌─ XOR ──────────────────────┐
│  BITWISE EXCLUS OR         │
│  Source A          I:1.0   │
│                            │
│  Source B          I:3.0   │
│                            │
│  Destination       O:4.0   │
└────────────────────────────┘
```

Source A
I:1.0

| 0 | 0 | 0 | 0 | 0 | 0 | 0 | 0 | 1 | 0 | 1 | 0 | 1 | 0 | 1 | 0 |

Source B
I:3.0

| 0 | 0 | 0 | 0 | 0 | 0 | 0 | 0 | 1 | 1 | 1 | 0 | 1 | 0 | 1 | 1 |

Destination
O:4.0

| 0 | 0 | 0 | 0 | 0 | 0 | 0 | 0 | 0 | 1 | 0 | 0 | 0 | 0 | 0 | 1 |

Figure 4-26   Word-level XOR instruction.
Source: Image Used with Permission of Rockwell Automation, Inc.

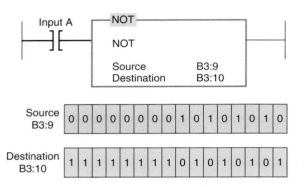

```
 Input A     ┌─ NOT ────────────────┐
──] [────────│  NOT                 │
             │                      │
             │  Source       B3:9   │
             │  Destination  B3:10  │
             └──────────────────────┘
```

Source
B3:9

| 0 | 0 | 0 | 0 | 0 | 0 | 0 | 0 | 1 | 0 | 1 | 0 | 1 | 0 | 1 | 0 |

Destination
B3:10

| 1 | 1 | 1 | 1 | 1 | 1 | 1 | 1 | 0 | 1 | 0 | 1 | 0 | 1 | 0 | 1 |

Figure 4-27   Word-level NOT operation.

Figure 4-27 illustrates the operation of a word-level NOT instruction. This instruction inverts the bits from the source word to the destination word. The bit pattern in B3:10 is the result of the instruction being true and is the inverse of the bit pattern in B3:9.

For 32-bit PLCs, such as the Allen-Bradley Control-Logix controller, the source and destination may be a SINT (one-byte integer), INT (two-byte integer), DINT (four-byte integer), or REAL (four-byte floating decimal point) value.

1. Explain the binary principle.
2. What is a logic gate?
3. Draw the logic symbol, construct a truth table, and state the Boolean equation for each of the following:
   a. Two-input AND gate
   b. NOT function
   c. Three-input OR gate
   d. XOR function
4. Express each of the following equations as a ladder logic program:
   a. $Y = (A + B)CD$
   b. $Y = A\overline{B}C + \overline{D} + E$
   c. $Y = [(\overline{A} + \overline{B})C] + DE$
   d. $Y = (\overline{ABC}) + (D\overline{E}F)$

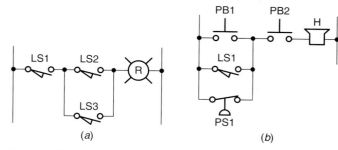

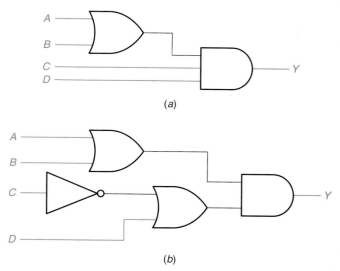

Figure 4-28   Question 5 relay ladder diagrams.

   d. $Y = \overline{A}(B + CD)$
   e. $Y = \overline{AB} + C$
   f. $Y = (ABC + D)(E\overline{F})$

5. Write the ladder logic program, draw the logic gate circuit, and state the Boolean equation for the two relay ladder diagrams in Figure 4-28.
6. Develop a logic gate circuit for each of the following Boolean expressions using AND, OR, and NOT gates:
   a. $Y = ABC + D$
   b. $Y = AB + CD$
   c. $Y = (A + B)(\overline{C} + D)$

7. State the logic instruction you would use when you want to:
   a. Know when one or both matching bits in two different words are 1.
   b. Reverse the state of bits in a word.
   c. Know when matching bits in two different words are both 1.
   d. Know when one or the other bit of matching bits, but not both, in two different words is 1.

1. It is required to have a pilot light come on when all of the following circuit requirements are met:
   • All four circuit pressure switches must be closed.
   • At least two out of three circuit limit switches must be closed.
   • The reset switch must not be closed.

   Using AND, OR, and NOT gates, design a logic circuit that will solve this hypothetical problem.
2. Write the Boolean equation for each of the logic gate circuits in Figure 4-29a–f.
3. The logic circuit of Figure 4-30 is used to activate an alarm when its output Y is logic **HIGH** or 1. Draw a truth table for the circuit showing the resulting output for all 16 of the possible input conditions.

Figure 4-29   Logic gate circuits for Problem 2.

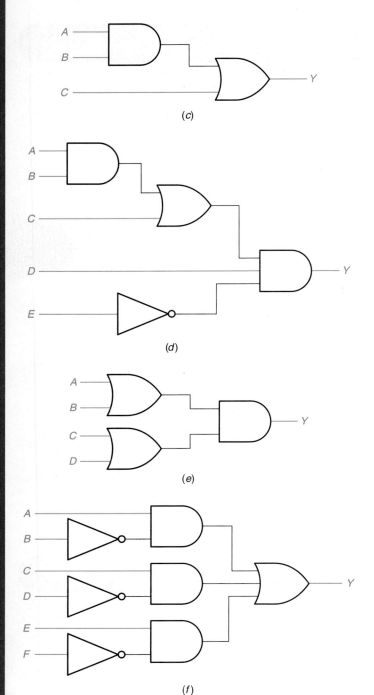

(c)

(d)

(e)

(f)

Figure 4-29 (*Continued*)

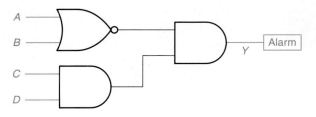

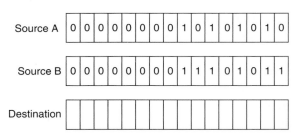

Figure 4-30 Logic circuit for Problem 3.

| Source A | 0 | 0 | 0 | 0 | 0 | 0 | 0 | 0 | 0 | 1 | 0 | 1 | 0 | 1 | 0 | 1 | 0 |

| Source B | 0 | 0 | 0 | 0 | 0 | 0 | 0 | 0 | 0 | 1 | 1 | 1 | 0 | 1 | 0 | 1 | 1 |

| Destination | | | | | | | | | | | | | | | | | | |

Figure 4-31 Data for Problem 4.

4. What will be the data stored in the destination address of Figure 4-31 for each of the following logical operations?
   a. AND operation
   b. OR operation
   c. XOR operation

5. Write the Boolean expression and draw the gate logic diagram and typical PLC ladder logic diagram for a control system wherein a fan is to run only when all of the following conditions are met:
   - Input A is OFF
   - Input B is ON or input C is ON, or both B and C are ON
   - Inputs D and E are both ON
   - One or more of inputs F, G, or H are ON

5

Basics of PLC Programming

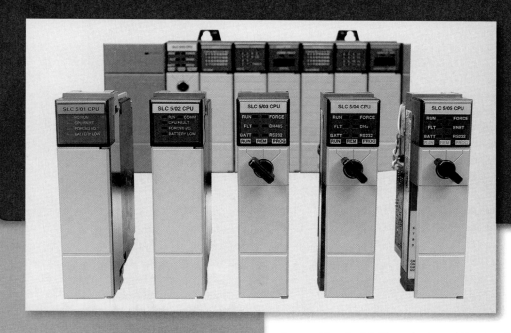

Image Used with Permission of Rockwell Automation, Inc.

Chapter Objectives

After completing this chapter, you will be able to:

5.1   Define and identify the functions of a PLC memory map

5.2   Describe input and output image table files and types of data files

5.3   Describe the PLC program scan sequence

5.4   Understand how ladder diagram language, Boolean language, and function chart programming language are used to communicate information to the PLC

5.5   Define and identify the function of internal relay instructions

5.6   Identify the common operating modes found in PLCs

5.7   Write and enter ladder logic programs

Each input and output PLC module terminal is identified by a unique address. In PLCs, the internal symbol for any input is a contact. Similarly, in most cases, the internal PLC symbol for all outputs is a coil. This chapter shows how these contact/coil functions are used to program a PLC for circuit operation. This chapter covers only the basic set of instructions that perform functions similar to relay functions. You will also learn more about the program scan cycle and the scan time of a PLC.

5.1 Processor Memory Organization

While the fundamental concepts of PLC programming are common to all manufacturers, differences in memory organization, I/O addressing, and instruction set mean that PLC programs are never perfectly interchangeable among different makers. Even within the same product line of a single manufacturer, different models may not be directly compatible.

The memory map or structure for a PLC processor consists of several areas, some of these having specific roles. Allen-Bradley PLCs have two different memory structures identified by the terms *rack-based* systems and *tag-based* systems. The memory organization for rack-based systems will be covered in this chapter and that for tag-based systems in a later chapter.

Memory organization takes into account the way a PLC divides the available memory into different sections. The memory space can be divided into two broad categories: *program files* and *data files*. Individual sections, their order, and the sections' length will vary and may be fixed or variable, depending on the manufacturer and model.

Program files are the part of the processor memory that stores the user ladder logic program. The program accounts for most of the total memory of a given PLC system. It contains the ladder logic that controls the machine operation. This logic consists of instructions that are programmed in a ladder logic format. Most instructions require one word of memory.

The data files store the information needed to carry out the user program. This includes information such as the status of input and output devices, timer and counter values, data storage, and so on. Contents of the data table can be divided into two categories: status data and numbers or codes. Status is ON/OFF type of information represented by 1s and 0s, stored in unique bit locations. Number or code information is represented by groups of bits that are stored in unique byte or word locations.

The memory organizations of the rack-based Allen-Bradley PLC-5 and SLC 500 controllers are very similar. Figure 5-1 shows the program and data file organization for the SLC 500 controller. The contents of each file are as follows.

Program Files

Program files are the areas of processor memory where ladder logic programming is stored. They may include:

- **System functions (file 0)**—This file is always included and contains various system-related information and user-programmed information such as processor type, I/O configuration, processor file name, and password.

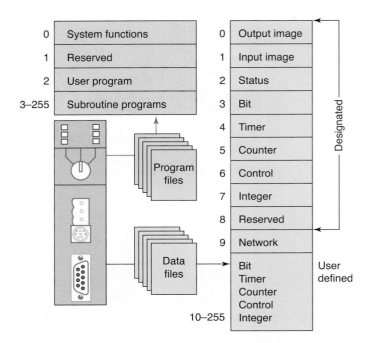

Figure 5-1 Program and data file organization for the SLC 500 controller.

- **Reserved (file 1)**—This file is reserved by the processor and is not accessible to the user.
- **Main ladder program (file 2)**—This file is always included and contains user-programmed instructions that define how the controller is to operate.
- **Subroutine ladder program (files 3–255)**—These files are user-created and are activated according to subroutine instructions residing in the main ladder program file.

Data Files

The data file portion of the processor's memory stores input and output status, processor status, the status of various bits, and numerical data. All this information is accessed via the ladder logic program. These files are organized by the type of data they contain and may include:

- **Output (file 0)**—This file stores the state of the output terminals for the controller.
- **Input (file 1)**—This file stores the status of the input terminals for the controller.
- **Status (file 2)**—This file stores controller operation information and is useful for troubleshooting controller and program operation.
- **Bit (file 3)**—This file is used for internal relay logic storage.
- **Timer (file 4)**—This file stores the timer accumulated and preset values and status bits.

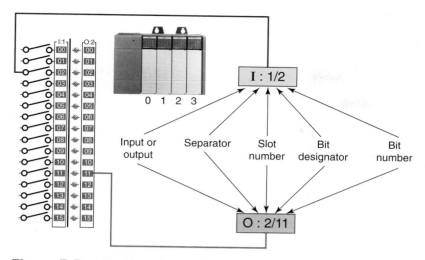

Figure 5-2 I/O address format for the SLC family of PLCs.
Source: Image Used with Permission of Rockwell Automation, Inc.

- **Counter (file 5)**—This file stores the counter accumulated and preset values and status bits.
- **Control (file 6)**—This file stores the length, pointer position, and status bit for specific instructions such as shift registers and sequencers.
- **Integer (file 7)**—This file is used to store numerical values or bit information.
- **Reserved (file 8)**—This file is not accessible to the user.
- **Network communications (file 9)**—This file is used for network communications if installed or used like files 10–255.
- **User-defined (files 10–255)**—These files are user-defined as bit, timer, counter, control, and/or integer data storage.

The I/O address format for the SLC family of PLCs is shown in Figure 5-2. The format consists of the following three parts:

Part 1: I for input, and a colon to separate the module type from the slot.

O for output and a colon to separate the module type from the slot.

Part 2: The module slot number and a forward slash to separate the slot from the terminal screw.

Part 3: The screw terminal number.

There are about 1000 program files for an Allen-Bradley PLC-5 controller. These program files may be set up in two ways: either (1) standard ladder logic programming, with the main program in program file 2 and program files 3 through 999 assigned, as needed, to subroutines; or (2) in sequential function charts in which files 2 through 999 are assigned steps or transitions, as required. With the

processor set up for standard ladder logic, the main program will always be in program file 2, and program files 3 through 999 will be subroutines. In either case, the processor can store and execute only one program at a time.

Figure 5-3 shows a typical data file memory organization for an Allen-Bradley PLC-5 controller. Each data file is made up of numerous *elements*. Each element may be one, two, or three words in length. Timer, counter, and control elements are three words in length; floating-point elements are two words in length; and all other elements are a single word in length. A *word* consists of 16 bits, or binary digits. The processor operates on two different data types: integer and floating point. All data types, except the floating-point files, are treated as integers or whole numbers. All element and bit addresses in the output and input data files are numbered octally. Element and bit addresses in all other data files are numbered decimally.

The PLC-5 and SLC 500 store all data in global data tables and are based on 16-bit operations. You access these data by specifying the address of the data you want. Typical addressing formats for the PLC-5 controller are as follows:

- The addresses in the output data file and the input data file are potential locations for either input modules or output modules mounted in the I/O chassis:
  - The address O:012/15 is in the output image table file, rack 1, I/O group 2, bit 15.
  - The address I:013/17 is in the input image table file, rack 1, I/O group 3, bit 17.
- The *status data file* contains information about the processor status:
  - The address S:015 addresses word 15 of the status file.
  - The address S:027/09 addresses bit 9 in word 27 of the status file.

Address range		Size, in elements
O:000 O:037	Output image file	32
I:000 I:037	Input image file	32
S:000 S:031	Processor status	32
B3:000 B3:999	Bit file	1–1000
T4:000 T4:999	Timer file	1–1000
C5:000 C5:999	Counter file	1–1000
R6:000 R6:999	Control file	1–1000
N7:000 N7:999	Integer file	1–1000
F8:000 F8:999	Floating-point file	1–1000
	Files to be assigned file nos. 9–999	1–1000 per file

Figure 5-3 Data file memory organization for an Allen-Bradley PLC-5 controller.
Source: Image Used with Permission of Rockwell Automation, Inc.

- The **bit data file** stores bit status. It frequently serves for storage when using internal outputs, sequencers, bit-shift instructions, and logical instructions:
  - The address B3:400 addresses word 400 of the bit file. The file number (3) must be included as part of the address. Note that the input, output, and status data files are the only files that do not require the file number designator because there can only be one input data, one output data, and one status data file.
  - Word 2, bit 15 is addressed as B3/47 because bit numbers are always measured from the beginning of the file. Remember that here, bits are numbered decimally (not octally, as the word representing the rack and slot).
- The **timer file** stores the timer status and timer data. A timer element consists of three words: the control word, preset word, and accumulated word. The addressing of the timer control word is the assigned timer number. Timers in file 4 are numbered starting with T4:0 and running through T4:999. The addresses for the three timer words in timer T4:0 are:

Control word:	T4:0
Preset word:	T4:0.PRE
Accumulated word:	T4:0.ACC

The enable-bit address in the control word is T4:0/EN, the timer-timing-bit address is T4:0/TT, and the done-bit address is T4:0/DN.

- The **counter file** stores the counter status and counter data. A counter element consists of three words: the control word, preset word, and accumulated word. The addressing of the counter control is the assigned counter number. Counters in file 5 are numbered beginning with C5:0 and running through C5:999. The addresses for the three counter words in counter C5:0 are:

Control word:	C5:0
Preset word:	C5:0.PRE
Accumulated word:	C5:0.ACC

The count-up-enable-bit address in the control word is C5:0/CU, the count-down-enable-bit address is C5:0/CD, the done-bit address is C5:0/DN, the overflow address is C5:0/OV, and the underflow address is C5:0/UN.

- The **control file** stores the control element's status and data, and it is used to control various file instructions. The control element consists of three words: the control word, length word, and position word. The addressing of the control's control word is the assigned control number. Control elements in control file 6 are numbered beginning with R6:0 and running through R6:999. The addresses for the three words in control element R6:0 are:

Control word:	R6:0
Length:	R6:0.LEN
Position:	R6:0.POS

There are numerous control bits in the control word, and their function depends on the instruction in which the control element is used.

- The *integer file* stores integer data values, with a range from −32,768 through 32,767. Stored values are displayed in decimal form. The integer element is a single-word (16-bit) element. As many as 1000 integer elements, addressed from N7:000 through N7:999, can be stored.
  - The address N7:100 addresses word 100 of the integer file.
  - Bit addressing is decimal, from 0 through 15. For example, bit 12 in word 15 is addressed N7:015/12.

- The *floating-point file* element can store values in the range from ±1.1754944e-38 to ±3.4028237e+38. The floating-point element is a two-word (32-bit) element. As many as 1000 elements, addressed from F8:000 through F8:999, can be stored. Individual words or bits cannot be addressed in the floating-point file.

- Data files 9 through 999 may be assigned to different data types, as required. When assigned to a certain type, a file is then reserved for that type and cannot be used for any other type. Additional input, output, or status files cannot be created.

The bit file, integer file, or floating-point file can be used to store status or data. Which of these you use depends on the intended use of the data. If you are dealing with status rather than data, the bit file is preferable. If you are using very large or very small numbers and require a decimal point, the floating-point file is preferable. The floating-point data type may have a restriction, however, because it may not interface well with external devices or with internal instructions such as counters and timers, which use only 16-bit words. In such a situation, it may be necessary to use the integer file type.

The *input image table file* is that part of the program memory allocated to storing the on/off status of connected discrete inputs. Figure 5-4 shows the connection of an open and closed switch to the input image table file

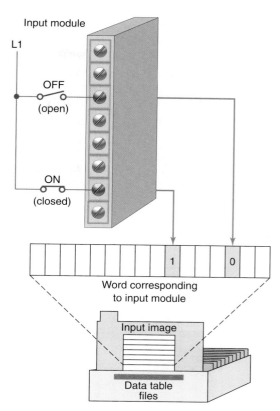

Figure 5-4 Connection of an open and closed switch to the input image table file through the input module.

through the input module. Its operation can be summarized as follows.

- For the switch that is closed, the processor detects a voltage at the input terminal and records that information by storing a binary 1 in its bit location.
- For the switch that is open, the processor detects no voltage at the input terminal and records that information by storing a binary 0 in its bit location.
- Each connected input has a bit in the input image table file that corresponds exactly to the terminal to which the input is connected.
- The input image table file is changed to reflect the current status of the switch during the I/O scan phase of operation.
- If the input is on (switch closed), its corresponding bit in the table is set to 1.
- If the input is off (switch open), the corresponding bit is cleared, or reset to 0.
- The processor continually reads the current input status and updates the input image table file.

The *output image table file* is that part of the program memory allocated to storing the actual on/off status of connected discrete outputs. Figure 5-5 shows a typical

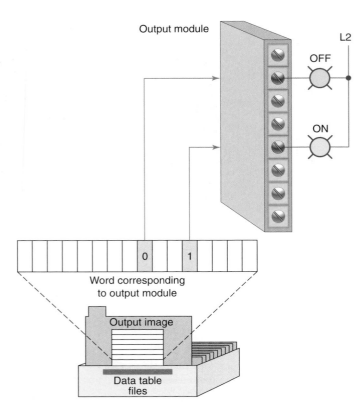

Figure 5-5 Connections of pilot lights to the output image table file through the output module.

connection of two pilot lights to the output image table file through the output module. Its operation can be summarized as follows.

- The status of each light (ON/OFF) is controlled by the user program and is indicated by the presence of 1 (ON) and 0 (OFF).

- Each connected output has a bit in the output image table file that corresponds exactly to the terminal to which the output is connected.
- If the program calls for a specific output to be ON, its corresponding bit in the table is set to 1.
- If the program calls for the output to be OFF, its corresponding bit in the table is set to 0.
- The processor continually activates or deactivates the output status according to the output table file status.

Typically, micro PLCs have a fixed number of inputs and outputs. Figure 5-6 shows the MicroLogix controller from the Allen-Bradley MicroLogix 1000 family of controllers. The controller has 20 discrete inputs with predefined addresses I/0 through I/19 and 12 discrete outputs with predefined addresses O/1 through O/11. Some units also contain analog inputs and outputs embedded into the base unit or available through add-on modules.

5.2 Program Scan

When a PLC executes a program, it must know—in real time—when external devices controlling a process are changing. During each operating cycle, the processor reads all the inputs, takes these values, and energizes or de-energizes the outputs according to the user program. This process is known as a *program scan cycle*. Figure 5-7 illustrates a single PLC operating cycle consisting of the *input scan, program scan, output scan,* and housekeeping duties. Because the inputs can change at any time, it constantly repeats this cycle as long as the PLC is in the RUN mode.

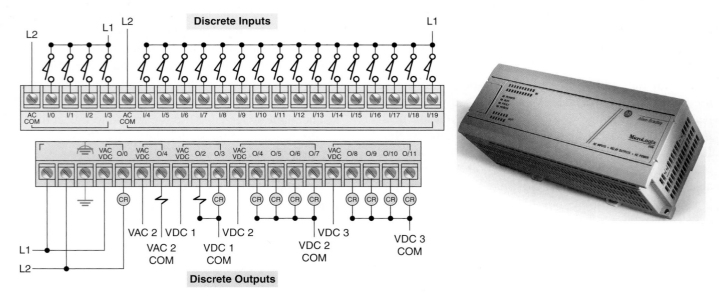

Figure 5-6 Typical micro PLC with predefined addresses.
Source: Image Used with Permission of Rockwell Automation, Inc.

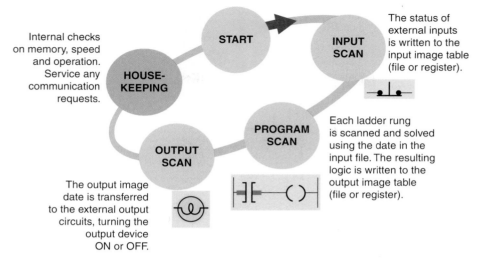

Figure 5-7 PLC program scan cycle.

The time it takes to complete a scan cycle is called the scan cycle time and indicates how fast the controller can react to changes in inputs. The time required to make a single scan can vary from about 1 millisecond to 20 milliseconds. If a controller has to react to an input signal that changes states twice during the scan time, it is possible that the PLC will never be able to detect this change. For example, if it takes 8 ms for the CPU to scan a program, and an input contact is opening and closing every 4 ms, the program may not respond to the contact changing state. The CPU will detect a change if it occurs during the update of the input image table file, but the CPU will not respond to every change. The scan time is a function of the following:

- The speed of the processor module
- The length of the ladder program
- The type of instructions executed
- The actual ladder true/false conditions

The actual scan time is calculated and stored in the PLC's memory. The PLC computes the scan time each time the END instruction is executed. Scan time data can be monitored via the PLC programming. Typical scan time data include the maximum scan time and the last scan time.

The scan is normally a continuous and sequential process of reading the status of inputs, evaluating the control logic, and updating the outputs. Figure 5-8 shows an overview of the data flow during the scan process. For each rung executed, the PLC processor will:

- Examine the status of the input image table bits.
- Solve the ladder logic in order to determine logical continuity.

- Update the appropriate output image table bits, if necessary.
- Copy the output image table status to all of the output terminals. Power is applied to the output device if the output image table bit has been previously set to a 1.
- Copy the status of all of the input terminals to the input image table. If an input is active (i.e., there is electrical continuity), the corresponding bit in the input image table will be set to a 1.

Figure 5-9 illustrates the scan process applied to a simple single rung program. The operation of the scan process can be summarized as follows:

- If the input device connected to address I:3/6 is closed, the input module circuitry senses *electrical continuity* and a 1 (ON) condition is entered into the input image table bit I:3/6.

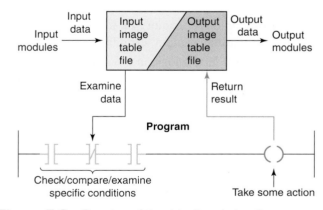

Figure 5-8 Overview of the data flow during the scan process.

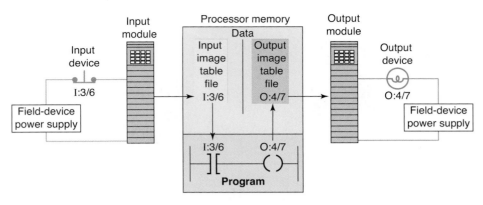

Figure 5-9 Scan process applied to a single rung program.

- During the program scan, the processor examines bit I:3/6 for a 1 (ON) condition.
- In this case, because input I:3/6 is 1, the rung is said to be TRUE or have *logic continuity*.
- The processor then sets the output image table bit O:4/7 to 1.
- The processor turns on output O:4/7 during the next I/O scan, and the output device (light) wired to this terminal becomes energized.
- This process is repeated as long as the processor is in the RUN mode.
- If the input device opens, electrical continuity is lost, and a 0 would be placed in the input image table. As a result, the rung is said to be FALSE due to loss of logic continuity.
- The processor would then set the output image table bit O:4/7 to 0, causing the output device to turn off.

Ladder programs process inputs at the beginning of a scan and outputs at the end of a scan, as illustrated in Figure 5-10. For each rung executed, the PLC processor will:

Step 1 Update the input image table by sensing the voltage of the input terminals. Based on the absence or presence of a voltage, a 0 or a 1 is stored into the memory bit location designated for a particular input terminal.

Step 2 Solve the ladder logic in order to determine logical continuity. The processor scans the ladder program and evaluates the logical continuity of each rung by referring to the input image table to see if the input conditions are met. If the conditions controlling an output are met, the processor immediately writes a 1 in its memory location, indicating that the output will be turned ON; conversely, if the conditions are not met a 0 indicating that the device will be turned OFF is written into its memory location.

Step 3 The final step of the scan process is to update the actual states of the output devices by transferring the output table results to the output module, thereby switching the connected output devices ON (1) or OFF (0). If the status of any input devices changes when the processor is in step 2 or 3, the output condition will not react to them until the next processor scan.

Each instruction entered into a program requires a certain amount of time for the instruction to be executed. The amount of time required depends on the instruction. For example, it takes less time for a processor to read the status of an input contact than it does to read the accumulated value of a timer or counter. The time taken to scan

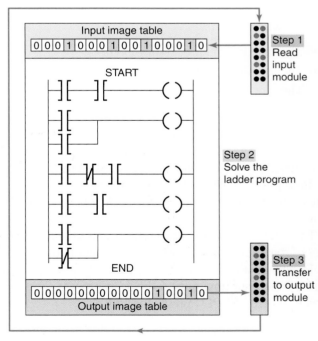

Figure 5-10 Scan process applied to a multiple rung program.

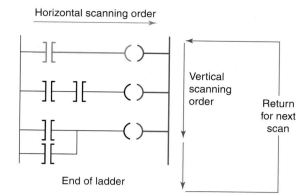

Figure 5-11 Scanning can be vertical or horizontal.

the user program is also dependent on the clock frequency of the microprocessor system. The higher the clock frequency, the faster is the scan rate.

There are two basic scan patterns that different PLC manufacturers use to accomplish the scan function (Figure 5-11). Allen-Bradley PLCs use the *horizontal* scan by rung method. In this system, the processor examines input and output instructions from the first command, top left in the program, horizontally, rung by rung. Modicon PLCs use the *vertical* scan by column method. In this system, the processor examines input and output instructions from the top left command entered in the ladder diagram, vertically, column by column and page by page. Pages are executed in sequence. Both methods are appropriate; however, misunderstanding the way the PLC scans a program can cause programming bugs.

5.3 PLC Programming Languages

The term *PLC programming language* refers to the method by which the user communicates information to the PLC. The standard IEC 61131 (Figure 5-12) was established to standardize the multiple languages associated with PLC programming by defining the following five standard languages:

- **Ladder Diagram (LD)**—a graphical depiction of a process with rungs of logic, similar to the relay ladder logic schemes that were replaced by PLCs.
- **Function Block Diagram (FBD)**—a graphical depiction of process flow using simple and complex interconnecting blocks.
- **Sequential Function Chart (SFC)**—a graphical depiction of interconnecting steps, actions, and transitions.
- **Instruction List (IL)**—a low-level, text-based language that uses mnemonic instructions.
- **Structured Text (ST)**—a high-level, text-based language such as BASIC, C, or PASCAL specifically developed for industrial control applications.

Ladder diagram language is the most commonly used PLC language and is designed to mimic relay logic. The ladder diagram is popular for those who prefer to define control actions in terms of relay contacts and coils, and other functions as block instructions. Figure 5-13 shows a comparison of ladder diagram programming and instruction list programming. Figure 5-13a shows

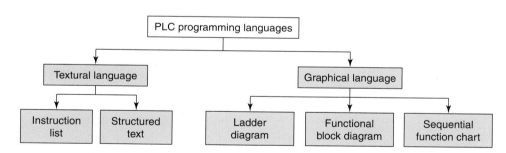

Figure 5-12 Standard IEC 61131 languages associated with PLC programming.

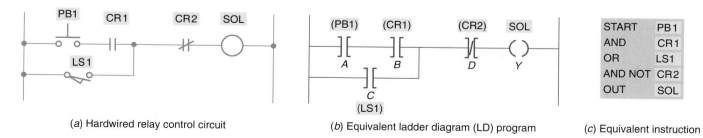

(a) Hardwired relay control circuit

(b) Equivalent ladder diagram (LD) program

(c) Equivalent instruction list (IL) program

Figure 5-13 Comparison of ladder diagram and instruction list programming.

the original relay hardwired control circuit. Figure 5-13b shows the equivalent logic ladder diagram programmed into a controller. Note how closely the ladder diagram program closely resembles the hardwired relay circuit. The input/output addressing is generally different for each PLC manufacturer. Figure 5-13c show how the original hardwired circuit could be programmed using the instruction list programming language. Note that the instructional list consists of a series of instructions that refer to the basic AND, OR, and NOT logic gate functions.

Functional block diagram programming uses instructions that are programmed as blocks wired together on screen to accomplish certain functions. Typical types of function blocks include logic, timers, and counters. Functional block diagrams are similar in layout to electrical/electronic block diagrams used to simplify complex systems by showing blocks of functionality. The primary concept behind a functional block diagram is data flow. Function blocks are linked together to complete a circuit that satisfies a control requirement. Data flow on a path from inputs, through function blocks or instructions, and then to outputs.

The use of function blocks for programming of programmable logic controllers (PLCs) is gaining wider acceptance. Rather than the classic contact and coil representation of ladder diagram or relay ladder logic programming, function blocks present a graphical image to the programmer with underlying algorithms already defined. The programmer simply completes needed information within the block to complete that phase of the program. Figure 5-14 shows function block diagram equivalents to ladder logic contacts.

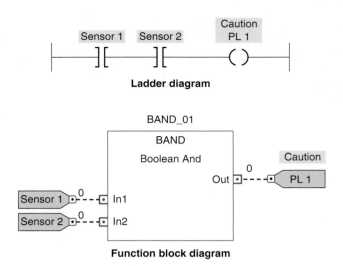

Figure 5-15 PLC ladder and equivalent function block diagram.

Figure 5-15 illustrates how ladder diagram and functional block diagram programming could be used to produce the same logical output. For this application, the objective is to turn on caution pilot light PL 1 whenever both sensor switch 1 and sensor switch 2 are closed. The ladder logic consists of a single rung across the power rails. This rung contains the two input sensor instructions programmed in series with the pilot light output instruction. The function block solution consists of a logic *Boolean And* function block with two input references tags for the sensors and a single output reference tag for the pilot light. Note there are no power rails in the function block diagram.

Sequential function chart programming language is similar to a flowchart of your process. SFC programming is designed to accommodate the programming of more advanced processes. This type of program can be split into steps with multiple operations happening in parallel branches. The basic elements of a sequential function chart program are shown in Figure 5-16.

Structured text is a high level text language primarily used to implement complex procedures that cannot be easily expressed with graphical languages. Structured text uses statements to define what to execute. Figure 5-17 illustrates how structured text and ladder diagram programming could be used to produce the same logical output. For this application, the objective is to energize SOL 1 whenever either one of the two following circuit conditions exists:

- Sensor 1 and Sensor 2 switches are both closed.
- Sensor 3 and Sensor 4 switches are both closed and Sensor 5 switch is open.

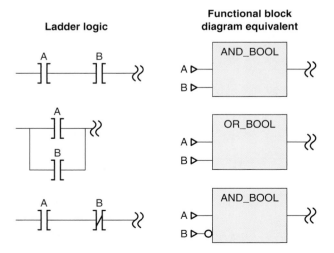

Figure 5-14 Function block diagram equivalents to ladder logic contacts.

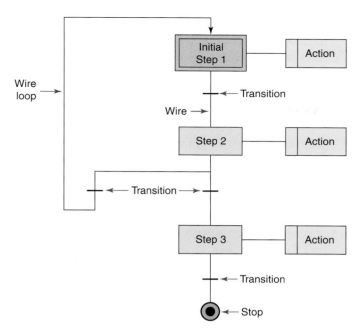

Figure 5-16 Major elements of a sequential function chart program.

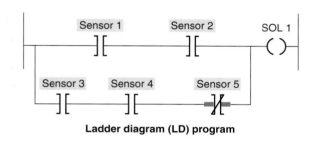

Ladder diagram (LD) program

```
IF Sensor_1 AND Sensor_2 THEN
        SOL_1 := 1;
ELSEIF Sensor_3 AND Sensor_4 AND NOT Sensor_5 THEN
        SOL_1 := 1;
END_IF;
```

Structured text (ST) program

Figure 5-17 PLC ladder and equivalent structured text program.

5.4 Relay-Type Instructions

The ladder diagram language is basically a *symbolic* set of instructions used to create the controller program. These ladder instruction symbols are arranged to obtain the desired control logic that is to be entered into the memory of the PLC. Because the instruction set is composed of contact symbols, ladder diagram language is also referred to as *contact symbology.*

Representations of contacts and coils are the basic symbols of the logic ladder diagram instruction set. The three fundamental symbols that are used to translate relay control

logic to contact symbolic logic are Examine If Closed (XIC), Examine If Open (XIO), and Output Energize (OTE). Each of these instructions relates to a single bit of PLC memory that is specified by the instruction's address.

The symbol for the *Examine If Closed (XIC)* instruction is shown in Figure 5-18. The XIC instruction, which is also called the Examine-on instruction, looks and operates like a normally open relay contact. Associated with each XIC instruction is a memory bit linked to the status of an input device or an internal logical condition in a rung. This instruction asks the PLC's processor to examine if the contact is *closed.* It does this by examining the bit at the memory location specified by the address in the following manner:

- The memory bit is set to 1 or 0 depending on the status of the input (physical) device or internal (logical) relay address associated with that bit.
- A 1 corresponds to a true status or on condition.
- A 0 corresponds to a false status or off condition.
- When the Examine-on instruction is associated with a physical input, the instruction will be set to 1 when a physical input is present (voltage is applied to the input terminal), and 0 when there is no physical input present (no voltage applied to the input terminal).
- When the Examine-on instruction is associated by address with an internal relay, then the status of the

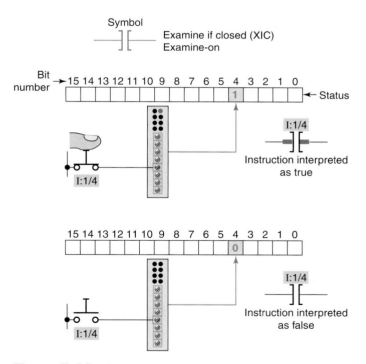

Figure 5-18 Examine If Closed (XIC) instruction.

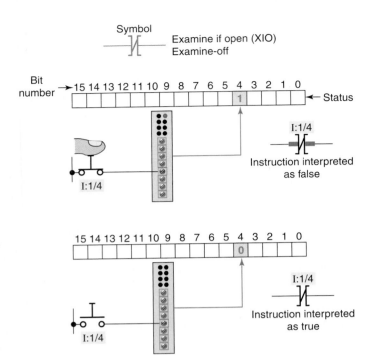

Symbol
Examine if open (XIO)
Examine-off

Bit number → 15 14 13 12 11 10 9 8 7 6 5 4 3 2 1 0 ← Status

I:1/4

I:1/4

Instruction interpreted as false

15 14 13 12 11 10 9 8 7 6 5 4 3 2 1 0

I:1/4

I:1/4

Instruction interpreted as true

Figure 5-19 Examine If Open (XIO) instruction.

bit is dependent on the logical status of the internal bit with the same address as the instruction.

- If the instruction memory bit is a 1 (true) this instruction will allow rung continuity through itself, like a closed relay contact.
- If the instruction memory bit is a 0 (false) this instruction will not allow rung continuity through itself and will assume a normally open state just like an open relay contact.

The symbol for the *Examine If Open (XIO)* instruction is shown in Figure 5-19. The XIO instruction, which is also called the Examine-off instruction, looks and operates like a normally closed relay contact. Associated with each XIO instruction is a memory bit linked to the status of an input device or an internal logical condition in a rung. This instruction asks the PLC's processor to examine if the contact is *open*. It does this by examining the bit at the memory location specified by the address in the following manner:

- As with any other input the memory bit is set to 1 or 0 depending on the status of the input (physical) device or internal (logical) relay address associated with that bit.
- A 1 corresponds to a true status or on condition.
- A 0 corresponds to a false status or off condition.
- When the Examine-off instruction is used to examine a physical input, then the instruction will be

interpreted as false when there is a physical input (voltage) present (the bit is 1) and will be interpreted as true when there is no physical input present (the bit is 0).

- If the Examine-off instruction were associated by address with an internal relay, then the status of the bit would be dependent on the logical status of the internal bit with the same address as the instruction.
- Like the Examine-on instruction, the status of the instruction (true or false) determines if the instruction will allow rung continuity through itself, like a closed relay contact.
- The memory bit always follows the status (true = 1 or false = 0) of the input address or internal address assigned to it. The interpretation of that bit, however, is determined by which instruction is used to examine it.
- Examine-on instructions always interpret a 1 status as true and a 0 status as false, while Examine-off instructions interpret a 1 status as false and a 0 status as true.

The symbol for the *Output Energize (OTE)* instruction is shown in Figure 5-20. The OTE instruction looks and operates like a relay coil and is associated with a memory bit. This instruction signals the PLC to energize (switch on) or de-energize (switch off) the output. The processor makes this instruction true (analogous to energizing a coil) when there is a logical path of true XIC and XIO instructions in the rung. The operation of the Output Energize instruction can be summarized as follows:

- The status bit of the addressed Output Energize instruction is set to 1 to energize the output and to 0 to de-energize the output.
- If a true logic path is established with the input instructions in the rung, the OTE instruction is energized and the output device wired to its terminal is energized.
- If a true logic path cannot be established or rung conditions go false, the OTE instruction is de-energized and the output device wired to it is switched off.

Sometimes beginner programmers used to thinking in terms of hardwired relay control circuits tend to use the same type of contact (NO or NC) in the ladder logic program that corresponds to the type of field switch wired to the discrete input. While this is true in many instances, it is not the best way to think of the concept. A better approach is to separate the action of the field device from

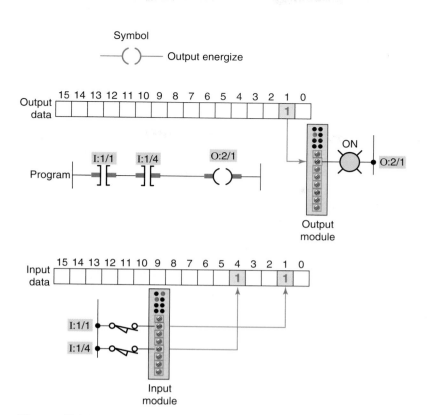

Symbol

—( )— Output energize

Figure 5-20 Output Energize (OTE) instruction.

the action of the PLC bits as illustrated in Figure 5-21. A signal present makes the NO bit (1) true; a signal absent makes the NO bit (0) false. The reverse is true for an NC bit. A signal present makes the NC bit (1) false; a signal absent makes the NO bit (0) true.

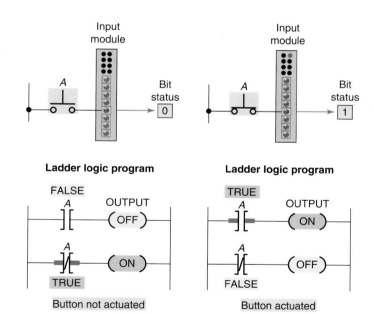

Figure 5-21 Separating the action of the field device and PLC bit.

The main function of the ladder logic diagram program is to control outputs based on input conditions, as illustrated in Figure 5-22. This control is accomplished through the use of what is referred to as a ladder rung. In general, a rung consists of a set of input conditions, represented by contact instructions, and an output instruction at the end of the rung, represented by the coil symbol. Each contact or coil symbol is referenced with an address that identifies what is being evaluated and what is being controlled. The same contact instruction can be used throughout the program whenever that condition needs to be evaluated. The number of ladder logic relays and input and output instructions is limited only by memory size. Most PLCs allow more than one output per rung.

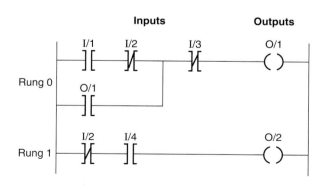

Figure 5-22 Ladder logic diagram rungs.

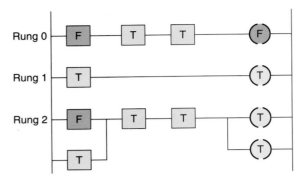

Figure 5-23 Logical continuity.

For an output to be activated or energized, at least one left-to-right true logical path must exist, as illustrated in Figure 5-23. A complete closed path is referred to as having logical continuity. When logical continuity exists in at least one path, the rung condition and Output Energize instruction are said to be true. The rung condition and OTE instruction are false if no logical continuity path has been established. During controller operation, the processor evaluates the rung logic and changes the state of the outputs according to the logical continuity of rungs.

5.5 Instruction Addressing

To complete the entry of a relay-type instruction, you must assign an address to each instruction. This address indicates what PLC input is connected to what input device and what PLC output will drive what output device.

The addressing of real inputs and outputs, as well as internals, depends on the PLC model used. Addressing formats can vary from one PLC family to another as well as for different manufacturers. These addresses can be represented in decimal, octal, or hexadecimal depending on the number system used by the PLC. The address identifies the function of an instruction and links it to a particular bit in the data table portion of the memory. Figure 5-24 shows the addressing format for an Allen-Bradley SLC 500 controller. Addresses contain the slot number of the module where input or output devices are connected. Addresses are formatted as file type, slot number, and bit.

The assignment of an I/O address can be included in the I/O connection diagram, as shown in Figure 5-25. Inputs and outputs are typically represented by squares and diamonds, respectively.

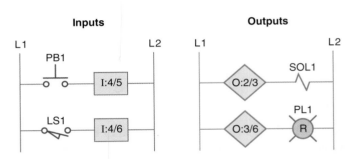

Figure 5-25 I/O connection diagram.

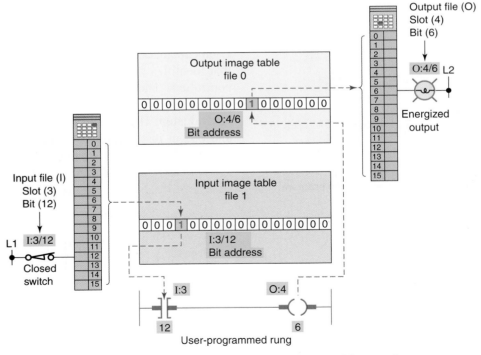

Figure 5-24 Addressing format for an Allen-Bradley SLC 500 controller.

5.6 Branch Instructions

Branch instructions are used to create parallel paths of input condition instructions. This allows more than one combination of input conditions (OR logic) to establish logic continuity in a rung. Figure 5-26 illustrates a typical branch instruction. The rung will be true if either instruction A or B is true.

Input branching by formation of parallel branches can be used in your application program to allow more than one combination of input conditions. If at least one of these parallel branches forms a true logic path, the rung logic is true and the output will be energized. If none of the parallel branches complete a logical path, logic rung continuity is not established and the output will not be de-energized. In the example shown in Figure 5-27, either A and B, or C provides logical continuity and energizes output D.

On most PLC models, branches can be established at both input and output portions of a rung. With output branching, you can program parallel outputs on a rung to allow a true logic path to control multiple outputs, as illustrated in Figure 5-28. When there is a true logic rung path, all parallel outputs become true. In the example shown, either A or B provides a true logical path to all three output instructions: C, D, and E.

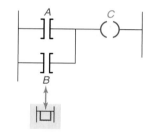

Figure 5-26   Typical branch instruction.

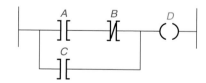

Figure 5-27   Parallel input branches.

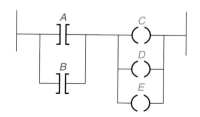

Figure 5-28   Parallel output branches.

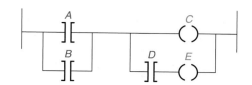

Figure 5-29   Parallel output branching with conditions.

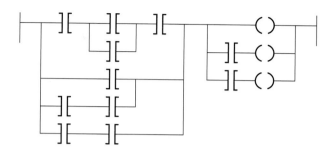

Figure 5-30   Nested input and output branches.

Additional input logic instructions (conditions) can be programmed in the output branches to enhance conditional control of the outputs. When there is a true logic path, including extra input conditions on an output branch, that branch becomes true. In the example shown in Figure 5-29, either A and D or B and D provide a true logic path to E.

Input and output branches can be *nested* to avoid redundant instructions and to speed up processor scan time. Figure 5-30 illustrates nested input and output branches. A nested branch starts or ends within another branch.

In some PLC models, the programming of a branch circuit within a branch circuit or a *nested* branch cannot be done directly. It is possible, however, to program a logically equivalent branching condition. Figure 5-31 shows an example of a circuit that contains a nested contact D. To obtain the required logic, the circuit would be programmed as shown in Figure 5-32. The duplication of contact C eliminates the nested contact D. Nested branching can be converted into non-nested branches by repeating instructions to make parallel equivalents.

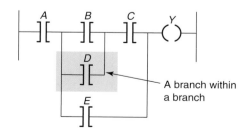

Figure 5-31   Nested contact program.

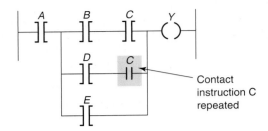

Figure 5-32 Program required to eliminate nested contact.

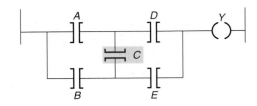

Boolean equation: $Y = (AD) + (BCD) + (BE) + (ACE)$

Figure 5-34 Program with vertical contact.

Some PLC manufacturers have virtually no limitations on allowable series elements, parallel branches, or outputs. For others, there may be limitations to the number of series contact instructions that can be included in one rung of a ladder diagram as well as limitations to the number of parallel branches. Also, there is an additional limitation with some PLCs: only one output per rung and the output must be located at the end of the rung. The only limitation on the number of rungs is memory size. Figure 5-33 shows the matrix limitation diagram for a typical PLC. A maximum of seven parallel lines and 10 series contacts per rung is possible.

Another limitation to branch circuit programming is that the PLC will not allow for programming of vertical contacts. A typical example of this limitation is contact C of the user program drawn in Figure 5-34. To obtain the required logic, the circuit would be reprogrammed as shown in Figure 5-35.

The processor examines the ladder logic rung for logic continuity from left to right *only*. The processor never allows for flow from right to left. This situation presents a problem for user program circuits similar to that shown in Figure 5-36. If programmed as shown, contact combination *FDBC* would be ignored. To obtain the required logic, the circuit would be reprogrammed as shown in Figure 5-37.

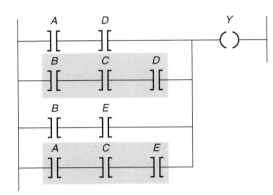

Figure 5-35 Reprogrammed to eliminate vertical contact.

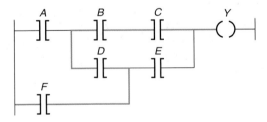

Boolean equation: $Y = (ABC) + (ADE) + (FE) + (FDBC)$

Figure 5-36 Original circuit.

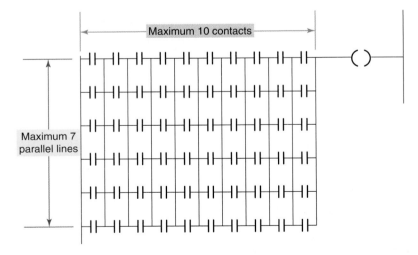

Figure 5-33 PLC matrix limitation diagram.

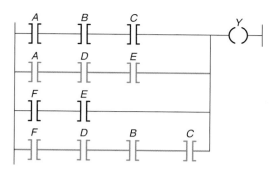

Figure 5-37 Reprogrammed circuit.

5.7 Internal Relay Instructions

Most PLCs have an area of the memory allocated for what are known as *internal storage bits*. These storage bits are also called *internal outputs, internal coils, internal control relays,* or simply *internal bits.* Internal outputs are on/off signals generated by programmed logic. Unlike a discrete output, an internal output does not directly control an output field device. The internal output operates just like any output that is controlled by programmed logic; however, the output is used strictly for internal purposes.

The advantage of using internal outputs is that there are many situations in which an output instruction is required in a program but no physical connection to a field device is needed. If there are no physical outputs wired to a bit address, the address can be used as an internal storage point. Internal storage bits or points can be programmed by the user to perform relay functions without occupying a physical output. In this way internal outputs can minimize output module point requirements whenever practical.

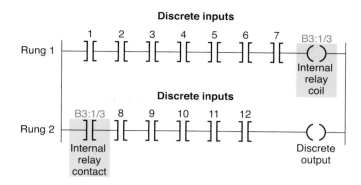

Figure 5-39 Programmed internal relay control.

Internal outputs are single-bit storage locations in memory and are addressed as such. SLC 500 controllers use bit file B3 for storage and addressing of internal output bits. The addressing for bit B3:1/3 illustrated in Figure 5-38 consists of the file number followed by word and bit numbers.

An internal control relay can be used when a program requires more series contacts than the rung allows. Figure 5-39 shows a circuit that allows for only 7 series contacts when 12 are actually required for the programmed logic. To solve this problem, the contacts are split into two rungs. Rung 1 contains seven of the required contacts and is programmed to control internal relay coil B3:1/3. The address of the first programmed contact on Rung 2 is B3:1/3 followed by the remaining five contacts and the discrete output. When the logic controlling the internal output is true, the referenced bit B3:1/3 is turned on or set to 1. The advantage of an internal storage bit in this manner is that it does not waste space in a physical output.

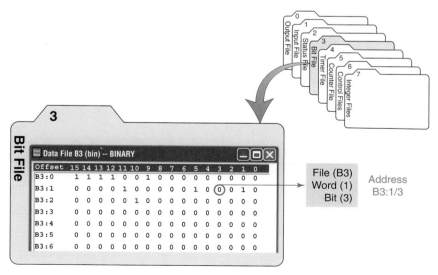

Figure 5-38 SLC 500 controllers use bit file B3 for internal bit addressing.

Hardwired circuit

PB1

PB2

PL

User program providing the same results

PB_1 PB_2 PL

Figure 5-40 Simple program that uses the Examine If Closed (XIC) instruction.

5.8 Programming Examine If Closed and Examine If Open Instructions

A simple program using the Examine If Closed (XIC) instruction is shown in Figure 5-40. This figure shows a hardwired circuit and a user program that provides the same results. You will note that *both the NO and the NC* pushbuttons are represented by the Examine If Closed

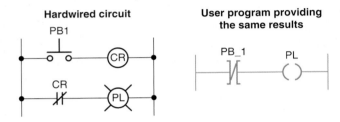

Hardwired circuit

PB1

CR

CR PL

User program providing the same results

PB_1 PL

Figure 5-41 Simple program that uses the Examine If Open (XIO) instruction.

symbol. This is because the normal state of an input (NO or NC) does not matter to the controller. What does matter is that if contacts need to close to energize the output, then the Examine If Closed instruction is used. Since both PB1 and PB2 must be closed to energize the pilot light, the Examine If Closed instruction is used for both.

A simple program using the Examine If Open (XIO) instruction is shown in Figure 5-41. Both the hardwired circuit and user program are shown. In the hardwired circuit, when the pushbutton is *open* relay coil CR is de-energized and its NO contact closes to switch the pilot light on. When the pushbutton is *closed,* relay coil CR is energized and its NC contact opens to switch the pilot light off. The pushbutton is represented in the user program by an Examine If Open instruction. This is because the rung must be true when the external pushbutton is open and false when the pushbutton is closed. Using an Examine If Open instruction to represent the pushbutton satisfies these requirements. The NO or NC mechanical action of the pushbutton is not a consideration. It is important to remember that the user program is not an electrical circuit but a *logic* circuit. In effect, we are interested in logic continuity when establishing an output.

Figure 5-42 shows a simple program using both the XIC and XIO instructions. The logic states (0 or 1) indicate whether an instruction is true or false and is the basis

If the data table bit is	The status of the instruction is		
	XIC EXAMINE IF CLOSED ⊣⊢	XIO EXAMINE IF OPEN ⊣/⊢	OTE OUTPUT ENERGIZE ⊸( )⊸
Logic 0	False	True	False
Logic 1	True	False	True

Input instructions

XIC XIO

Output instruction

OTE

	Instruction outcome		
Time	XIC	XIO	OTE
t_1 (initial)	False	True	False
t_2	True	True	Goes true
t_3	True	False	Goes false
t_4	False	False	Remains false

Input bit status		
XIC	XIO	OTE
0	0	0
1	0	1
1	1	0
0	1	0

Figure 5-42 Simple program using both the XIC and XIO instructions.

of controller operation. The figure summarizes the on/off state of the output as determined by the changing states of the inputs in the rung. The time aspect relates to the repeated scans of the program, wherein the input table is updated with the most current status bits.

5.9 Entering the Ladder Diagram

Most of today's PLC programming packages operate in the Windows environment. For example, Allen-Bradley's RSLogix software packages are Windows programming packages used to develop ladder logic programs. This software, in various versions, can be used to program the PLC-5, SLC 500, ControlLogix, and MicroLogic family of processors. An added feature is that RSLogix programs are compatible with programs that have been previously created with DOS-based programming packages. You can import projects that were developed with DOS products or export to them from RSLogix.

Entering the ladder diagram, or actual programming, is usually accomplished with a computer keyboard or hand-held programming device. Because hardware and programming techniques vary with each manufacturer, it is necessary to refer to the programming manual for a specific PLC to determine how the instructions are entered.

One method of entering a program is through a hand-held keyboard. Keyboards usually have relay symbol and special function keys along with numeric keys for addressing. Some also have alphanumeric keys (letters and numbers) for other special programming functions. In hand-held units, the keyboard is small and the keys have multiple functions. Multiple-function keys work like second-function keys on calculators.

A personal computer is most often used today as the programmer. The computer is adapted to the particular PLC model through the use of the relevant programmable controller software.

Figure 5-43 shows the RSLogix SLC 500 main window. Different screens, toolbars, and dialog boxes are used to navigate through the Windows environment. It is important that you understand the purpose of the various screens, toolbars, and windows to make the most effective use of the software. This information is available from the software reference manual for the particular PLC family and will become more familiar to you as you develop programs using the software.

Figure 5-44 shows a typical instruction toolbar with bit instructions selected. To place an instruction on a rung, click its icon on the toolbar and simply drag the instruction straight off the toolbar onto the rung of the ladder. Drop points are shown on the ladder to help position the instruction. In addition, instructions can also be dragged

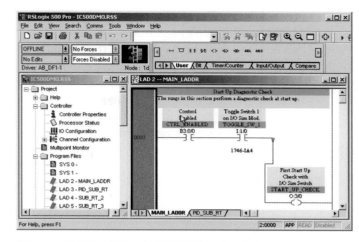

Figure 5-43 RSLogix SLC 500 main window.
Source: Image Used with Permission of Rockwell Automation, Inc.

from other rungs in the project. There are several different methods that you can use to address instructions. You can enter an address by manually typing it in or by dragging the address from data files or other instructions.

Some of the windows you will need to use when working with RSLogix 500 software include:

- **Main Window**—This window opens each time you create a new project or open an existing one. Some of the features associated with this window include the following:
  - Window Title Bar—The title bar is located at the topmost strip of the window and displays the name of the program as well as that of the opened file.
  - Menu Bar—The menu bar is located below the title bar. The menu contains key words associated with menus that are opened by clicking on the key word.
  - Windows Toolbar—The Windows toolbar buttons execute standard Windows commands when you click on them.
  - Program/Processor Status Toolbar—This toolbar contains four drop-down lists that identify the current processor operating mode, current online edit status, and whether forces are present and enabled.
  - Project Window—This window displays the file folders listed in the project tree.

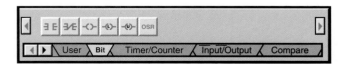

Figure 5-44 Typical instruction toolbar with bit instructions selected.

- Project Tree—The project tree is a visual representation of all folders and their associated files contained in the current project. From the project tree, you can open files, create files, modify file parameters, copy files, hide or unhide files, delete files, and rename files.
- Result Window—This window displays the results of either a search or a verify operation. The verify operation is used to check the ladder program for errors.
- Active Tab—This tab identifies which program is currently active.
- Status Bar—This bar contains information relevant to the current file.
- Split Bar—The split bar is used to split the ladder window to display two different program files or groups of ladder rungs.
- Tabbed Instruction Toolbar—This toolbar displays the instruction set as a group of tabbed categories.
- Instruction Palette—This tool contains all the available instructions displayed in one table to make the selection of instructions easier.
- Ladder Window—This window displays the currently open ladder program file and is used to develop and edit ladder programs.
- Ladder Window Properties—This window allows you to change the display of your ladder program and its associated addressing and documentation.
- **Select Processor Type**—The programming software needs to know what processor is being used in conjunction with the user program. The Select Processor Type screen (Figure 5-45) contains a list of the different processors that the RSLogix software can program. You simply scroll down the list until you find the processor you are using and select it.

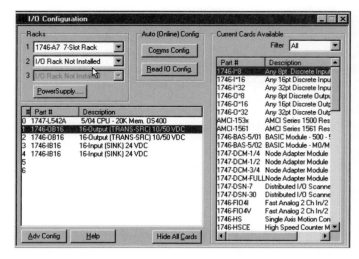

Figure 5-46 I/O configuration screen.

- **I/O Configuration**—The I/O Configuration screen (Figure 5-46) lets you click or drag-and-drop a module from an all-inclusive list to assign it to a slot in your configuration.
- **Data Files**—Data File screens contain data that are used in conjunction with ladder program instructions and include input and output files as well as timer, counter, integer, and bit files. Figure 5-47 shows an example of the bit file B3, which is used for internal relays. Note that all the addresses from this file start with B3.

Relay ladder logic is a graphical programming language designed to closely represent the appearance of a wired relay system. It offers considerable advantages for PLC control. Not only is it reasonably intuitive, especially for users with relay experience, but it is also particularly effective in an online mode when the PLC is actually performing control. Operation of the logic is apparent from the highlighting of rungs of the various instructions on-screen,

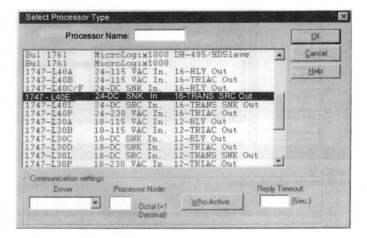

Figure 5-45 Select processor type screen.

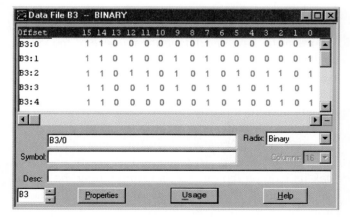

Figure 5-47 Data bit file B3 screen.

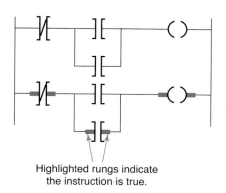

Highlighted rungs indicate
the instruction is true.

Figure 5-48 Monitoring a ladder logic program.

which identifies the logic state of contacts in real time (Figure 5-48) and which rungs have logic continuity.

For most PLC systems, each Examine If Closed and Examine If Open contact, each output, and each branch Start/End instruction requires one word of user memory. You can refer to the SLC 500 Controller Properties to see the number of instruction words used and the number left as the program is being developed.

5.10 Modes of Operation

A processor has basically two modes of operation: the *program mode* and some variation of the *run mode*. The number of different operating modes and the method of accessing them varies with the manufacturer. Figure 5-49 shows a typical three-position keyswitch used to select different processor modes of operation.

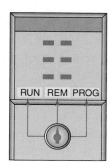

Figure 5-49 Three-position keyswitch used to select different processor modes of operation.

Some common operating modes are explained in the following paragraphs.

Program Mode   The program mode is used to enter a new program, edit or update an existing program, upload files, download files, document (print out) programs, or change any software configuration file in the program. When the PLC is switched into the program mode, all outputs from the PLC are forced off regardless of their rung logic status, and the ladder I/O scan sequence is halted.

Run Mode   The run mode is used to execute the user program. Input devices are monitored and output devices are energized accordingly. After all instructions have been entered in a new program or all changes made to an existing program, the processor is put in the run mode.

Test Mode   The test mode is used to operate or monitor the user program without energizing any outputs. The processor still reads inputs, executes the ladder program, and updates the output status table files, but without energizing the output circuits. This feature is often used after developing or editing a program to test the program execution before allowing the PLC to operate real-world outputs. Variations of the test mode can include the *single-step test mode,* which directs the processor to execute a selected single rung or group of rungs; the *single-scan test mode,* which executes a single processor operating scan or cycle; and the *continuous-scan test mode,* which directs the processor to continuously run the program for checking or troubleshooting.

Remote Mode   Some processors have a three-position switch to change the processor operating mode. In the Run position, all logic is solved and the I/O is enabled. In the Program position, all logic solving is stopped and the I/O is disabled. The Remote position allows the PLC to be remotely changed between program and run mode by a personal computer connected to the PLC processor. The remote mode may be beneficial when the controller is in a location that is not easily accessible.

1. What does the memory map for a typical PLC processor consist of?

2. Compare the function of the PLC program and data files.

3. In what manner are data files organized?

4. List eight different types of data files use by an SLC 500 controller.

5. **a.** What information is stored in the input image table file?
   b. In what form is this information stored?

6. **a.** What information is stored in the output image table file?
   b. In what form is this information stored?

7. Outline the sequence of events involved in a PLC scan cycle.

8. List four factors that enter into the length of the scan time.

9. Compare the way horizontal and vertical scan patterns examine input and output instructions.

10. List the five standard PLC languages as defined by the International Standard for Programmable Controllers, and give a brief description of each.

11. Draw the symbol and state the equivalent instruction for each of the following: NO contact, NC contact, and coil.

12. Answer the following with regard to the Examine If Closed instruction:
    a. What is another common name for this instruction?
    b. What is this instruction asking the processor to examine?
    c. Under what condition is the status bit associated with this instruction 0?
    d. Under what condition is the status bit associated with this instruction 1?
    e. Under what condition is this instruction logically true?
    f. What state does this instruction assume when it is false?

13. Answer the following with regard to the Examine If Open instruction:
    a. What is another common name for this instruction?
    b. What is this instruction asking the processor to examine?
    c. Under what condition is the status bit associated with this instruction 0?

d. Under what condition is the status bit associated with this instruction 1?
e. Under what condition is this instruction logically true?
f. What state does this instruction assume when it is false?

14. Answer the following with regard to the Output Energize instruction:
    a. What part of an electromagnetic relay does this instruction look and act like?
    b. What is this instruction asking the processor to do?
    c. Under what condition is the status bit associated with this instruction 0?
    d. Under what condition is the status bit associated with this instruction 1?

15. A normally closed pushbutton is connected to a PLC discrete input. Does this mean it must be represented by a normally closed contact in the ladder logic program? Explain why or why not.

16. Answer the following with regard to a ladder logic rung:
    a. Describe the basic makeup of a ladder logic rung.
    b. How are the contacts and coil of a rung identified?
    c. When is the ladder rung considered as having logic continuity?

17. What does the address assigned to an instruction indicate?

18. When are input branch instructions used as part of a ladder logic program?

19. Identify two matrix limitations that may apply to certain PLCs.

20. In what way does an internal output differ from a discrete output.

21. A normally open limit switch is to be programmed to control a solenoid. What determines whether an Examine-on or Examine-off contact instruction is used?

22. Explain the purpose of Windows based programming software such as RSLogix.

23. Briefly describe each of the following PLC modes of operation:
    a. Program
    b. Test
    c. Run

1. Assign each of the following discrete input and output addresses based on the SLC 500 format.
   a. Limit switch connected to terminal screw 4 of the module in slot 1 of the chassis.
   b. Pressure switch connected to terminal screw 2 of the module in slot 3 of the chassis.
   c. Pushbutton connected to terminal screw 0 of the module in slot 6 of the chassis.
   d. Pilot light connected to terminal screw 13 of the module in slot 2 of the chassis.
   e. Motor starter coil connected to terminal screw 6 of the module in slot 4 of the chassis.
   f. Solenoid connected to terminal screw 8 of the module in slot 5 of the chassis.

2. Redraw the program shown in Figure 5-50 corrected to solve the problem of a nested contact.

3. Redraw the program shown in Figure 5-51 corrected to solve the problem of a nested vertical programmed contact.

4. Redraw the program shown in Figure 5-52 corrected to solve the problem of some logic ignored.

5. Redraw the program shown in Figure 5-53 corrected to solve the problem of too many series contacts (only four allowed).

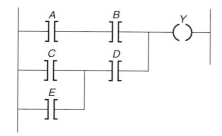

Figure 5-50 Program for Problem 2.

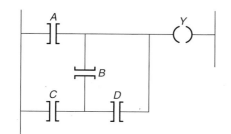

Figure 5-51 Program for Problem 3.

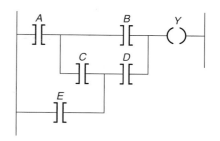

Figure 5-52 Program for Problem 4.

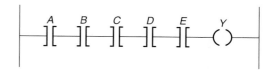

Figure 5-53 Program for Problem 5.

6. Draw the equivalent ladder logic program used to implement the hardwired circuit drawn in Figure 5-54, wired using:
   a. A limit switch with a single NO contact connected to the PLC discrete input module
   b. A limit switch with a single NC contact connected to the PLC discrete input module

7. Assuming the hardwired circuit drawn in Figure 5-55 is to be implemented using a PLC program, identify
   a. All input field devices
   b. All output field devices
   c. All devices that could be programmed using internal relay instructions

8. What instruction would you select for each of the following discrete input field devices to accomplish the desired task? (State the reason for your answer.)
   a. Turn on a light when a conveyor motor is running in reverse. The input field device is a set of

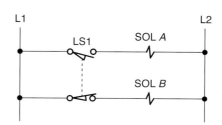

Figure 5-54 Hardwired circuit for Problem 6.

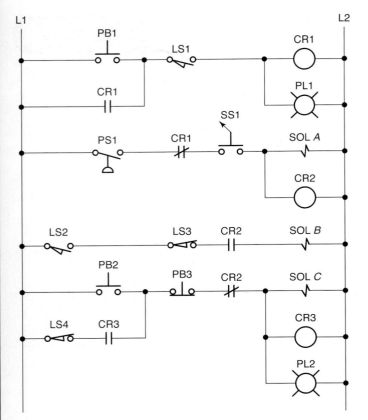

Figure 5-55 Hardwired circuit for Problem 7.

contacts on the conveyor start relay that close when the motor is running forward and open when it is running in reverse.

b. When a pushbutton is pressed, it operates a solenoid. The input field device is a normally open pushbutton.

c. Stop a motor from running when a pushbutton is pressed. The input field device is a normally closed pushbutton.

d. When a limit switch is closed, it triggers an instruction ON. The input field device is a limit switch that stores a 1 in a data table bit when closed.

9. Write the ladder logic program needed to implement each of the following (assume inputs *A*, *B*, and *C* are all normally open toggle switches):

a. When input *A* is closed, turn ON and hold ON outputs *X* and *Y* until *A* opens.

b. When input *A* is closed and either input *B* or *C* is open, turn ON output *Y*; otherwise, it should be OFF.

c. When input *A* is closed or open, turn ON output *Y*.

d. When input *A* is closed, turn ON output *X* and turn OFF output *Y*.

6

Developing Fundamental PLC Wiring Diagrams and Ladder Logic Programs

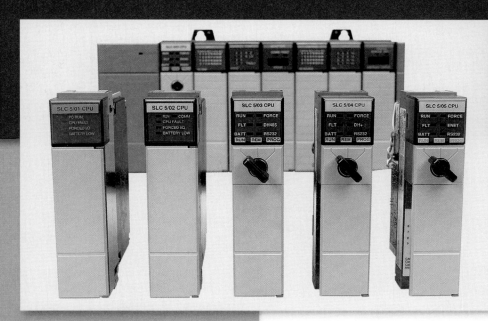

Image Used with Permission of Rockwell Automation, Inc.

Chapter Objectives

After completing this chapter, you will be able to:

6.1 Identify the functions of electromagnetic control relays, contactors, and motor starters

6.2 Identify switches commonly found in PLC installations

6.3 Explain the operation of sensors commonly found in PLC installations

6.4 Explain the operation of output control devices commonly found in PLC installations

6.5 Describe the operation of an electromagnetic latching relay and the PLC-programmed LATCH/UNLATCH instruction

6.6 Compare sequential and combination control processes

6.7 Convert fundamental relay ladder diagrams to PLC ladder logic programs

6.8 Write PLC programs directly from a narrative description

For ease of understanding, ladder logic programs can be compared to relay schematics. This chapter gives examples of how traditional relay schematics are converted into PLC ladder logic programs. You will learn more about the wide variety of field devices commonly used in connection with the I/O modules.

6.1 Electromagnetic Control Relays

The PLC's original purpose was the replacement of electromagnetic relays with a solid-state switching system that could be programmed. Although the PLC has replaced much of the relay control logic, electromagnetic relays are still used as auxiliary devices to switch I/O field devices. The programmable controller is designed to replace the physically small control relays that make logic decisions but are not designed to handle heavy current or high voltage (Figure 6-1). In addition, an understanding of electromagnetic relay operation and terminology is important for correctly converting relay schematic diagrams to ladder logic programs.

An electrical relay is a magnetic switch. It uses electromagnetism to switch contacts. A relay will usually have only one coil but may have any number of different contacts. Figure 6-2 illustrates the operation of a typical

control relay. With no current flow through the coil (de-energized), the armature is held away from the core of the coil by spring tension. When the coil is energized, it produces an electromagnetic field. Action of this field, in turn, causes the physical movement of the armature. Movement of the armature causes the contact points of the relay to open or close. The coil and contacts are insulated from each other; therefore, under normal conditions, no electric circuit will exist between them.

The symbol used to represent a control relay is shown in Figure 6-3. The contacts are represented by a pair of short parallel lines and are identified with the coil by means of the letters. The letter M frequently indicates a motor starter, while CR is used for control relays. *Normally open (NO) contacts* are defined as those contacts that are open when no current flows through the coil but that *close* as soon as the coil conducts a current or is energized. *Normally closed (NC) contacts* are *closed* when the coil is de-energized and open when the coil is energized. Each contact is usually drawn as it would appear with the coil de-energized.

A typical control relay used to control two pilot lights is shown in Figure 6-4. The operation of the circuit can be summarized as follows:

- With the switch open, coil CR is de-energized.
- The circuit to the green pilot light is completed through the normally closed contact, so this light will be on.
- At the same time, the circuit to the red pilot light is opened through the normally open contact, so this light will be off.

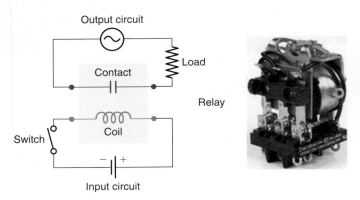

Figure 6-1 Electromechanical control relay.
Source: Photo courtesy Tyco Electronics, **www.tycoelectronics.com**.

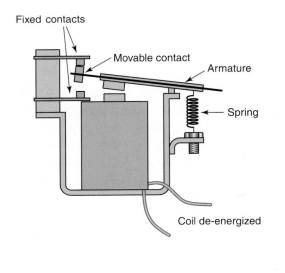
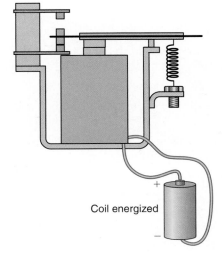

Figure 6-2 Relay operation.

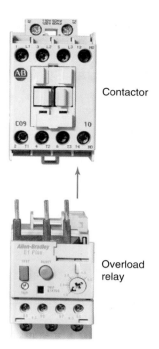

Contactor

Overload relay

Figure 6-7 Motor starter is a contactor with an attached overload relay.

Source: Image Used with Permission of Rockwell Automation, Inc.

- Control contact M (across START button) closes to seal in the coil circuit when the START button is released. This contact is part of the *control* circuit and, as such, is only required to handle the small amount of current needed to energize the coil.

- An overload (OL) relay is provided to protect the motor against current overloads. The normally closed relay contact OL opens automatically when

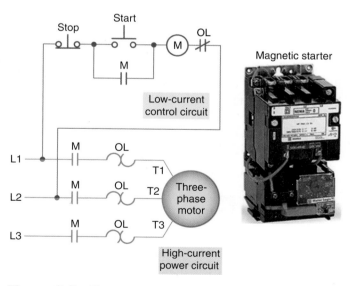

Figure 6-8 Three-phase magnetic motor starter.

Source: This material and associated copyrights are proprietary to, and used with the permission of Schneider Electric.

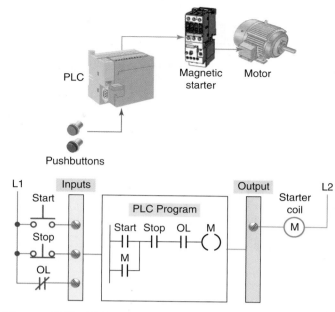

Figure 6-9 PLC control of a motor.

an overload current is sensed to de-energize the M coil and stop the motor.

Motor starters are available in various standard National Electric Manufacturers Association (NEMA) sizes and ratings. When a PLC needs to control a large motor, it must work in conjunction with a starter as illustrated in Figure 6-9. The power requirements for the starter coil must be within the power rating of the output module of the PLC. Note that the control logic is determined and executed by the program within the PLC and not by the hardwired arrangement of the input control devices.

6.4 Manually Operated Switches

Manually operated switches are controlled by hand. These include toggle switches, pushbutton switches, knife switches, and selector switches.

Pushbutton switches are the most common form of manual control. A pushbutton operates by opening or closing contacts when pressed. Figure 6-10 shows commonly used types of pushbutton switches, which include:

- *Normally open (NO) pushbutton,* which makes a circuit when it is pressed and returns to its open position when the button is released.

- *Normally closed (NC) pushbutton,* which opens the circuit when it is pressed and returns to the closed position when the button is released.

- *Break-before-make pushbutton* in which the top section contacts are NC and the bottom section contacts are NO. When the button is pressed, the top contacts open before the bottom contacts are closed.

Developing Fundamental PLC Wiring Diagrams and Ladder Logic Programs   **Chapter 6**   **99**

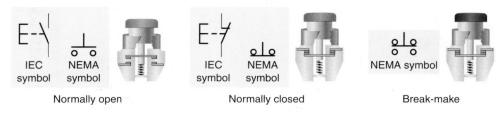

IEC symbol | NEMA symbol | IEC symbol | NEMA symbol | NEMA symbol

Normally open — Normally closed — Break-make

Figure 6-10 Commonly used types of pushbutton switches.

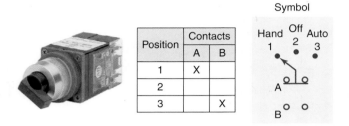

Position	Contacts	
	A	B
1	X	
2		
3		X

Symbol

Hand 1 — Off 2 — Auto 3

A

B

Figure 6-11 Three-position selector switch.
Source: Image Used with Permission of Rockwell Automation, Inc.

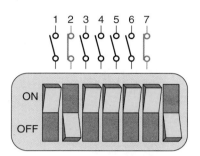

1 2 3 4 5 6 7

ON
OFF

Figure 6-12 DIP switch.

The *selector switch* is another common manually operated switch. The main difference between a pushbutton and selector switch is the operator mechanism. A selector switch operator is rotated (instead of pushed) to open and close contacts of the attached contact block. Figure 6-11 shows a three-position selector switch. Switch positions are established by turning the operator knob right or left. Selector switches may have two or more selector positions, with either maintained contact position or spring return to give momentary contact operation.

Dual in-line package (DIP) switches are small switch assemblies designed for mounting on printed circuit board modules (Figure 6-12). The pins or terminals on the bottom of the DIP switch are the same size and spacing as an integrated circuit (IC) chip. The individual switches may be of the toggle, rocker, or slide kind. DIP switches use binary (on/off) settings to set the parameters for a particular module. For example, the input voltage range

on a particular input module may be selected by means of DIP switches located on the back of the module.

6.5 Mechanically Operated Switches

A *mechanically operated switch* is controlled automatically by factors such as pressure, position, or temperature. The *limit switch,* shown in Figure 6-13, is a very common industrial control device. Limit switches are designed to operate only when a predetermined limit is reached, and they are usually actuated by contact with an object such as a cam. These devices take the place of a human operator. They are often used in the control circuits of machine processes to govern the starting, stopping, or reversal of motors.

The *temperature switch,* or *thermostat,* shown in Figure 6-14 is used to sense temperature changes. Although there are many types available, they are all actuated by some specific environmental temperature change.

Cam (on machine)

Operating force

Operator

Enclosure containing contact mechanism

NEMA symbols

Normally open limit switch

Normally closed limit switch

IEC symbols

Normally open

Normally closed

Figure 6-13 Mechanically operated limit switch.
Source: Photo courtesy Eaton Corporation.

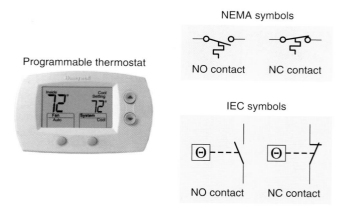

Figure 6-14 Temperature switch.
Source: Photo courtesy Honeywell, **www.honeywell.com**.

Temperature switches open or close when a designated temperature is reached. Industrial applications for these devices include maintaining the desired temperature range of air, gases, liquids, or solids.

Pressure switches, such as that shown in Figure 6-15, are used to control the pressure of liquids and gases. Although many different types are available, they are all basically designed to actuate (open or close) their contacts when a specified pressure is reached. Pressure switches can be pneumatically (air) or hydraulically (liquid) operated switches. Generally, bellows or a diaphragm presses up against a small microswitch and causes it to open or close.

Level switches are used to sense liquid levels in vessels and provide automatic control for motors that transfer liquids from sumps or into tanks. They are also used to open

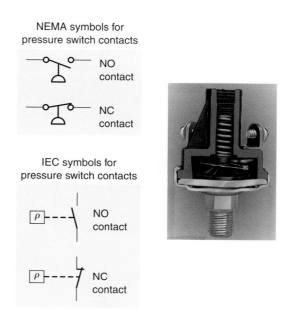

Figure 6-15 Pressure switch.
Source: Photo courtesy Honeywell, **www.honeywell.com**.

Figure 6-16 Float type level switch.
Source: Courtesy Dwyer Instruments.

or close piping solenoid valves to control fluids. The float switch shown in Figure 6-16 is a type of level switch. This switch is weighted so that as the liquid rises the switch float and turns upside down, actuating its internal contacts.

6.6 Sensors

Sensors are used for detecting, and often measuring, the magnitude of something. They convert mechanical, magnetic, thermal, optical, and chemical variations into electric voltages and currents. Sensors are usually categorized by what they measure, and they play an important role in modern manufacturing process control.

Proximity Sensor

Proximity sensors or switches, such as that shown in Figure 6-17, are pilot devices that detect the presence of an object (usually called the target) *without physical contact.* These solid-state electronic devices are completely encapsulated to protect against excessive vibration, liquids, chemicals, and corrosive agents found in the industrial environment. Proximity sensors are used when:

- The object being detected is too small, lightweight, or soft to operate a mechanical switch.
- Rapid response and high switching rates are required, as in counting or ejection control applications.
- An object has to be sensed through nonmetallic barriers such as glass, plastic, and paper cartons.
- Hostile environments demand improved sealing properties, preventing proper operation of mechanical switches.

Figure 6-17 Proximity sensor.
Source: Photo courtesy Turck, Inc., **www.turck.com**.

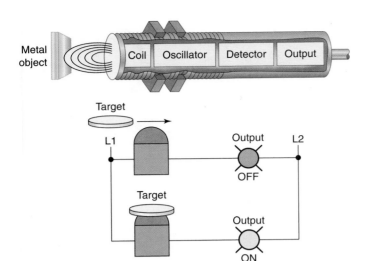

Figure 6-18 Inductive proximity sensor.

- Long life and reliable service are required.
- A fast electronic control system requires a bounce-free input signal.

Proximity sensors operate on different principles, depending on the type of matter being detected. When an application calls for noncontact metallic target sensing, an *inductive-type proximity sensor* is used. Inductive proximity sensors are used to detect both ferrous metals (containing iron) and nonferrous metals (such as copper, aluminum, and brass).

Inductive proximity sensors operate under the electrical principle of inductance, where a fluctuating current induces an electromotive force (emf) in a target object. The block diagram for an inductive proximity sensor is shown in Figure 6-18 and its operation can be summarized as follows:

- The oscillator circuit generates a high-frequency electromagnetic field that radiates from the end of the sensor.
- When a metal object enters the field, eddy currents are induced in the surface of the object.
- The eddy currents on the object absorb some of the radiated energy from the sensor, resulting in a loss of energy and change of strength of the oscillator.
- The sensor's detection circuit monitors the oscillator's strength and triggers a solid-state output at a specific level.
- Once the metal object leaves the sensing area, the oscillator returns to its initial value.

Most sensor applications operate either at 24V DC or at 120V AC. The method of connecting a proximity

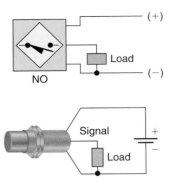

Figure 6-19 Typical three-wire DC sensor connection.

sensor varies with the type of sensor and its application. Figure 6-19 shows a typical three-wire DC sensor connection. The three-wire DC proximity sensor has the positive and negative line leads connected directly to it. When the sensor is actuated, the circuit will connect the signal wire to the positive side of the line if operating normally open. If operating normally closed, the circuit will disconnect the signal wire from the positive side of the line.

Figure 6-20 shows a typical two-wire proximity sensor connection intended to be connected in series with the load. They are manufactured for either AC or DC supply voltages. In the off state, enough current must flow through the circuit to keep the sensor active. This off state current is called leakage current and typically may range from 1 to 2 mA. When the switch is actuated, it will conduct the normal load circuit current.

Figure 6-21 shows the proximity sensor sensing range. Hysteresis is the distance between the operating point when the target approaches the proximity sensor face and the release point when the target is moving away from the sensor face. The object must be closer to turn the sensor on rather than to turn it off. If the target is moving toward the sensor, it will have to move to a closer point. Once the

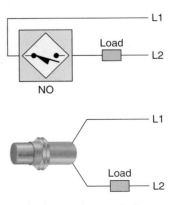

Figure 6-20 Typical two-wire proximity sensor connection.

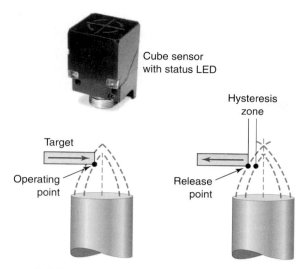

Figure 6-21 Proximity sensor sensing range.
Source: Photo courtesy Eaton Corporation, **www.eaton.com**.

sensor turns on, it will remain on until the target moves to the release point. Hysteresis is needed to keep proximity sensors from chattering when subjected to shock and vibration, slow-moving targets, or minor disturbances such as electrical noise and temperature drift. Most proximity sensors come equipped with an LED status indicator to verify the output switching action.

As a result of solid-state switching of the output, a small leakage current flows through the sensor even when the output is turned off. Similarly, when the sensor is on, a small voltage drop is lost across its output terminals. To operate properly, a proximity sensor should be powered continuously. Figure 6-22 illustrates the use of a bleeder resistor connected to allow enough current for the sensor to operate but not enough to turn on the input of the PLC.

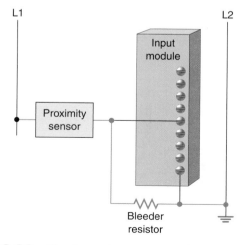

Figure 6-22 Bleeder resistor connected to continuously power a proximity sensor.

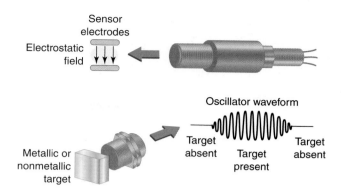

Figure 6-23 Capacitive proximity sensor.

Capacitive proximity sensors are similar to inductive proximity sensors. The main differences between the two types are that capacitive proximity sensors produce an electrostatic field instead of an electromagnetic field and are actuated by both conductive and nonconductive materials.

Figure 6-23 illustrates the operation of a capacitive sensor. A capacitive sensor contains a high-frequency oscillator along with a sensing surface formed by two metal electrodes. When the target nears the sensing surface, it enters the electrostatic field of the electrodes and changes the capacitance of the oscillator. As a result, the oscillator circuit begins oscillating and changes the output state of the sensor when it reaches certain amplitude. As the target moves away from the sensor, the oscillator's amplitude decreases, switching the sensor back to its original state.

Capacitive proximity sensors will sense metal objects as well as nonmetallic materials such as paper, glass, liquids, and cloth. They typically have a short sensing range of about 1 inch, regardless of the type of material being sensed. The larger the dielectric constant of a target, the easier it is for the capacitive sensor to detect. This makes possible the detection of materials inside nonmetallic containers as illustrated in Figure 6-24. In this example, the liquid has a much higher dielectric constant than the cardboard container, which gives the sensor the ability to

Figure 6-24 Capacitive proximity sensor liquid detection.
Source: Photo courtesy Omron Industrial Automation, **www.ia.omron.com**.

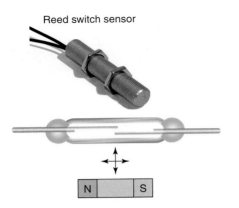

Figure 6-25 Magnetic reed switch.
Source: Courtesy of Reed Switch Developments Corp., used with permission.

see through the container and detect the liquid. In the process shown, detected empty containers are automatically diverted via the push rod.

Inductive proximity switches may be actuated only by a metal and are insensitive to humidity, dust, dirt, and the like. Capacitive proximity switches, however, can be actuated by any dirt in their environment. For general applications, the capacitive proximity switches are not really an alternative but a supplement to the inductive proximity switches. They are a supplement when there is no metal available for the actuation (e.g., for woodworking machines and for determining the exact level of liquids or powders).

Magnetic Reed Switch

A *magnetic reed switch* is composed of two flat contact tabs that are hermetically sealed (airtight) in a glass tube filled with protective gas, as illustrated in Figure 6-25. When a magnetic force is generated parallel to the reed switch, the reeds become flux carriers in the magnetic circuit. The overlapping ends of the reeds become opposite magnetic poles, which attract each other. If the magnetic force between the poles is strong enough to overcome the restoring force of the reeds, the reeds will be drawn together to actuate the switch. Because the contacts are sealed, they are unaffected by dust, humidity, and fumes; thus, their life expectancy is quite high.

Light Sensors

The photovoltaic cell and the photoconductive cell, illustrated in Figure 6-26, are two examples of light sensors. The *photovoltaic* or *solar cell* reacts to light by converting the light energy directly into electric energy. The *photoconductive cell* (also called a *photoresistive cell*) reacts to light by change in the resistance of the cell.

A *photoelectric sensor* is an optical control device that operates by detecting a visible or invisible beam of light and responding to a change in the received light intensity. Photoelectric sensors are composed of two basic components: a transmitter (light source) and a receiver (sensor), as shown in Figure 6-27. These two components may or may not be housed in separate units. The

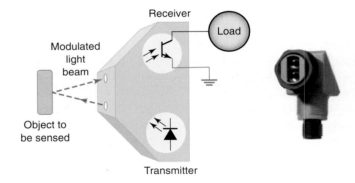

Figure 6-27 Photoelectric sensor.
Source: Photo courtesy SICK, Inc., **www.sick.com**.

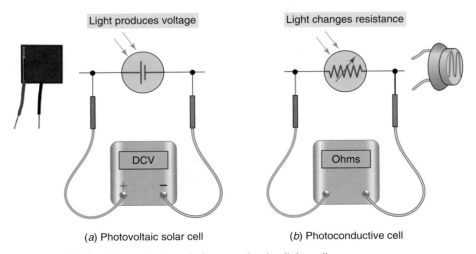

(*a*) Photovoltaic solar cell   (*b*) Photoconductive cell

Figure 6-26 Photovoltaic and photoconductive light cells.

basic operation of a photoelectric sensor can be summarized as follows:

- The transmitter contains a light source, usually an LED along with an oscillator.
- The oscillator modulates or turns the LED on and off at a high rate of speed.
- The transmitter sends this modulated light beam to the receiver.
- The receiver decodes the light beam and switches the output device, which interfaces with the load.
- The receiver is tuned to its emitter's modulation frequency and will only amplify the light signal that pulses at the specific frequency.
- Most sensors allow adjustment of how much light will cause the output of the sensor to change state.
- Response time is related to the frequency of the light pulses. Response times may become important when an application calls for the detection of very small objects, objects moving at a high rate of speed, or both.

The scan technique refers to the method used by photoelectric sensors to detect an object. The *through-beam* scan technique (also called direct scan) places the transmitter and receiver in direct line with each other, as illustrated in Figure 6-28. Because the light beam travels in only one direction, through-beam scanning provides long-range sensing. Quite often, a garage door opener has a through-beam photoelectric sensor mounted near the floor, across the width of the door. For this application the sensor senses that nothing is in the path of the door when it is closing.

In a *retroreflective scan,* the transmitter and receiver are housed in the same enclosure. This arrangement requires the use of a separate reflector or reflective tape mounted across from the sensor to return light back to the receiver. The retroreflective scan is designed to respond to objects that interrupt the beam normally maintained between the transmitter and receiver, as illustrated

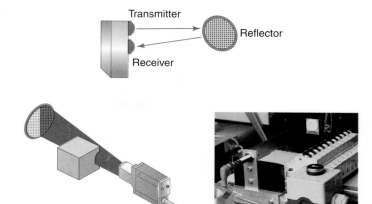

Figure 6-29 Retroreflective scan.
Source: Photo courtesy ifm efector, **www.ifm.com/us**.

in Figure 6-29. In contrast to a through-beam application, retroreflective sensors are used for medium-range applications.

Fiber optics is not a scan technique, but another method for transmitting light. *Fiber optic sensors* use a flexible cable containing tiny fibers that channel light from emitter to receiver, as illustrated in Figure 6-30. Fiber optic sensor systems are completely immune to all forms of electrical interference. The fact that an optical fiber does not contain any moving parts and carries only light means that there is no possibility of a spark. This means that it can be safely used even in the most hazardous sensing environments such as a refinery for producing gases, grain bins, mining, pharmaceutical manufacturing, and chemical processing.

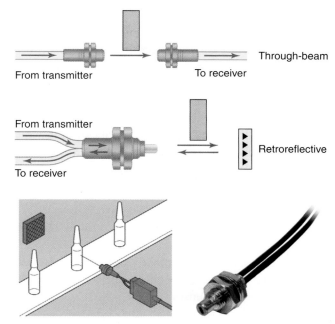

Figure 6-30 Fiber optic sensors.
Source: Photo courtesy Omron Industrial Automation, **www.ia.omron.com**.

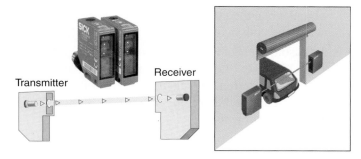

Figure 6-28 Through-beam scan.
Source: Photo courtesy SICK, Inc., **www.sick.com**.

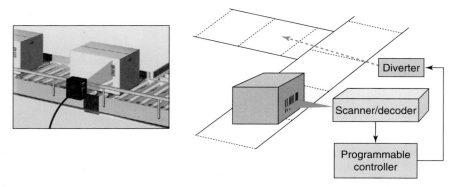

Figure 6-31 PLC bar code application.
Source: Courtesy Keyence Canada, Inc.

Bar code technology is widely implemented in industry to enter data quickly and accurately. *Bar code scanners* are the eyes of the data collection system. A light source within the scanner illuminates the bar code symbol; those bars absorb light, and spaces reflect light. A photodetector collects this light in the form of an electronic-signal pattern representing the printed symbol. The decoder receives the signal from the scanner and converts these data into the character data representation of the symbol's code. Figure 6-31 illustrates a typical PLC application which involves a bar code module reading the bar code on boxes as they move along a conveyor line. The PLC is then used to divert the boxes to the appropriate product lines according to the data read from the bar code.

Ultrasonic Sensors

An *ultrasonic sensor* operates by sending high-frequency sound waves toward the target and measuring the time it takes for the pulses to bounce back. The time taken for this echo to return to the sensor is directly proportional to the distance or height of the object because sound has a constant velocity.

Figure 6-32 illustrates a practical application in which the returning echo signal is electronically converted to a 4- to 20-mA output, which supplies a monitored flow rate to external control devices. The operation of this process can be summarized as follows:

- The 4-20 mA represents the sensor's measurement span.
- The 4-mA set point is typically placed near the bottom of the empty tank, or the greatest measurement distance from the sensor.
- The 20-mA set point is typically placed near the top of the full tank, or the shortest measurement distance from the sensor.

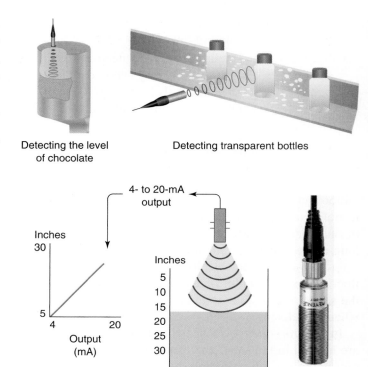

Detecting the level of chocolate

Detecting transparent bottles

Level detection

Figure 6-32 Ultrasonic sensor.
Source: Courtesy Keyence Canada, Inc.

- The sensor will proportionately generate a 4-mA signal when the tank is empty and a 20-mA signal when the tank is full.
- Ultrasonic sensors can detect solids, fluids, granular objects, and textiles. In addition, they enable the detection of different objects irrespective of color and transparency and therefore are ideal for monitoring transparent objects.

Strain/Weight Sensors

A *strain gauge* converts a mechanical strain into an electric signal. Strain gauges are based on the principle that

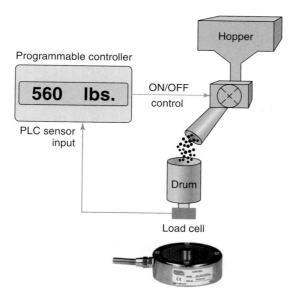

Figure 6-33 Strain gauge load cell.
Source: Courtesy RDP Group.

the resistance of a conductor varies with length and cross-sectional area. The force applied to the gauge causes the gauge to bend. This bending action also distorts the physical size of the gauge, which in turn changes its *resistance*. This resistance change is fed to a bridge circuit that detects small changes in the gauge's resistance. *Strain gauge load cells* are usually made with steel and sensitive strain gauges. As the load cell is loaded, the metal elongates or compresses very slightly. The strain gauge detects this movement and translates it to a varying voltage signal. Many sizes and shapes of load cells are available, and they range in sensitivity from grams to millions of pounds. Strain gauge–based load cells are used extensively for industrial weighing applications similar to the one illustrated in Figure 6-33.

Temperature Sensors

The thermocouple is the most widely used temperature sensor. Thermocouples operate on the principle that when two dissimilar metals are joined, a predictable DC voltage will be generated that relates to the difference in temperature between the hot junction and the cold junction (Figure 6-34). The hot junction (measuring junction) is the joined end of a thermocouple that is exposed to the process where the temperature measurement is desired. The cold junction (reference junction) is the end of a thermocouple that is kept at a constant temperature to provide a reference point. For example, a K-type thermocouple, when heated to a temperature of 300°C at the hot junction, will produce 12.2 mV at the cold junction. Because of their ruggedness and wide temperature range, thermocouples are used in industry to monitor and control oven and furnace temperatures.

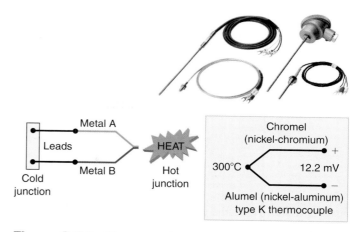

Figure 6-34 Thermocouple temperature sensor.
Source: Photo courtesy Omron Industrial Automation, **www.ia.omron.com**.

Flow Measurement

Many industrial processes depend on accurate measurement of fluid flow. Although there are several ways to measure fluid flow, the usual approach is to convert the kinetic energy that the fluid has into some other measurable form.

Turbine-type flowmeters are a popular means of measurement and control of liquid products in industrial, chemical, and petroleum operations. Turbine flowmeters, like windmills, utilize their angular velocity (rotation speed) to indicate the flow velocity. The operation of a turbine flowmeter is illustrated in Figure 6-35. Its basic construction consists of a bladed turbine rotor installed in a flow tube. The bladed rotor rotates on its axis in proportion to the rate of the liquid flow through the tube. A magnetic pickup sensor is positioned as close to the rotor as practical. Fluid passing through the flow tube causes the rotor to rotate, which generates pulses in the pickup coil. The frequency of the pulses is then transmitted to readout electronics and displayed as gallons per minute.

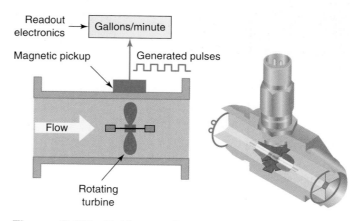

Figure 6-35 Turbine type flowmeter.

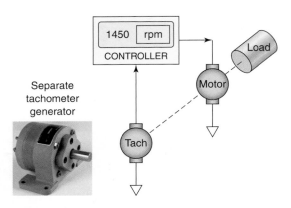

Figure 6-36 Tachometer generator feedback.
Source: Courtesy ATC Digitec.

Velocity and Position Sensors

Tachometer generators provide a convenient means of converting rotational speed into an analog voltage signal that can be used for motor speed indication and control applications. A tachometer generator is a small AC or DC generator that develops an output voltage (proportional to its rpm) whose phase or polarity depends on the rotor's direction of rotation. The DC tachometer generator usually has permanent magnetic field excitation. The AC tachometer generator field is excited by a constant AC supply. In either case, the rotor of the tachometer is mechanically connected, directly or indirectly, to the load.

Figure 6-36 illustrates motor speed control applications in which a tachometer generator is used to provide a feedback voltage to the motor controller that is proportional to motor speed. The control motor and tachometer generator may be contained in the same or separate housings.

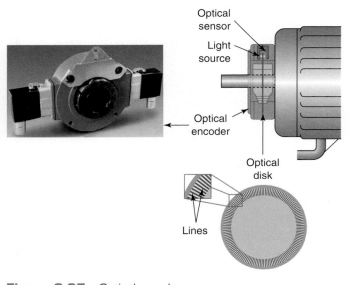

Figure 6-37 Optical encoder.
Source: Photo courtesy Avtron, **www.avtron.com**.

An *encoder* is used to convert linear or rotary motion into a binary digital signal. Encoders are used in applications where positions have to be precisely determined. The optical encoder illustrated in Figure 6-37 uses a light source shining on an optical disk with lines or slots that interrupt the beam of light to an optical sensor. An electronic circuit counts the interruptions of the beam and generates the encoder's digital output pulses.

6.7 Output Control Devices

A variety of output control devices can be operated by the PLC output to control traditional industrial processes. These devices include pilot lights, control relays, motor starters, alarms, heaters, solenoids, solenoid valves, small motors, and horns. Similar electrical symbols are used to represent these devices both on relay schematics and PLC output connection diagrams. Figure 6-38 shows common electrical symbols used

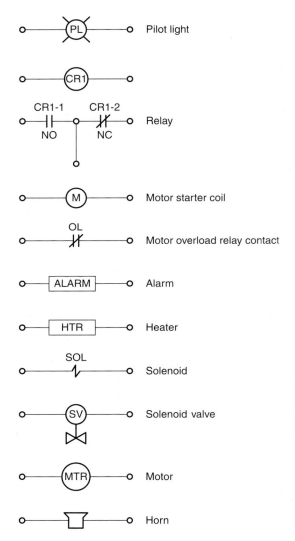

Figure 6-38 Symbols for output control devices.

for various output devices. Although these symbols are generally acceptable, some differences among manufacturers do exist.

An *actuator,* in the electrical sense, is any device that converts an electrical signal into mechanical movement. An electromechanical solenoid is an actuator that uses electrical energy to magnetically cause mechanical control action. A solenoid consists of a coil, frame, and plunger (or armature, as it is sometimes called). Figure 6-39 shows the basic construction and operation of a solenoid. Its operation can be summarized as follows:

- The coil and frame form the fixed part.
- When the coil is energized, it produces a magnetic field that attracts the plunger, pulling it into the frame and thus creating mechanical motion.
- When the coil is de-energized the plunger returns to its normal position through gravity or assistance from spring assemblies within the solenoid.
- The frame and plunger of an AC-operated solenoid are constructed with laminated pieces instead of a solid piece of iron to limit eddy currents induced by the magnetic field.

Solenoid valves are electromechanical devices that work by passing an electrical current through a solenoid, thereby changing the state of the valve. Normally, there is a mechanical element, which is often a spring, that holds the valve in its default position. A solenoid valve is a combination of a solenoid coil operator and valve, which controls the flow of liquids, gases, steam, and other media. When electrically energized, they open, shut off, or direct the flow of media.

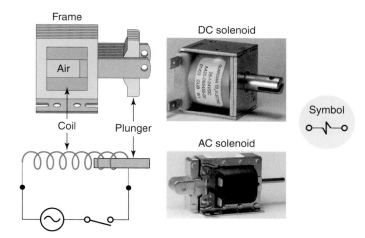

Figure 6-39 Solenoid construction and operation.
Source: Photos courtesy Guardian Electric, **www.guardian-electric.com**.

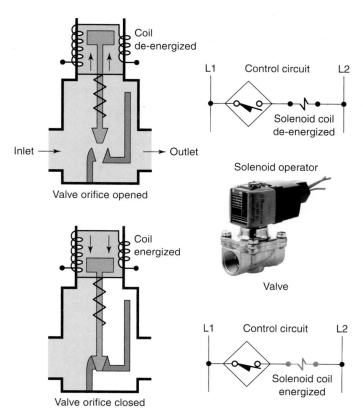

Figure 6-40 Solenoid valve construction and operation.
Source: Photo courtesy ASCO Valve Inc., **www.ascovalve.com**.

Figure 6-40 illustrates the construction and principle of operation of a typical fluid solenoid valve. Its operation can be summarized as follows:

- The valve body contains an orifice in which a disk or plug is positioned to restrict or allow flow.
- Flow through the orifice is either restricted or allowed depending on whether the solenoid coil is energized or de-energized.
- When the coil is energized, the core is drawn into the solenoid coil to open the valve.
- The spring returns the valve to its original closed position when the current coil is de-energized.
- A valve must be installed with direction of flow in accordance with the arrow cast on the side of the valve body.

Stepper motors operate differently than standard types, which rotate continuously when voltage is applied to their terminals. The shaft of a stepper motor rotates in discrete increments when electrical command pulses are applied to it in the proper sequence. Every revolution is divided into a number of steps, and the motor must be sent a voltage pulse for each step. The amount of rotation is directly proportional to the number of pulses, and the speed of rotation is relative to the frequency of those pulses. A 1-degree-per-step motor will require 360 pulses to move through one revolution; the degrees per step are known as the *resolution*. When stopped, a stepper motor inherently holds its position. Stepper systems are used most often in "open-loop" control systems, where the controller tells the motor only how many steps to move and how fast to move, but does not have any way of knowing what position the motor is at.

The movement created by each pulse is precise and repeatable, which is why stepper motors are so effective for load-positioning applications. Conversion of rotary to linear motion inside a linear actuator is accomplished through a threaded nut and lead screw. Generally, stepper motors produce less than 1 hp and are therefore frequently used in low-power position control applications. Figure 6-41 shows a stepper motor/drive unit along with typical rotary and linear applications.

All *servo motors* operate in closed-loop mode, whereas most stepper motors operate in open-loop mode. Closed-loop and open-loop control schemes are illustrated in Figure 6-42. *Open loop* is control without feedback, for example, when the controller tells the stepper motor how many steps to move and how fast to move, but does not verify where the motor is. *Closed-loop* control compares speed or position feedback with the commanded speed or position and generates a modified command to make the error smaller. The error is the difference between the required speed or position and the actual speed or position.

Figure 6-43 illustrates a closed-loop servo motor system. The motor controller directs operation of the servo motor by sending speed or position command signals to the amplifier, which drives the servo motor. A feedback device such as an encoder for position and a tachometer

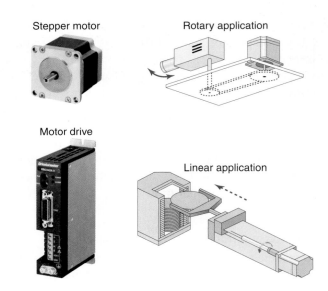

Figure 6-41 Stepper motor/drive unit.
Source: Photos courtesy Oriental Motor, **www.orientalmotor.com**.

for speed are either incorporated within the servo motor or are remotely mounted, often on the load itself. These provide the servo motor's position and speed feedback information that the controller compares to its programmed motion profile and uses to alter its position or speed.

6.8 Seal-In Circuits

Seal-in, or *holding,* circuits are very common in both relay logic and PLC logic. Essentially, a seal-in circuit is a method of maintaining current flow after a momentary switch has been pressed and released. In these types of circuits, the seal-in contact is usually in parallel with the momentary device.

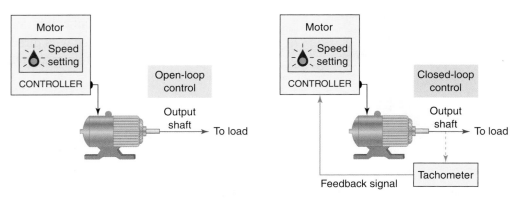

Figure 6-42 Open- and closed-loop motor control systems.

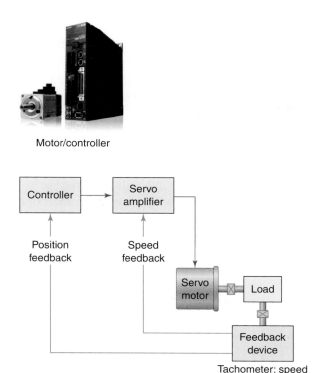

Motor/controller

Figure 6-43 Closed-loop servo motor system.
Source: Photos courtesy Omron Industrial Automation, **www.ia.omron.com**.

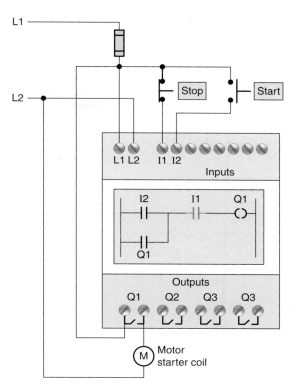

Figure 6-45 Motor seal-in circuit implemented using an Allen-Bradley Pico controller.

The motor stop/start circuit shown in Figure 6-44 is a typical example of a seal-in circuit. The hardwired circuit consists of a normally closed stop button in series with a normally open start button. The seal-in auxiliary contact of the starter is connected in parallel with the start button to keep the starter coil energized when the start button is released. When this circuit is programmed into a PLC, both the start and stop buttons are examined for a closed condition because both buttons must be closed to cause the motor starter to operate.

Figure 6-45 shows a PLC wiring diagram of the motor seal-in circuit implemented using an Allen-Bradley Pico controller. The controller is programmed using ladder logic. Each programming element can be entered directly

via the Pico display. This controller also lets you program the circuit from a personal computer using PicoSoft programming software.

6.9 Latching Relays

Electromagnetic latching relays are designed to hold the relay closed after power has been removed from the coil. Latching relays are used where it is necessary for contacts to stay open and/or closed even though the coil is energized only momentarily. Figure 6-46 shows a latching relay that uses two coils. The *latch* coil is momentarily energized to set the latch and hold the relay in the latched position. The *unlatch* or release coil is momentarily

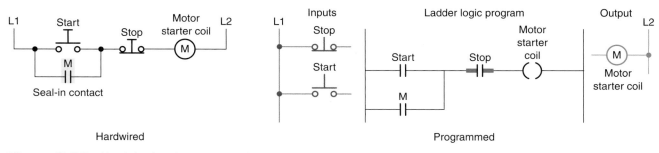

Figure 6-44 Hardwired and programmed seal-in circuit.

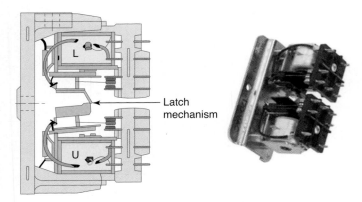

Figure 6-46 Two-coil mechanical latching relay.
Source: Courtesy Relay Service Company.

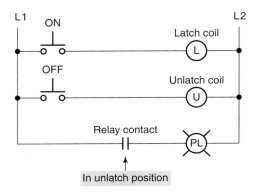

Figure 6-47 Hardwired control circuit for an electromagnetic latching relay.

energized to disengage the mechanical latch and return the relay to the unlatched position.

Figure 6-47 shows a hardwired control circuit for an electromagnetic latching relay. The operation of the circuit can be summarized as follows:

- The contact is shown with the relay in the *unlatched* position.
- In this state the circuit to the pilot light is open and so the light is off.

- When the ON button is *momentarily* actuated, the latch coil is energized to set the relay to its latched position.
- The contacts close, completing the circuit to the pilot light, and so the light is switched on.
- The relay coil does *not* have to be continuously energized to hold the contacts closed and keep the light on.
- The only way to switch the lamp off is to actuate the OFF button, which will energize the unlatch coil and return the contacts to their open, unlatched state.
- In cases of power loss, the relay will remain in its original latched or unlatched state when power is restored.

An electromagnetic latching relay function can be programmed on a PLC to work like its real-world counterparts. The instruction set for the SLC 500 includes a set of output instructions that duplicates the operation of the mechanical latch. A description of the output latch (OTL) and output unlatch (OTU) instruction is given in Figure 6-48. The OTL and OTU instructions differ from the OTE instruction in that they must be used together. Both the latch and unlatch outputs must have the same address. The OTL (latch) instruction can only turn a bit on and the OTU (unlatch) instruction can only turn a bit off.

The operation of the output latch and output unlatch coil instruction is illustrated in the ladder program of Figure 6-49. The operation of the program can be summarized as follows:

- Both the latch (L) and the unlatch (U) coil have the *same* address (O:2/5).
- When the on pushbutton (I:1/0) is momentarily actuated, the latch rung becomes true and the latch status bit (O:2/5) is set to 1, and so the light output is switched on.

Command	Name	Symbol	Description
OTL	Output latch	–(L)–	OTL sets the bit to "1" when the rung becomes true and retains its state when the rung loses continuity or a power cycle occurs.
OTU	Output unlatch	–(U)–	OTU resets the bit to "0" when the rung becomes true and retains it.

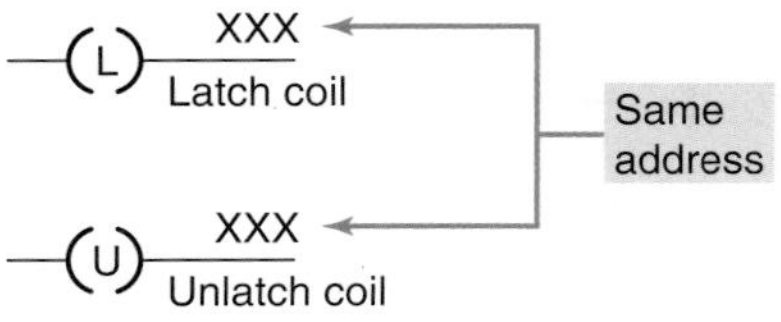

Figure 6-48 Output latch and output unlatch instruction.

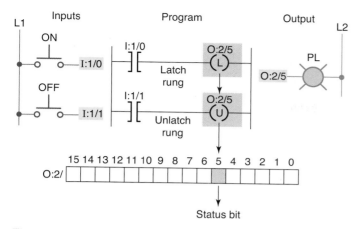

Figure 6-49 Output latch and output unlatch operation.

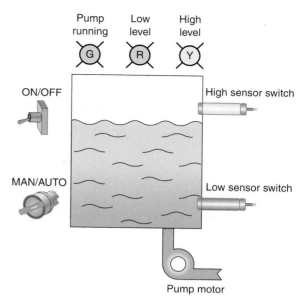

Figure 6-50 Process used to control the level of water in a storage tank.

- The status bit *will remain set to 1* when the pushbutton is released and logical continuity of the latch rung is lost.
- When the off pushbutton (I:1/1) is momentarily actuated, the unlatch rung becomes true and the latch status bit (O:2/5) is reset back to 0 and so the light is switched off.
- The status bit *will remain reset to 0* when the pushbutton is released and logical continuity of the latch rung is lost.

Output latch is an output instruction with a bit-level address. When the instruction is true, it sets a bit in the output image file. It is a retentive instruction because the bit remains set when the latch instruction goes false. In most applications it is used with an unlatch instruction. The output unlatch instruction is also an output instruction with a bit-level address. When the instruction is true, it resets a bit in the output image file. It, too, is a retentive instruction because the bit remains reset when the instruction goes false.

The process shown in Figure 6-50 is to be used to control the level of water in a storage tank by turning a discharge pump on or off. The modes of operation are to be programmed as follows:

OFF Position—The water pump will *stop* if it is running and will *not* start if it is stopped.

Manual Mode—The pump will start if the water in the tank is at any level except low.

Automatic Mode—If the level of water in the tank *reaches a high point*, the water pump will *start* so that water can be removed from the tank, thus lowering the level.
- When the water level *reaches a low point*, the pump will *stop*.

Status Indicating Lights—Water pump running light (green)
- Low water level status light (red)
- High water level status light (yellow)

Figure 6-51 shows a program that can be used to implement control of the water level in the storage tank. The latch and unlatch instructions form part of the program. The operation of the program can be summarized as follows:

- An internal storage bit is used for the latch and address rather than an actual discrete output address. Both have the same addresses.
- The rung 1 Examine-on instruction addressed to the off/on switch prevents the pump motor from starting under any condition when in the off (open) state.
- In the MAN mode, the rung 1 Examine-on instruction addressed to the low sensor switch allows the pump motor to operate only when the low level sensor switch is closed.
- In the AUTO mode, whenever the high sensor switch is momentarily closed the Examine-on instruction of rung 1 addressed to it will energize the latch coil. The pump will begin running and continue to operate until the unlatch coil is energized by the rung 3 Examine-off instruction addressed to the low sensor switch.
- The pump running status light is controlled by the rung 4 Examine-on instruction addressed to the motor output.

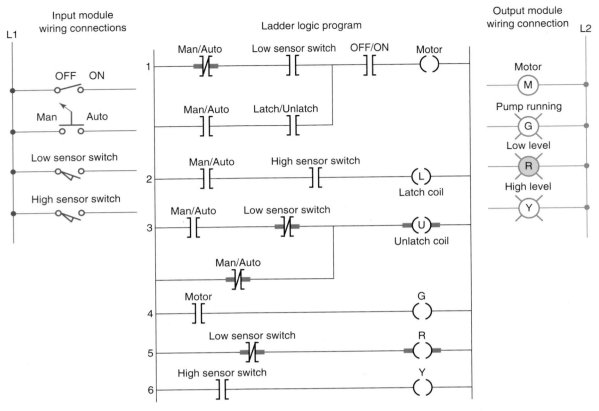

Figure 6-51 Program used to implement control of the water level in the storage tank.

- The low level status light is controlled by the rung 5 Examine-off instruction addressed to the low sensor switch.
- The high level status light is controlled by the rung 6 Examine-on instruction addressed to the high sensor switch.

Figure 6-52 shows a typical I/O module wiring diagram and addressing format for the water level control program implemented using an Allen-Bradley modular SLC 500 controller. The chassis power supply has a relatively small power rating and is used to supply DC power to all devices physically mounted in the backplane of the PLC rack. In this application a 24 VDC field power supply is used for the input devices and a 120 VAC field power supply for the output devices. This allows a low-voltage 24-volt control signal to control 240-volt output devices. SLC 500 controllers use a rack/slot-based address system where the slot location of the I/O modules in the rack establishes the PLC address. The addresses for the field devices of this particular application are shown below:

FIELD DEVICE	ADDRESS	Signifies
OFF/ON Switch	I:2/0	The input module in slot 2 and screw terminal 0
MAN/AUTO Switch	I:2/4	The input module in slot 2 and screw terminal 4
LOW SENSOR SWITCH	I:2/8	The input module in slot 2 and screw terminal 8
HIGH SENSOR SWITCH	I:2/12	The input module in slot 2 and screw terminal 12
MOTOR	O:3/1	The output module in slot 3 and screw terminal 1
PUMP RUNNING Light	O:3/5	The output module in slot 3 and screw terminal 5
LOW LEVEL Light	O:3/9	The output module in slot 3 and screw terminal 9
HIGH LEVEL Light	O:3/13	The output module in slot 3 and screw terminal 13
	B3:0/0	Internal retentive bit instruction that does not drive a real-word device

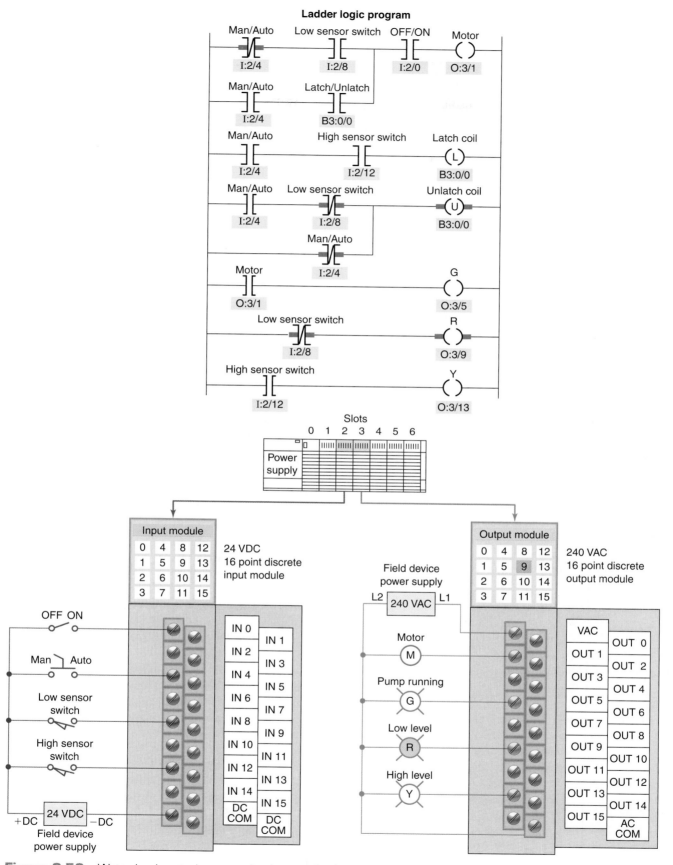

Figure 6-52 Water level control program implemented using an Allen-Bradley modular SLC 500 controller.

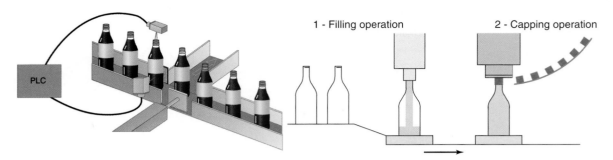

Figure 6-53 Sequential control process.
Source: Photo courtesy Omron Industrial Automation, **www.ia.omron.com**.

6.10 Converting Relay Schematics into PLC Ladder Programs

The best approach to developing a PLC program from a relay schematic is to understand first the operation of each relay ladder rung. As each relay ladder rung is understood, an equivalent PLC rung can be generated. This process will require access to the relay schematic, documentation of the various input and output devices used, and possibly a process flow diagram of the operation.

Most control processes require the completion of several operations to produce the required output. Manufacturing, machining, assembling, packaging, finishing, or transporting of products requires the precise coordination of tasks.

A *sequential* control process is required for processes that demand that certain operations be performed in a specific order. Figure 6-53 illustrates part of a bottle filling process. In the filling and capping operations, the tasks are (1) fill bottle and (2) press on cap. These tasks must be performed in the proper order. Obviously we could not fill the bottle after the cap is pressed on. This process, therefore, requires sequential control.

Combination controls require that certain operations be performed without regard to the order in which they are performed. Figure 6-54 illustrates another part of the same bottle filling process. Here, the tasks are (1) place label 1 on bottle and (2) place label 2 on bottle. The order in which the tasks are performed does not really matter. In fact, however, many industrial processes that are not inherently sequential in nature are performed in a sequential manner for the most efficient order of operations.

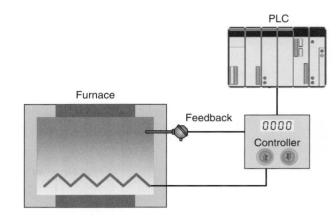

Figure 6-55 Automatic control process.

Automatic control involves maintaining a desired set point at an output. One example is maintaining a certain set-point temperature in a furnace as illustrated in Figure 6-55. If there is deviation from that set point, an error is determined by comparing the output against the set point and using this error to make a correction. This requires feedback from the output to the control for the input.

The converting of a simple sequential process can be examined with reference to the process flow diagram illustrated in Figure 6-56. The sequential task is as follows:

1. Start button is pressed.
2. Table motor is started.

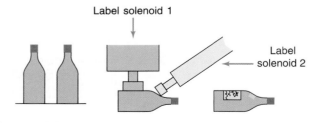

Figure 6-54 Combination control process.

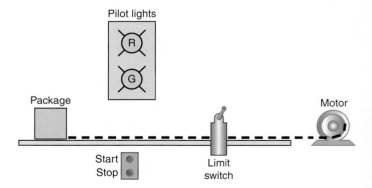

Figure 6-56 Sequential process flow diagram.

3. Package moves to the position of the limit switch and automatically stops.

Other auxiliary features include:

- A stop button that will stop the table, for any reason, before the package reaches the limit switch position
- A red pilot light to indicate the table is stopped
- A green pilot light to indicate the table is running

A relay schematic for the sequential process is shown in Figure 6-57. The operation of this hardwired circuit can be summarized as follows:

- Start button is actuated; CR is energized if stop button and limit switch are not actuated.
- Contact CR-1 closes, sealing in CR when the start button is released.
- Contact CR-2 opens, switching the red pilot light from on to off.
- Contact CR-3 closes, switching the green pilot light from off to on.
- Contact CR-4 closes to energize the motor starter coil, starting the motor and moving the package toward the limit switch.
- Limit switch is actuated, de-energizing relay coil CR.
- Contact CR-1 opens, opening the seal-in circuit.
- Contact CR-2 closes, switching the red pilot light from off to on.
- Contact CR-3 opens, switching the green pilot light from on to off.
- Contact CR-4 opens, de-energizing the motor starter coil to stop the motor and end the sequence.

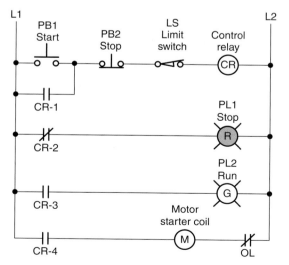

Figure 6-57 Relay schematic for the sequential process.

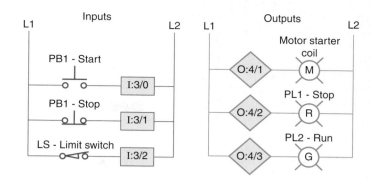

Figure 6-58 I/O connection diagram.

Figure 6-58 shows an I/O connection diagram for a programmed version of the sequential process. Each input and output device is represented by its symbol and associated address. These addresses will indicate what PLC input is connected to what input device and what PLC output will drive what output device. The address code, of course, will depend on the PLC model used. This example uses SLC 500 addressing for the process. Note that the electromagnetic control relay CR is *not* needed because its function is replaced by an *internal* PLC control relay.

The hardwired relay schematic for the sequential process can be converted to the PLC ladder logic program shown in Figure 6-59. In converting the process to a program the operation of each rung must be understood. The pushbuttons PB1, PB2 as well as limit switch LS are all programmed using the examine-closed (–] [–) instruction to produce the desired logic control. Also, internal relay B3:1/0 is used to replace control relay CR. To obtain the desired control logic, all internal relay contacts are programmed using the PLC contact instruction that matches the coil de-energized state. The internal relay implemented in software requires one coil address the contacts of which can be examined for an ON or OFF condition as many times as you like.

There is more than one method to correctly design the ladder logic program for a given control process. In some cases one arrangement may be more efficient in terms of the amount of memory used and the time required to scan the program. Figure 6-60 illustrates an example of an arrangement of series instructions of a rung programmed for optimum scan time. The series instructions are programmed from the most likely to be *false* (far left) to the least likely to be *false* (far right). Once the processor sees a false input instruction in series, the processor stops checking the rung at the false condition and sets the output false.

Figure 6-61 illustrates an example of an arrangement of parallel instructions of a rung programmed for optimum scan time. The parallel path that is most often *true* is

Ladder logic program

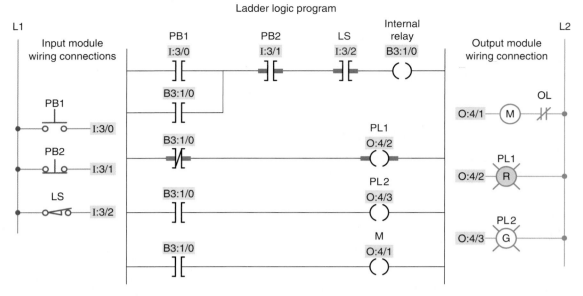

Figure 6-59 Sequential process PLC ladder logic program.

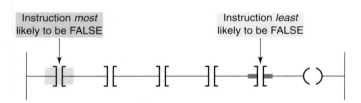

Figure 6-60 Series instructions programmed for optimum scan time.

placed on the top of the rung. The processor will not look at the others unless the top path is *false*.

Figure 6-62 shows a hardwired jog control circuit that incorporates a jog control relay. The operation of the circuit can be summarized as follows:

- Pressing the start pushbutton completes a circuit for the CR coil, closing the CR1 and CR2 contacts.
- The CR1 contact completes the circuit for the M coil, starting the motor.

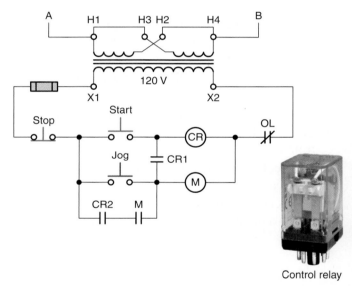

Figure 6-62 Jog circuit with control relay.
Source: Photo courtesy IDEC Corporation, **www.IDEC.com/usa**, RR Relay.

- The M maintaining contact closes; this maintains the circuit for the M coil.
- Pressing the jog button energizes the M coil only, starting the motor. Both CR contacts remain open, and the CR coil is de-energized. The M coil will not remain energized when the jog push button is released.

Figure 6-63 shows a PLC program equivalent of the hardwired relay jog circuit. Note that the function of the control relay is now accomplished using an internal PLC instruction (B3:1/0).

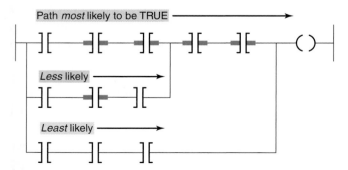

Figure 6-61 Parallel instructions programmed for optimum scan time.

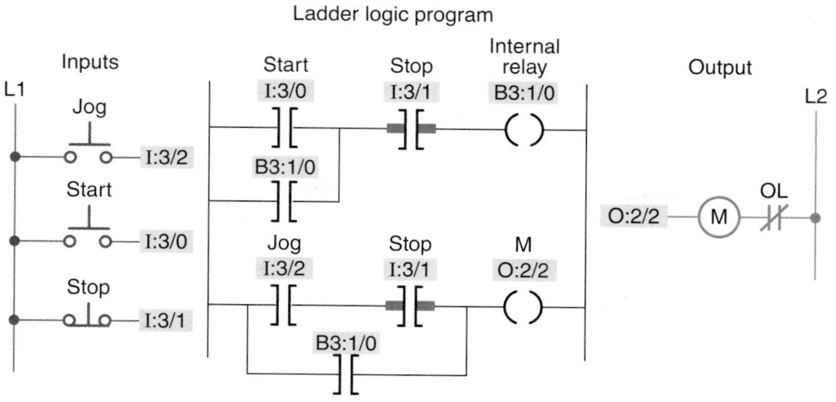

Ladder logic program

Figure 6-63 PLC program equivalent of the hardwired relay jog circuit.

6.11 Writing a Ladder Logic Program Directly from a Narrative Description

In most cases, it is possible to prepare a ladder logic program directly from the narrative description of a control process. Some of the steps in planning a program are as follows:

- Define the process to be controlled.
- Draw a sketch of the process, including all sensors and manual controls needed to carry out the control sequence.

- List the sequence of operational steps in as much detail as possible.
- Write the ladder logic program to be used as a basis for the PLC program.
- Consider different scenarios where the process sequence may go astray and make adjustments as needed.
- Consider the safety of operating personnel and make adjustments as needed.

The following are examples of ladder logic programs derived from narrative descriptions of control processes.

EXAMPLE 6-1

Figure 6-64 shows the sketch of a drilling process that requires the drill press to turn on only if there is a part present and the operator has one hand on each of the start switches. This precaution will ensure that the operator's hands are not in the way of the drill.

The sequence of operation requires that switches 1 and 2 and the part sensor all be activated to make the drill motor operate. Figure 6-65 shows the ladder logic program required for the process implemented using an SLC 500 controller.

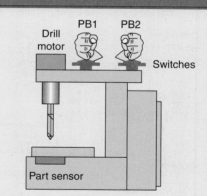

Figure 6-64 Sketch of the drilling process.

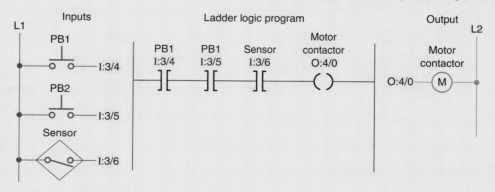

Figure 6-65 Drilling process PLC program.

EXAMPLE 6-2

A motorized overhead garage door is to be operated automatically to preset open and closed positions. The field devices include one of each of the following:

- Reversing *motor contactor* for the up and down directions.
- Normally *closed down limit switch* to sense when the door is fully closed.
- Normally closed *up limit switch* to sense when the door is fully opened.
- Normally open *door up button* for the up direction.
- Normally open *door down button* for the down direction.
- Normally closed *door stop button* for stopping the door.
- Red *door ajar light* to signal when the door is partially open.

- Green *door open light* to signal when the door is fully open.
- Yellow *door closed light* to signal when the door is fully closed.

The sequence of operation requires that:

- When the up button is pushed, the up motor contactor energizes and the door travels upward until the up limit switch is actuated.
- When the down button is pushed, the down motor contactor energizes and the door travels down until the down limit switch is actuated.
- When the stop button is pushed, the motor stops. The motor must be stopped before it can change direction.

Figure 6-66 shows the ladder logic program required for the operation implemented using an SLC 500 controller.

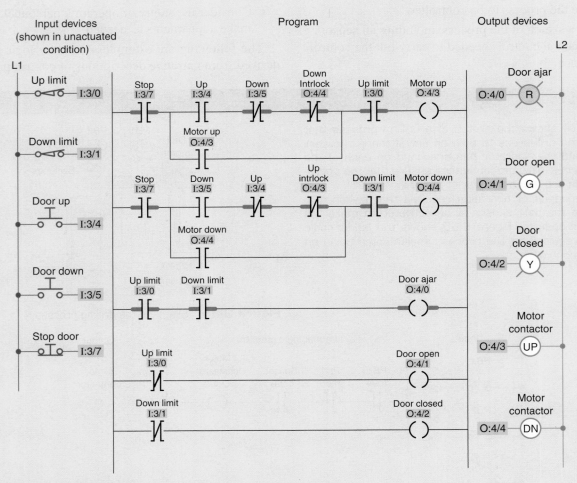

Figure 6-66 Motorized overhead garage door PLC program.

EXAMPLE 6-3

Figure 6-67 shows the sketch of a continuous filling operation. This process requires that boxes moving on a conveyor be automatically positioned and filled.

The sequence of operation for the continuous filling operation is as follows:

- Start the conveyor when the start button is momentarily pressed.
- Stop the conveyor when the stop button is momentarily pressed.
- Energize the run status light when the process is operating.
- Energize the standby status light when the process is stopped.
- Stop the conveyor when the right edge of the box is first sensed by the photosensor.
- With the box in position and the conveyor stopped, open the solenoid valve and allow the box to fill. Filling should stop when the level sensor goes true.
- Energize the full light when the box is full. The full light should remain energized until the box is moved clear of the photosensor.

Figure 6-68 shows the ladder logic program required for the operation.

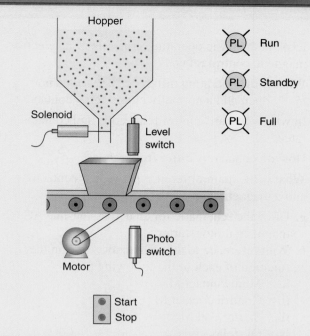

Figure 6-67 Sketch of the continuous filling operation.

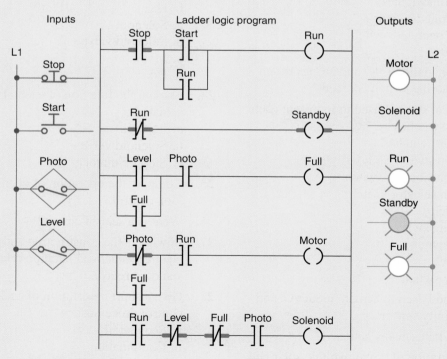

Figure 6-68 Continuous filling operation PLC program.

1. Explain the basic operating principle of an electromagnetic control relay.

2. What is the operating difference between a normally open and a normally closed relay contact?

3. In what ways are control relay coils and contacts rated?

4. How do contactors differ from relays?

5. What is the main difference between a contactor and a magnetic motor starter?

6. **a.** Draw the schematic for an across-the-line AC magnetic motor starter.
   b. With reference to this schematic, explain the function of each of the following parts:
   i. Main contact M
   ii. Control contact M
   iii. Starter coil M
   iv. OL relay coils
   v. OL relay contact

7. The current requirement for the control circuit of a magnetic starter is normally much smaller than that required by the power circuit. Why?

8. Compare the method of operation of each of the following types of switches:
   a. Manually operated switch
   b. Mechanically operated switch
   c. Proximity switch

9. What do the abbreviations NO and NC represent when used to describe switch contacts?

10. Draw the electrical symbol used to represent each of the following switches:
    a. NO pushbutton switch
    b. NC pushbutton switch
    c. Break-make pushbutton switch
    d. Three-position selector switch
    e. NO limit switch
    f. NC temperature switch
    g. NO pressure switch
    h. NC level switch
    i. NO proximity switch

11. Outline the method used to actuate inductive and capacitive proximity sensors.

12. How are reed switch sensors actuated?

13. Compare the operation of a photovoltaic solar cell with that of a photoconductive cell.

14. What are the two basic components of a photoelectric sensor?

15. Compare the operation of the reflective-type and through-beam photoelectric sensors.

16. Give an explanation of how a scanner and a decoder act in conjunction with each other to read a bar code.

17. How does an ultrasonic sensor operate?

18. Explain the principle of operation of a strain gauge.

19. Explain the principle of operation of a thermocouple.

20. What is the most common approach taken with regard to the measurement of fluid flow?

21. Explain how a tachometer is used to measure rotational speed.

22. How does an optical encoder work?

23. Draw an electrical symbol used to represent each of the following PLC output control devices:
    a. Pilot light
    b. Relay
    c. Motor starter coil
    d. OL relay contact
    e. Alarm
    f. Heater
    g. Solenoid
    h. Solenoid valve
    i. Motor
    j. Horn

24. Explain the function of each of the following actuators:
    a. Solenoid
    b. Solenoid valve
    c. Stepper motor

25. Compare the operation of open-loop and closed-loop control.

26. What is a seal-in circuit?

27. In what is the construction and operation of an electromechanical latching relay different from a standard relay?

28. Give a short description of each of the following control processes:
    a. Sequential
    b. Combination
    c. Automatic

1. Design and draw the schematic for a conventional hardwired relay circuit that will perform each of the following circuit functions when a normally closed pushbutton is pressed:
   - Switch a pilot light on
   - De-energize a solenoid
   - Start a motor running
   - Sound a horn

2. Design and draw the schematic for a conventional hardwired circuit that will perform the following circuit functions using two break-make pushbuttons:
   - Turn on light L1 when pushbutton PB1 is pressed.
   - Turn on light L2 when pushbutton PB2 is pressed.
   - Electrically interlock the pushbuttons so that L1 and L2 cannot both be turned on at the same time.

3. Study the ladder logic program in Figure 6-69, and answer the questions that follow:
   a. Under what condition will the latch rung 1 be true?
   b. Under what conditions will the unlatch rung 2 be true?
   c. Under what condition will rung 3 be true?
   d. When PL1 is on, the relay is in what state (latched or unlatched)?
   e. When PL2 is on, the relay is in what state (latched or unlatched)?
   f. If AC power is removed and then restored to the circuit, what pilot light will automatically come on when the power is restored?
   g. Assume the relay is in its latched state and all three inputs are false. What input change(s) must occur for the relay to switch into its unlatched state?
   h. If the examine if closed instructions at addresses I/1, I/2, and I/3 are all true, what state will the relay remain in (latched or unlatched)?

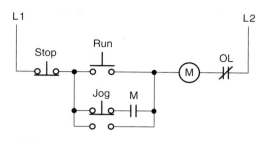

Figure 6-70 Hardwired control circuit for Problem 4.

4. Design a PLC program and prepare a typical I/O connection diagram and ladder logic program that will correctly execute the hardwired control circuit in Figure 6-70.
   Assume: Stop pushbutton used is an NO type.
   Run pushbutton used is an NO type.
   Jog pushbutton used has one set of NO contacts.
   OL contact is hardwired.

5. Design a PLC program and prepare a typical I/O connection diagram and ladder logic program that will correctly execute the hardwired control circuit in Figure 6-71.

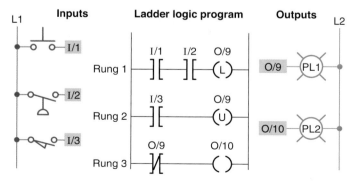

Figure 6-69 Ladder logic program for Problem 3.

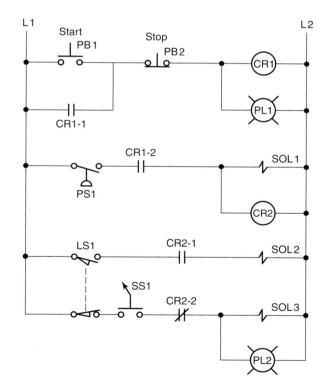

Figure 6-71 Hardwired control circuit for Problem 5.

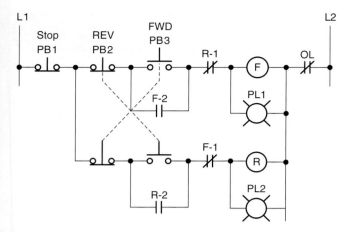

Figure 6-72 Hardwired control circuit for Problem 6.

Assume: PB1 pushbutton used is an NO type.
PB2 pushbutton used is an NC type.
PS1 pressure switch used is an NO type.
LS1 limit switch used has only one set of NC contacts.

6. Design a PLC program and prepare a typical I/O connection diagram and ladder logic program that will correctly execute the hardwired control circuit in Figure 6-72.
Assume: PB1 pushbutton used is an NC type.
PB2 and PB3 are each wired using one set of NO contacts.
OL contact is hardwired.

7. Design a PLC program and prepare a typical I/O connection diagram and ladder logic program for the following motor control specifications:
- A motor must be started and stopped from any one of three start/stop pushbutton stations.
- Each start/stop station contains one NO start pushbutton and one NC stop pushbutton.
- Motor OL contacts are to be hardwired.

8. Design a PLC program and prepare a typical I/O connection diagram and ladder logic program for the following motor control specifications:
- Three starters are to be wired so that each starter is operated from its own start/stop pushbutton station.

- A master stop station is to be included that will trip out all starters when pushed.
- Overload relay contacts are to be programmed so that an overload on any one of the starters will automatically drop all of the starters.
- *All* pushbuttons are to be wired using one set of NO contacts.

9. A temperature control system consists of four thermostats controlling three heating units. The thermostat contacts are set to close at 50°, 60°, 70°, and 80°F, respectively. The PLC ladder logic program is to be designed so that at a temperature below 50°F, three heaters are to be ON. Between 50° to 60°F, two heaters are to be ON. For 60° to 70°F, one heater is to be ON. Above 80°F, there is a safety shutoff for all three heaters in case one stays on because of a malfunction. A master switch is to be used to turn the system ON and OFF. Prepare a typical PLC program for this control process.

10. A pump is to be used to fill two storage tanks. The pump is manually started by the operator from a start/stop station. When the first tank is full, the control logic must be able to automatically stop flow to the first tank and direct flow to the second tank through the use of sensors and electric sole-noid valves. When the second tank is full, the pump must shut down automatically. Indicator lamps are to be included to signal when each tank is full.
a. Draw a sketch of the process.
b. Prepare a typical PLC program for this control process.

11. Write the optimum ladder logic rung for each of the following scenarios, and arrange the instructions for optimum performance:
a. If limit switches LS1 or LS2 or LS3 are on, or if LS5 and LS7 are on, turn on; otherwise, turn off. (Commonly, if LS5 and LS7 are on, the other conditions rarely occur.)
b. Turn on an output when switches SW6, SW7, and SW8 are all on, or when SW55 is on. (SW55 is an indication of an alarm state, so it is rarely on; SW7 is on most often, then SW8, then SW6.)

7

Programming Timers

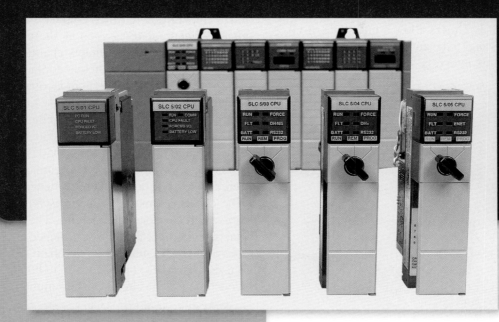

Chapter Objectives

After completing this chapter, you will be able to:

7.1 Describe the operation of pneumatic on-delay and off-delay timers

7.2 Describe PLC timer instruction and differentiate between a nonretentive and retentive timer

7.3 Convert fundamental timer relay schematic diagrams to PLC ladder logic programs

7.4 Analyze and interpret typical PLC timer ladder logic programs

7.5 Program the control of outputs using the timer instruction control bits

The most commonly used PLC instruction, after coils and contacts, is the timer. This chapter deals with how timers time intervals and the way in which they can control outputs. We discuss the basic PLC on-delay timer function, as well as other timing functions derived from it, and typical industrial timing tasks.

7.1 Mechanical Timing Relays

There are very few industrial control systems that do not need at least one or two timed functions. Mechanical timing relays are used to delay the opening or closing of contacts for circuit control. The operation of a mechanical timing relay is similar to that of a control relay, except that certain of its contacts are designed to operate at a preset time interval, after the coil is energized or de-energized. Typical types of mechanical and electronic timing relays are shown in Figure 7-1. Timers allow a multitude of operations in a control circuit to be automatically started and stopped at different time intervals.

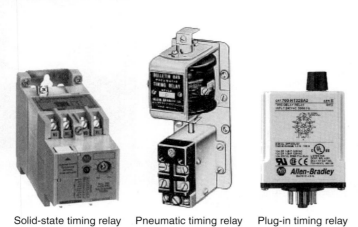

Solid-state timing relay   Pneumatic timing relay   Plug-in timing relay

Figure 7-1   Timing relays.
Source: Image Used with Permission of Rockwell Automation, Inc.

Figure 7-2 shows the construction of an on-delay pneumatic (air) timer. The time-delay function depends on the transfer of air through a restricted orifice. The time-delay period is adjusted by positioning the needle valve to vary the amount of orifice restriction. When the coil is energized, the timed contacts are delayed from opening or closing. However, when the coil is de-energized, the timed contacts return instantaneously to their normal state. This particular pneumatic timer has instantaneous contacts in addition to timed contacts. The instantaneous contacts change state as soon as the timer coil is powered while the delayed contacts change state at the end of the time delay. Instantaneous contacts are often used as holding or sealing contacts in a control circuit.

Mechanical timing relays provide time delay through two arrangements. The first arrangement, *on delay,* provides time delay when the relay coil is *energized.* The second arrangement, *off delay,* provides time delay when the relay coil is *de-energized.* Figure 7-3 illustrates the different relay symbols used for timed contacts.

The on-delay timer is sometimes referred to as DOE, which stands for delay on energize. The time delay of the contacts begins once the timer is switched on; hence the term *on-delay timing.* Figure 7-4 shows an on-delay timer circuit that uses a normally open, timed closed (NOTC) contact. The operation of the circuit can be summarized as follows:

- With S1 initially open, TD coil is de-energized so TD1 contacts are open and light L1 will be off.

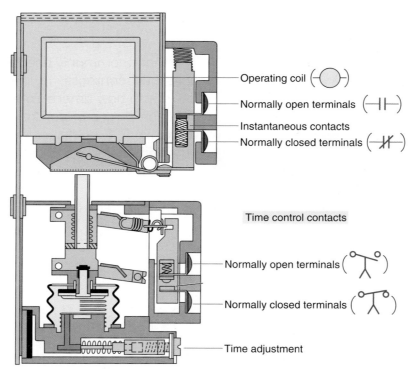

Figure 7-2   Pneumatic on-delay timer.

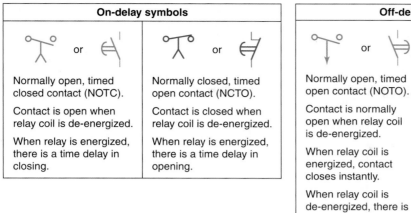

On-delay symbols	
or	or
Normally open, timed closed contact (NOTC).	Normally closed, timed open contact (NCTO).
Contact is open when relay coil is de-energized.	Contact is closed when relay coil is de-energized.
When relay is energized, there is a time delay in closing.	When relay is energized, there is a time delay in opening.

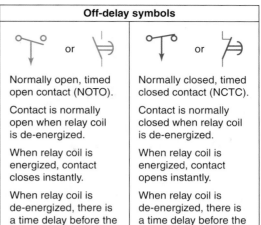

Off-delay symbols	
or	or
Normally open, timed open contact (NOTO).	Normally closed, timed closed contact (NCTC).
Contact is normally open when relay coil is de-energized.	Contact is normally closed when relay coil is de-energized.
When relay coil is energized, contact closes instantly.	When relay coil is energized, contact opens instantly.
When relay coil is de-energized, there is a time delay before the contact opens.	When relay coil is de-energized, there is a time delay before the contact closes.

Figure 7-3 Timed contact symbols.

- When S1 is closed TD coil is energized and the timing period starts. TD1 contacts are delayed from closing so L1 remains off.
- After the 10 s time-delay period has elapsed, TD1 contacts close and L1 is switched on.
- When S1 is opened, TD coil is de-energized and TD1 contacts open instantly to switch L1 off.

Figure 7-5 shows an on-delay timer circuit that uses a normally closed, timed open (NCTO) contact. The operation of the circuit can be summarized as follows:

- With S1 initially open, TD coil is de-energized so TD1 contacts are closed and light L1 will be on.
- When S1 is closed, TD coil is energized and the timing period starts. TD1 contacts are delayed from opening so L1 remains on.
- After the 10 s time-delay period has elapsed, TD1 contacts open and L1 is switched off.
- When S1 is opened, TD coil is de-energized and TD1 contacts close instantly to switch L1 on.

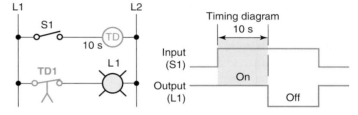

Figure 7-5 On-delay timer circuit that uses a normally closed, timed open (NCTO) contact.

Figure 7-6 shows an off-delay timer circuit that uses a normally open, timed open (NOTO) contact. The operation of the circuit can be summarized as follows:

- With S1 initially open, TD coil is de-energized so TD1 contacts are open and light L1 will be off.
- When S1 is closed, TD coil is energized and TD1 contacts close instantly to switch light L1 on.
- When S1 is opened, TD coil is de-energized and the timing period starts.
- After the 10 s time-delay period has elapsed, TD1contacts open to switch the light off.

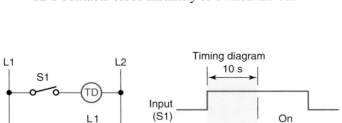

Figure 7-4 On-delay timer circuit that uses a normally open, timed closed (NOTC) contact.

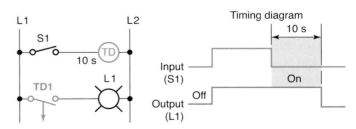

Figure 7-6 Off-delay timer circuit that uses a normally open, timed open (NOTO) contact.

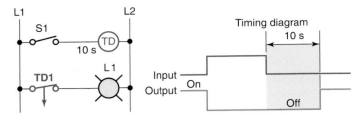

Figure 7-7 Off-delay timer circuit that uses a normally closed, timed closed (NCTC) contact.

Figure 7-7 shows an off-delay timer circuit that uses a normally closed, timed closed (NCTC) contact. The operation of the circuit can be summarized as follows:

- With S1 initially open, TD coil is de-energized so TD1 contacts are closed and light L1 will be on.
- When S1 is closed, TD coil is energized and TD1 contacts open instantly to switch light L1 off.
- When S1 is opened, TD coil is de-energized and the timing period starts. TD1 contacts are delayed from closing so L1 remains off.
- After the 10 s time-delay period has elapsed, TD1 contacts close to switch the light on.

7.2 Timer Instructions

PLC timers are instructions that provide the same functions as on-delay and off-delay mechanical and electronic timing relays. PLC timers offer several advantages over their mechanical and electronic counterparts. These include the fact that:

- Time settings can be easily changed.
- The number of them used in a circuit can be increased or decreased through the use of programming changes rather than wiring changes.
- Timer accuracy and repeatability are extremely high because its time delays are generated in the PLC processor.

In general, there are three different PLC timer types: the *on-delay timer (TON), off-delay timer (TOF),* and *retentive timer on (RTO).* The most common is the on-delay timer, which is the basic function. There are also many other timing configurations, all of which can be derived from one or more of the basic time-delay functions. Figure 7-8 shows the timer selection toolbar for the Allen-Bradley SLC 500 PLC and its associated RSLogix software. These timer commands can be summarized as follows:

TON (Timer On Delay)—Counts time-based intervals when the instruction is true.

TOF (Timer Off Delay)—Counts time-based intervals when the instruction is false.

Figure 7-8 Timer selection toolbar.

RTO (Retentive Timer On)—Counts time-based intervals when the instruction is true and retains the accumulated value when the instruction goes false or when power cycle occurs.

RES (Reset)—Resets a retentive timer's accumulated value to zero.

Several quantities are associated with the timer instruction:

- The *preset time* represents the time duration for the timing circuit. For example, if a time delay of 10 s is required, the timer will have a preset of 10 s.
- The *accumulated time* represents the amount of time that has elapsed from the moment the timing coil became energized.
- Every timer has a *time base.* Once the timing rung has continuity, the timer counts in time-based intervals and times until the preset value and accumulated value are equal or, depending on the type of controller, up to the maximum time interval of the timer. The intervals that the timers time out at are generally referred to as the time bases of the timer. Timers can be programmed with several different time bases: 1 s, 0.1 s, and 0.01 s are typical time bases. If a programmer entered 0.1 for the time base and 50 for the number of delay increments, the timer would have a 5-s delay (50 × 0.1 s = 5 s). The smaller the time base selected, the better the accuracy of the timer.

Although each manufacturer may represent timers differently on the ladder logic program, most timers operate in a similar manner. One of the first methods used depicts the timer instruction as a relay coil similar to that of a mechanical timing relay. Figure 7-9 shows a coil-formatted timer instruction. Its operation can be summarized as follows:

- The timer is assigned an address and is identified as a timer.
- Also included as part of the timer instruction is the time base of the timer, the timer's preset value or time-delay period, and the accumulated value or current time-delay period for the timer.

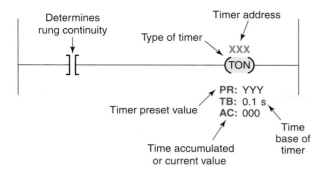

Figure 7-9 Coil-formatted timer instruction.

- When the timer rung has logic continuity, the timer begins counting time-based intervals and times until the accumulated value equals the preset value.
- When the accumulated time equals the preset time, the output is energized and the timed output contact associated with the output is closed. The timed contact can be used as many times as you wish throughout the program as an NO or NC contact.

Timers are most often represented by boxes in ladder logic. Figure 7-10 illustrates a generic block format for a retentive timer that requires two input lines. Its operation can be summarized as follows:

- The timer block has two input conditions associated with it, namely, the *control* and *reset*.
- The control line controls the actual timing operation of the timer. Whenever this line is true or power is supplied to this input, the timer will time. Removal of power from the control line input halts the further timing of the timer.
- The reset line resets the timer's accumulated value to zero.
- Some manufacturers require that *both* the control and reset lines be true for the timer to time; removal of power from the reset input resets the timer to zero.
- Other manufacturers' PLCs require power flow for the control input only and no power flow on the reset input for the timer to operate. For this type of timer operation, the timer is reset whenever the reset input is true.

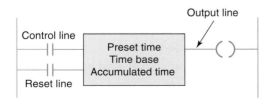

Figure 7-10 Block-formatted timer instruction.

- The timer instruction block contains information pertaining to the operation of the timer, including the preset time, the time base of the timer, and the current or accumulated time.
- All block-formatted timers provide at least one output signal from the timer. The timer continuously compares its current time with its preset time, and its output is false (logic 0) as long as the current time is less than the preset time. When the current time equals the preset time, the output changes to true (logic 1).

7.3 On-Delay Timer Instruction

Most timers are output instructions that are conditioned by input instructions. An *on-delay timer* is used when you want to program a time delay before an instruction becomes true. Figure 7-11 illustrates the principle of operation of an on-delay timer. Its operation can be summarized as follows:

- The on-delay timer operates such that when the rung containing the timer is true, the timer time-out period commences.
- At the end of the timer time-out period, an output is made true.
- The timed output becomes true sometime after the timer rung becomes true; hence, the timer is said to have an on-delay.
- The length of the time delay can be adjusted by changing the preset value.
- In addition, some PLCs allow the option of changing the time base, or resolution, of the timer. As the time base you select becomes smaller, the accuracy of the timer increases.

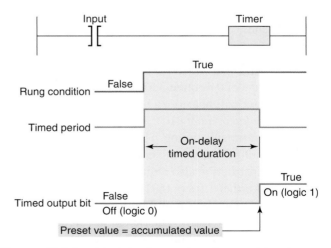

Figure 7-11 Principle of operation of an on-delay timer.

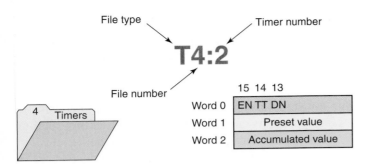

Figure 7-12 SLC 500 timer file.

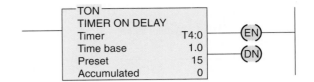

Figure 7-13 On-delay timer instruction.

The Allen-Bradley SLC 500 timer file is file 4 (Figure 7-12). Each timer is composed of three 16-bit words, collectively called a timer element. There can be up to 256 timer elements. Addresses for timer file 4, timer element number 2 (T4:2), are listed below.

T4 = timer file 4

:2 = timer element number 2 (0–255 timer elements per file)

T4:2/DN is the address for the done bit of the timer.

T4:2/TT is the address for the timer-timing bit of the timer.

T4:2/EN is the address for the enable bit of the timer.

The *control word* uses the following three control bits:

Enable (EN) bit—The enable bit is true (has a status of 1) whenever the timer instruction is true. When the timer instruction is false, the enable bit is false (has a status of 0).

Timer-timing (TT) bit—The timer-timing bit is true whenever the accumulated value of the timer is changing, which means the timer is timing. When the timer is not timing, the accumulated value is not changing, so the timer-timing bit is false.

Done (DN) bit—The done bit changes state whenever the accumulated value reaches the preset value. Its state depends on the type of timer being used.

The *preset value (PRE) word* is the set point of the timer, that is, the value up to which the timer will time. The preset word has a range of 0 through 32,767 and is stored in binary form. The preset will not store a negative number.

The *accumulated value (ACC) word* is the value that increments as the timer is timing. The accumulated value will stop incrementing when its value reaches the preset value.

The timer instruction also requires that you enter a *time base,* which is either 1.0 s or 0.01 s. The actual preset time interval is the time base multiplied by the value stored in the timer's preset word. The actual accumulated time interval is the time base multiplied by the value stored in the timer's accumulated word.

Figure 7-13 shows an example of the on-delay timer instruction used as part of the Allen-Bradley PLC-5 and SLC 500 controller instruction sets. The information to be entered includes:

Timer number—This number must come from the timer file. In the example shown, the timer number is T4:0, which represents timer file 4, timer 0 in that file. The timer address must be unique for this timer and may not be used for any other timer.

Time base—The time base (which is always expressed in seconds) may be either 1.0 s or 0.01 s. In the example shown, the time base is 1.0 s.

Preset value—In the example shown, the preset value is 15. The timer preset value can range from 0 through 32,767.

Accumulated value—In the example shown, the accumulated value is 0. The timer's accumulated value normally is entered as 0, although it is possible to enter a value from 0 through 32,767. Regardless of the value that is preloaded, the timer value will become 0 whenever the timer is reset.

The on-delay timer (TON) is the most commonly used timer. Figure 7-14 shows a PLC program that uses an on-delay timer. The operation of the program can be summarized as follows:

- The timer is activated by input switch A.
- The preset time for this timer is 10 s, at which time output D will be energized.
- When input switch is A is closed, the timer becomes true and the timer begins counting and counts until the accumulated time equals the preset value; the output D is then energized.
- If the switch is opened before the timer is timed out, the accumulated time is automatically reset to 0.
- This timer configuration is termed *nonretentive* because any loss of continuity to the timer causes the timer instruction to reset.
- This timing operation is that of an on-delay timer because output D is switched on 10 s after the switch has been actuated from the off to the on position.

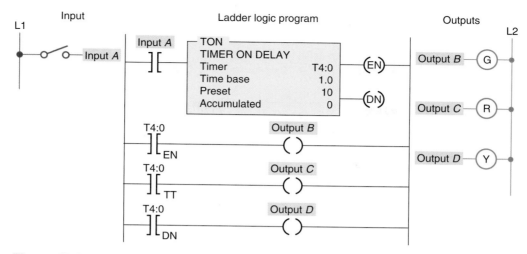

Figure 7-14 PLC on-delay timer program.

Figure 7-15 shows the timing diagram for the on-delay timer's control bits. The sequence of operation is as follows:

- The first true period of the timer rung shows the timer timing to 4 s and then going false.
- The timer resets, and both the timer-timing bit and the enable bit go false. The accumulated value also resets to 0.
- For the second true period input *A* remains true in excess of 10 s.
- When the accumulated value reaches 10 s, the done bit (DN) goes from false to true and the timer-timing bit (TT) goes from true to false.
- When input *A* goes false, the timer instruction goes false and also resets, at which time the control bits are all reset and the accumulated value resets to 0.

The timer table for an Allen-Bradley SLC 500 is shown in Figure 7-16. Addressing is done at three different levels: the element level, the word level, and the bit level.

The timer uses three words per element. Each element consists of a control word, a preset word, and an accumulated word. Each word has 16 bits, which are numbered from 0 to 15. When addressing to the bit level, the address always refers to the bit within the word:

EN = Bit 15 enable
TT = Bit 14 timer timing
DN = Bit 13 done

Timers may or may not have an instantaneous output (also known as the enable bit) signal associated with them. If an instantaneous output signal is required from a timer and it is not provided as part of the timer instruction, an equivalent instantaneous contact instruction can be programmed using an internally referenced relay coil. Figure 7-17 shows an application of this technique. The operation of the program can be summarized as follows:

- According to the hardwired relay circuit diagram, coil M is to be energized 5 s after the start pushbutton is pressed.

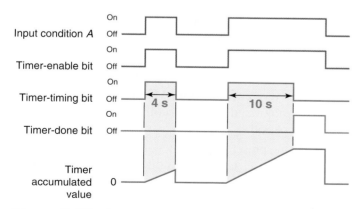

Figure 7-15 Timing diagram for an on-delay timer.

	/EN	/TT	/DN	.PRE	.ACC
T4:0	0	0	0	10	0
T4:1	0	0	0	0	0
T4:2	0	0	0	0	0
T4:3	0	0	0	0	0
T4:4	0	0	0	0	0
T4:5	0	0	0	0	0

Address: T4:0 Table: T4: Timer

Figure 7-16 SLC 500 timer table.

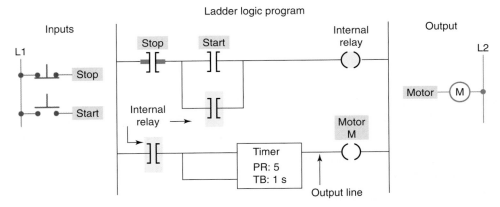

Hardwired relay circuit

Ladder logic program

Figure 7-17 Instantaneous contact instruction can be programmed using an internally referenced relay coil.

- Contact TD-1 is the instantaneous contact, and contact TD-2 is the timed contact.
- The ladder logic program shows that a contact instruction referenced to an internal relay is now used to operate the timer.
- The instantaneous contact is referenced to the internal relay coil, whereas the time-delay contact is referenced to the timer output coil.

Figure 7-18 shows an application for an on-delay timer that uses an NCTO contact. This circuit is used as a warning signal when moving equipment, such as a conveyor motor, is about to be started. The operation of the circuit can be summarized as follows:

- According to the hardwired relay circuit diagram, coil CR is energized when the start pushbutton PB1 is momentarily actuated.
- As a result, contact CR-1 closes to seal in CR coil, contact CR-2 closes to energize timer coil TD, and contact CR-3 closes to sound the horn.
- After a 10-s time-delay period, timer contact TD-1 opens to automatically switch the horn off.
- The ladder logic program shows how an equivalent circuit could be programmed using a PLC.

- The logic on the last rung is the same as the timer-timing bit and as such can be used with timers that do not have a timer-timing output.

Timers are often used as part of automatic sequential control systems. Figure 7-19 shows how a series of motors can be started automatically with only one start/stop control station. The operation of the circuit can be summarized as follows:

- According to the relay ladder schematic, lube-oil pump motor starter coil M1 is energized when the start pushbutton PB2 is momentarily actuated.
- As a result, M1-1 control contact closes to seal in M1, and the lube-oil pump motor starts.
- When the lube-oil pump builds up sufficient oil pressure, the lube-oil pressure switch PS1 closes.
- This in turn energizes coil M2 to start the main drive motor and energizes coil TD to begin the time-delay period.
- After the preset time-delay period of 15 s, TD-1 contact closes to energize coil M3 and start the feed motor.
- The ladder logic program shows how an equivalent circuit could be programmed using a PLC.

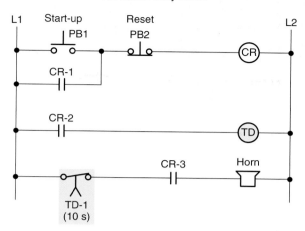

Hardwired relay circuit

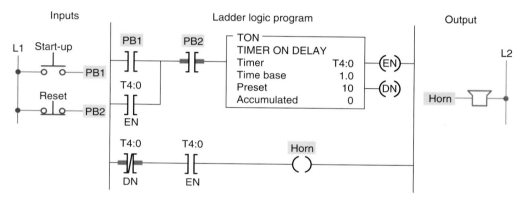

Figure 7-18 Conveyor warning signal circuit.

7.4 Off-Delay Timer Instruction

The *off-delay timer (TOF)* operation will keep the output energized for a time period after the rung containing the timer has gone false. Figure 7-20 illustrates the programming of an off-delay timer that uses the SLC 500 TOF timer instruction. If logic continuity is *lost,* the timer begins counting time-based intervals until the accumulated time equals the programmed preset value. The operation of the circuit can be summarized as follows:

- When the switch connected to input I:1/0 is first closed, timed output O:2/1 is set to 1 immediately and the lamp is switched on.
- If this switch is now opened, logic continuity is lost and the timer begins counting.
- After 15 s, when the accumulated time equals the preset time, the output is reset to 0 and the lamp switches off.
- If logic continuity is gained before the timer is timed out, the accumulated time is reset to 0. For

this reason, this timer is also classified as nonretentive.

Figure 7-21 illustrates the use of an off-delay timer instruction used to switch motors *off* sequentially at 5 second intervals. The operation of the program can be summarized as follows:

- Timer preset values for T4:1, T4:2, and T4:3 are set for 5 s, 10s, and 15 s, respectively.
- Closing the input switch SW immediately sets the done bit of each of the three off-delay timers to 1, immediately turning on motors M1, M2, and M3.
- If SW is then opened, logic continuity to all three timers is lost and each timer begins counting.
- Timer T4:1 times out after 5 s resetting its done bit to zero to de-energize motor M1.
- Timer T4:2 times out 5 s later resetting its done bit to zero to de-energize motor M2.
- Timer T4:3 times out 5 s later resetting its done bit to zero to de-energize motor M3.

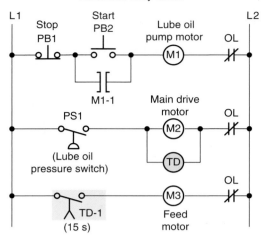

Hardwired relay circuit

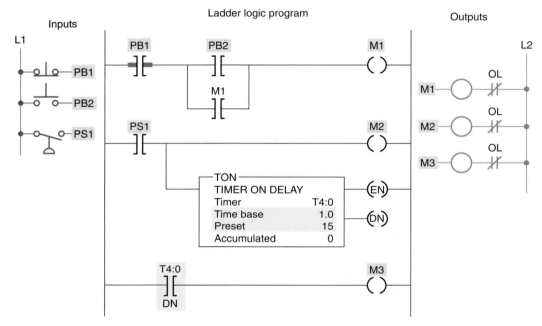

Figure 7-19 Automatic sequential control system.

Figure 7-22 shows how a hardwired off-delay timer relay circuit with both instantaneous and timed contacts. The operation of the circuit can be summarized as follows:

- When power is first applied (limit switch LS open), motor starter coil M1 is energized and the green pilot light is on.
- At the same time, motor starter coil M2 is de-energized, and the red pilot light is off.
- When limit switch LS closes, off-delay timer coil TD energizes.

- As a result, timed contact TD-1 opens to de-energize motor starter coil M1, timed contact TD-2 closes to energize motor starter coil M2, instantaneous contact TD-3 opens to switch the green light off, and instantaneous contact TD-4 closes to switch the red light on. The circuit remains in this state as long as limit switch LS1 is closed.
- When limit switch LS1 is opened, the off-delay timer coil TD de-energizes and the time-delay period is started.

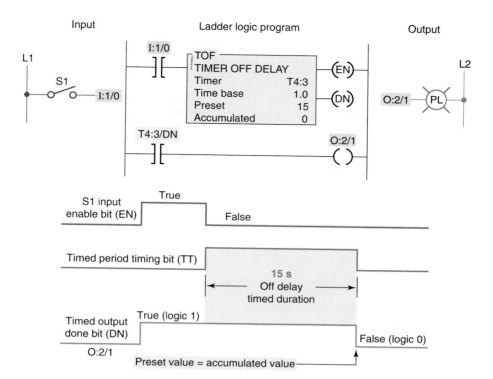

Figure 7-20 Off-delay programmed timer.

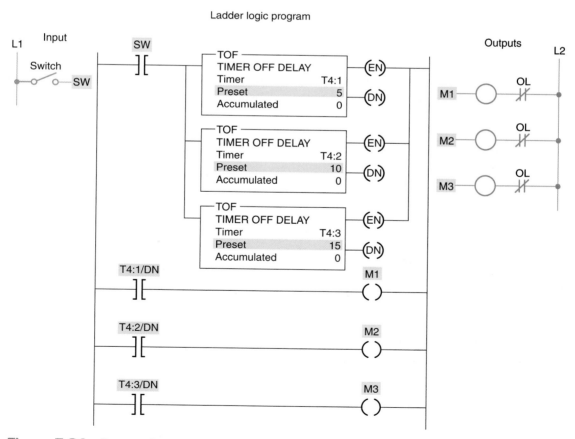

Figure 7-21 Program for switching motors off at 5 s intervals.

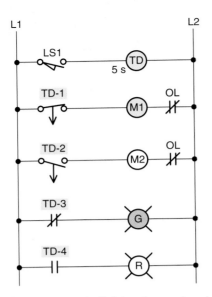

Figure 7-22 Hardwired off-delay timer relay circuit with both instantaneous and timed contacts.

- Instantaneous contact TD-3 closes to switch the green light on, and instantaneous contact TD-4 opens to switch the red light off.
- After a 5-s time-delay period, timed contact TD-1 closes to energize motor starter M1, and timed contact TD-2 opens to de-energize motor starter M2.

Figure 7-23 shows an equivalent PLC program of the hardwired off-delay timer relay circuit containing both instantaneous and timed contacts. The timer instruction carries out all of the functions of the original physical timer.

Figure 7-24 shows a program that uses both the on-delay and the off-delay timer instruction. The process involves pumping fluid from tank A to tank B. The operation of the process can be summarized as follows:

- Before starting, PS1 must be closed.
- When the start button is pushed, the pump starts. The button can then be released and the pump continues to operate.
- When the stop button is pushed, the pump stops.
- PS2 and PS3 must be closed 5 s after the pump starts. If either PS2 or PS3 opens, the pump will shut off and will not be able to start again for another 14 s.

7.5 Retentive Timer

A *retentive timer* accumulates time whenever the device receives power, and it maintains the current time should power be removed from the device. When the timer accumulates time equal to its preset value, the contacts of the device change state. Loss of power to the timer after reaching its preset value does not affect the state of the contacts. The retentive timer must be intentionally reset with a separate signal for the accumulated time to be reset and for the contacts of the device to return to its nonenergized state.

Figure 7-25 illustrates the action of a motor-driven, electromechanical retentive timer used in some appliances. The shaft-mounted cam is driven by a motor. Once power is applied, the motor starts turning the shaft and

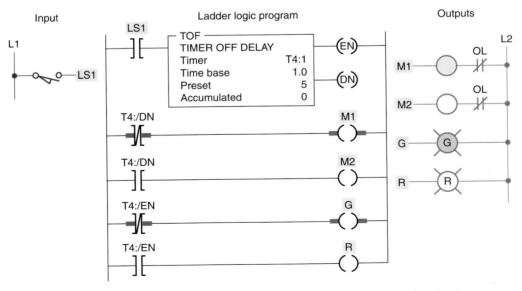

Figure 7-23 Equivalent PLC program of the hardwired off-delay timer relay circuit containing both instantaneous and timed contacts.

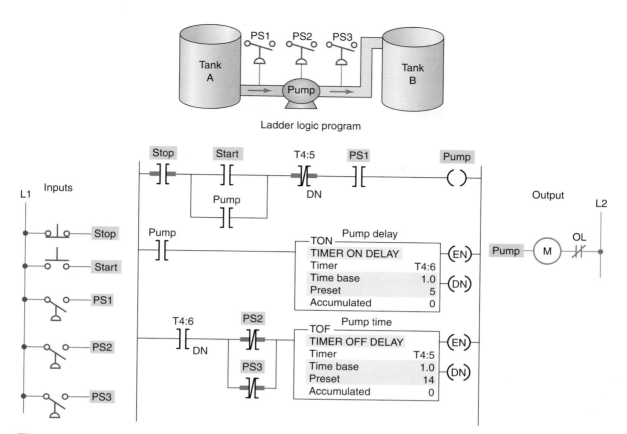

Figure 7-24 Fluid pumping process.

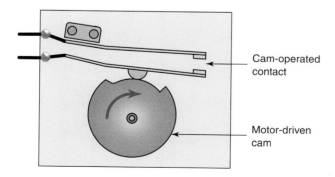

Figure 7-25 Electromechanical retentive timer.

cam. The positioning of the lobes of the cam and the gear reduction of the motor determine the time it takes for the motor to turn the cam far enough to activate the contacts. If power is removed from the motor, the shaft stops but *does not reset.*

A PLC retentive timer is used when you want to retain accumulated time values through power loss or the change in the rung state from true to false. The PLC-programmed retentive on-delay timer (RTO) is programmed in a manner similar to the nonretentive on-delay timer (TON), with one major exception—a retentive timer reset (RES) instruction. Unlike the TON, the RTO will hold its accumulated value when the timer rung goes false and will continue timing where it left off when the timer rung goes true again. This timer must be accompanied by a timer reset instruction to reset the accumulated value of the timer to 0. The RES instruction is the only automatic means of resetting the accumulated value of a retentive timer. The RES instruction has the same address as the timer it is to reset. Whenever the RES instruction is true, both the timer accumulated value and the timer done bit (DN) are reset to 0. Figure 7-26 shows a PLC program for a retentive on-delay timer. The operation of the program can be summarized as follows:

- The timer will start to time when time pushbutton PB1 is closed.
- If the pushbutton is closed for 3 seconds and then opened for 3 seconds, the timer accumulated value will remain at 3 seconds.
- When the time pushbutton is closed again, the timer picks up the time at 3 seconds and continues timing.

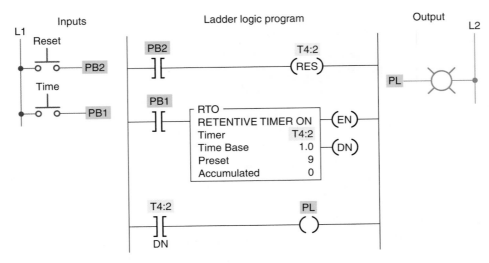

Figure 7-26 Retentive on-delay timer program.

- When the accumulated value (9) equals the preset value (9), the timer done bit T4:2/DN is set to 1 and the pilot light output PL is switched on.
- Whenever the momentary reset pushbutton is closed the timer accumulated value is reset to 0.

Figure 7-27 shows a timing chart for the retentive on-delay timer program. The timing operation can be summarized as follows:

- When the timing rung is true (PB1 closed) the timer will commence timing.

- If the timing rung goes false the timer will stop timing but will recommence timing for the stored accumulated value each time the rung goes true.
- When the reset PB2 is closed, the T4:2/DN bit is reset to 0 and turns the pilot light output off. The accumulated value is also reset and held at zero until the reset pushbutton is opened.

The program drawn in Figure 7-28 illustrates a practical application for an RTO. The purpose of the RTO timer is to detect whenever a piping system has sustained

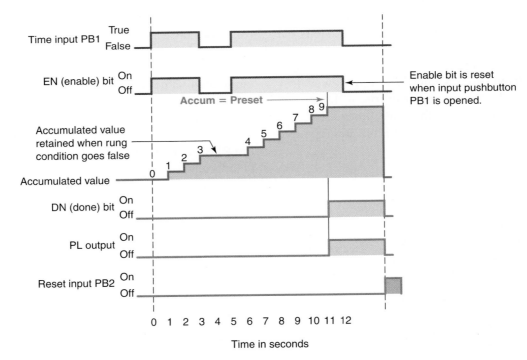

Figure 7-27 Retentive on-delay timer timing chart.

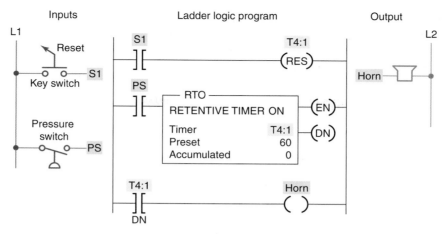

Figure 7-28 Retentive on-delay timer alarm program.

a *cumulative* overpressure condition for 60 s. At that point, a horn is sounded automatically to call attention to the malfunction. When they are alerted, maintenance personnel can silence the alarm by switching the key switch S1 to the reset (contact closed) position. After

the problem has been corrected, the alarm system can be reactivated by switching the key switch to open contact position.

Figure 7-29 shows a practical application that uses the on-delay, off-delay, and retentive on-delay

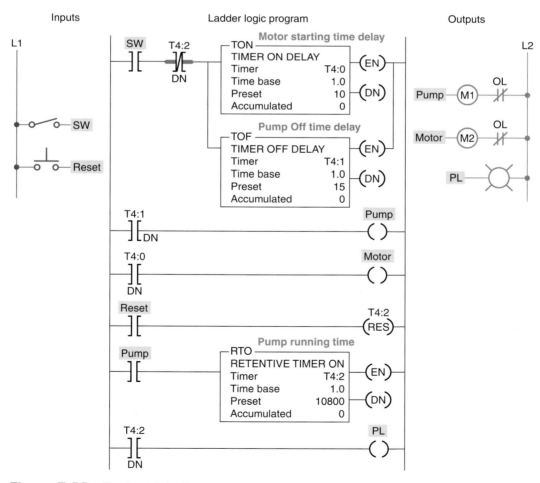

Figure 7-29 Bearing lubrication program.

instructions in the same program. In this industrial application, there is a machine with a large steel shaft supported by babbitted bearings. This shaft is coupled to a large electric motor. The bearings need lubrication, which is supplied by an oil pump driven by a small electric motor. The operation of the program can be summarized as follows:

- To start the machine, the operator turns SW on.
- Before the *motor* shaft starts to turn, the bearings are supplied with oil by the *pump* for 10 seconds.
- The bearings also receive oil when the machine is running.
- When the operator turns SW off to stop the machine, the oil pump continues to supply oil for 15 seconds.
- A retentive timer is used to track the total running time of the pump. When the total running time is 3 hours, the motor is shut down and a pilot light is turned on to indicate that the filter and oil need to be changed.
- A reset button is provided to reset the process after the filter and oil have been changed.

Retentive timers do not have to be timed out completely to be reset. Rather, such a timer can be reset at any time during its operation. Note that the reset input to the timer will override the control input of the timer even though the control input to the timer has logic continuity.

7.6 Cascading Timers

The programming of two or more timers together is called *cascading*. Timers can be interconnected, or cascaded, to satisfy a number of logic control functions.

Figure 7-30 shows how three motors can be started automatically in sequence with a 20 s time delay between each using two hardwired on-delay timers. The operation of the circuit can be summarized as follows:

- Motor starter coil M1 is energized when the momentary start pushbutton PB2 is actuated.
- As a result, motor 1 starts, contact M1-1 closes to seal in M1, and timer coil TD1 is energized to begin the first time-delay period.
- After the preset time period of 20 s, TD1-1 contact closes to energize motor starter coil M2.
- As a result, motor 2 starts and timer coil TD2 is energized to begin the second time-delay period.

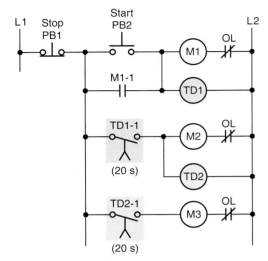

Figure 7-30 Hardwired sequential time-delayed motor-starting circuit.

- After the preset time period of 20 s, TD2-1 contact closes to energize motor starter coil M3, and so motor 3 starts.

Figure 7-31 shows an equivalent PLC program of the hardwired sequential time-delayed motor-starting circuit. Two programmed on-delay timers are cascaded together to obtain the same logic as the original hardwired timer relay circuit. Note that the output of timer T4:1 is used to control the input logic to timer T4:2.

Two timers can be interconnected to form an oscillator circuit. The oscillator logic is basically a timing circuit programmed to generate periodic output pulses of any duration. Figure 7-32 shows the program for an annunciator flasher circuit. Two internal timers form the oscillator circuit, which generates a timed, pulsed output. The oscillator circuit output is programmed in series with the alarm condition. If the alarm condition (temperature, pressure, or limit switch) is true, the appropriate output indicating light will flash. Note that any number of alarm conditions could be programmed using the same flasher circuit.

At times you may require a time-delay period longer than the maximum preset time allowed for the single timer instruction of the PLC being used. When this is the case, the problem can be solved by simply cascading timers, as illustrated in Figure 7-33. The operation of the program can be summarized as follows:

- The total time-delay period required is 42,000 s.
- The first timer, T4:1, is programmed for a preset time of 30,000 s and begins timing when input SW is closed.

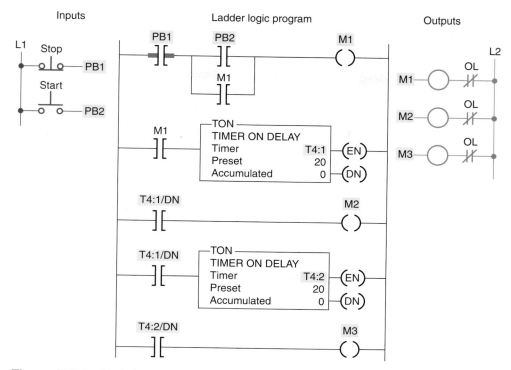

Figure 7-31 Equivalent PLC program of the sequential time-delayed motor-starting circuit.

- When T4:1 completes its time-delay period 30,000 s later, the T4:1/DN bit will be set to 1.
- This in turn activates the second timer, T4:2, which is preset for the remaining 12,000 s of the total 42,000-s time delay.

- Once T4:2 reaches its preset time, the T4:2/DN bit will be set to 1, which switches on the output PL, the pilot light, to indicate the completion of the full 42,000-s time delay.

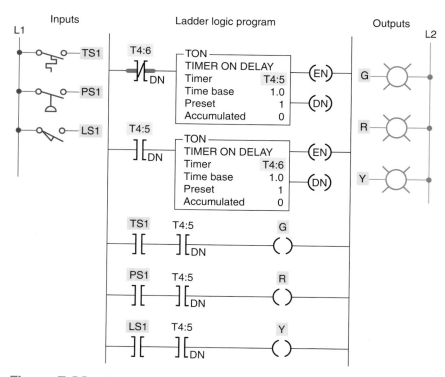

Figure 7-32 Annunciator flasher program.

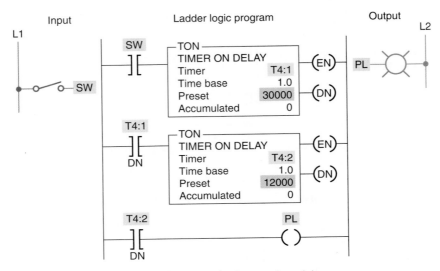

Figure 7-33 Cascading of timers for longer time delays.

- Opening input SW at any time will reset both timers and switch output PL off.

A typical application for PLC timers is the control of traffic lights. The ladder logic circuit of Figure 7-34 illustrates a control of a set of traffic lights in one direction. The operation of the program can be summarized as follows:

- Transition from red light to green light to amber light is accomplished by the interconnection of the three TON timer instructions.
- The input to timer T4:0 is controlled by the T4:2 done bit.
- The input to timer T4:1 is controlled by the T4:0 done bit.
- The input rung to timer T4:2 is controlled by the T4:1 done bit.
- The timed sequence of the lights is:
   - Red—30 s on
   - Green—25 s on
   - Amber—5 s on
- The sequence then repeats itself.

The chart shown in Figure 7-35 shows the timed sequence of the lights for two-directional control of traffic lights.

Figure 7-36 shows the original traffic light program modified to include three more lights that control traffic flow in two directions.

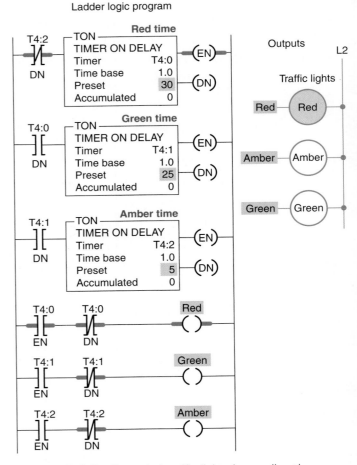

Figure 7-34 Control of traffic lights in one direction.

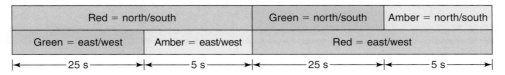

Red = north/south		Green = north/south	Amber = north/south
Green = east/west	Amber = east/west	Red = east/west	

|←——— 25 s ———→|←——— 5 s ———→|←——— 25 s ———→|←——— 5 s ———→|

Figure 7-35 Timing chart for two-directional control of traffic lights.

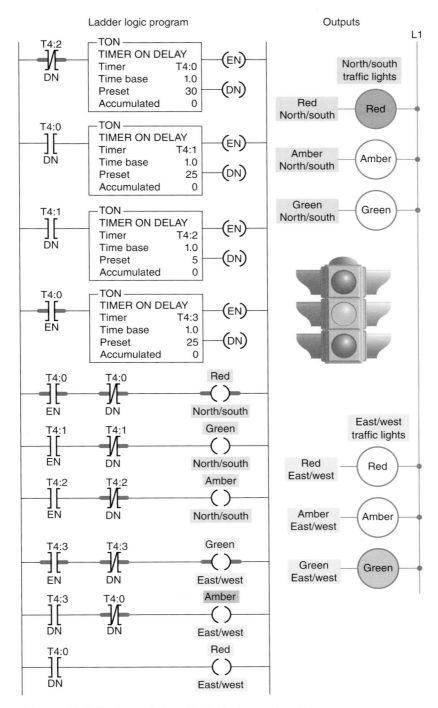

Figure 7-36 Control of traffic lights in two directions.

1. Explain the difference between the timed and instantaneous contacts of a mechanical timing relay.

2. Draw the symbol and explain the operation of each of the following timed contacts of a mechanical timing relay:
   a. On-delay timer—NOTC contact
   b. On-delay timer—NCTO contact
   c. Off-delay timer—NOTO contact
   d. Off-delay timer—NCTC contact

3. Name five pieces of information usually associated with a PLC timer instruction.

4. When is the output of a programmed timer energized?

5. a. What are the two methods commonly used to represent a timer instruction within a PLC's ladder logic program?
   b. Which method is preferred? Why?

6. a. Explain the difference between the operation of a nonretentive timer and that of a retentive timer.
   b. Explain how the accumulated count of programmed retentive and nonretentive timers is reset to zero.

7. State three advantages of using programmed PLC timers over mechanical timing relays.

8. For a TON timer:
   a. When is the enable bit of a timer instruction true?
   b. When is the timer-timing bit of a timer instruction true?
   c. When does the done bit of a timer change state?

9. For a TOF timer:
   a. When is the enable bit of a timer instruction true?
   b. When is the timer-timing bit of a timer instruction true?
   c. When does the done bit of a timer change state?

10. Explain what each of the following quantities associated with a PLC timer instruction represents:
    a. Preset time
    b. Accumulated time
    c. Time base

11. State the method used to reset the accumulated time of each of the following:
    a. TON timer
    b. TOF timer
    c. RTO timer

CHAPTER 7 PROBLEMS

1. a. With reference to the relay schematic diagram in Figure 7-37, state the status of each light (on or off) after each of the following sequential events:
   i.   Power is first applied and switch S1 is open.
   ii. Switch S1 has just closed.
   iii. Switch S1 has been closed for 5 s.
   iv. Switch S1 has just opened.
   v.   Switch S1 has been opened for 5 s.
   b. Design a PLC program and prepare a typical I/O connection diagram and ladder logic program that will execute this hardwired control circuit correctly.

2. Design a PLC program and prepare a typical I/O connection diagram and ladder logic program that will correctly execute the hardwired relay control circuit shown in Figure 7-38.

3. Study the ladder logic program in Figure 7-39 and answer the questions that follow:
   a. What type of timer has been programmed?
   b. What is the length of the time-delay period?

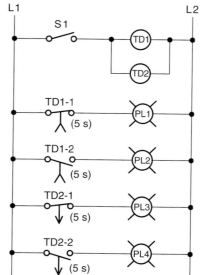

Figure 7-37 Relay schematic diagram for Problem 1.

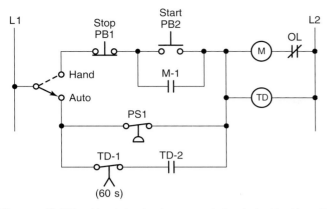

Figure 7-38 Hardwired relay control circuit for Problem 2.

c. What is the value of the accumulated time when power is first applied?
d. When does the timer start timing?
e. When does the timer stop timing and reset itself?
f. When input LS1 is first closed, which rungs are true and which are false?
g. When input LS1 is first closed, state the status (on or off) of each output.
h. When the timer's accumulated value equals the preset value, which rungs are true and which are false?
i. When the timer's accumulated value equals the preset value, state the status (on or off) of each output.

j. Suppose that rung 1 is true for 5 s and then power is lost. What will the accumulated value of the counter be when power is restored?

4. Study the ladder logic program in Figure 7-40 and answer the questions that follow:
   a. What type of timer has been programmed?
   b. What is the length of the time-delay period?
   c. What is the value of the accumulated time when power is first applied?
   d. When does the timer start timing?
   e. When does the timer stop timing and reset itself?
   f. When input LS1 is first closed, which rungs are true and which are false?
   g. When input LS1 is first closed, state the status (on or off) of each output.
   h. When the timer's accumulated value equals the preset value, which rungs are true and which are false?
   i. When the timer's accumulated value equals the preset value, state the status (on or off) of each output.
   j. Suppose that rung 1 is true for 5 s and then power is lost. What will the accumulated value of the counter be when power is restored?

5. Study the ladder logic program in Figure 7-41, and answer the questions that follow:
   a. What type of timer has been programmed?
   b. What is the length of the time-delay period?
   c. When does the timer start timing?

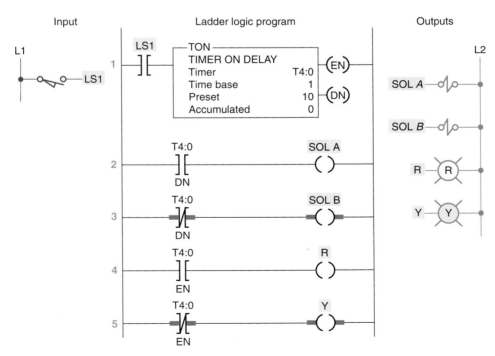

Figure 7-39 Ladder logic program for Problem 3.

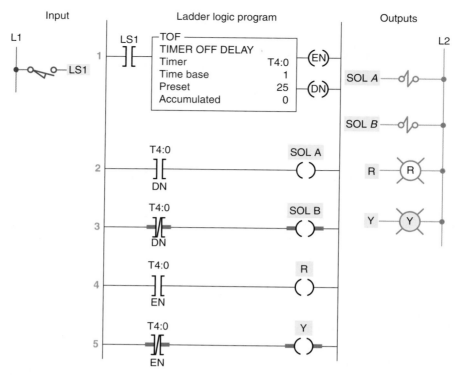

Figure 7-40 Ladder logic program for Problem 4.

d. When is the timer reset?
e. When will rung 3 be true?
f. When will rung 5 be true?
g. When will output PL4 be energized?

h. Assume that your accumulated time value is up to 020 and power to your system is lost. What will your accumulated time value be when power is restored?

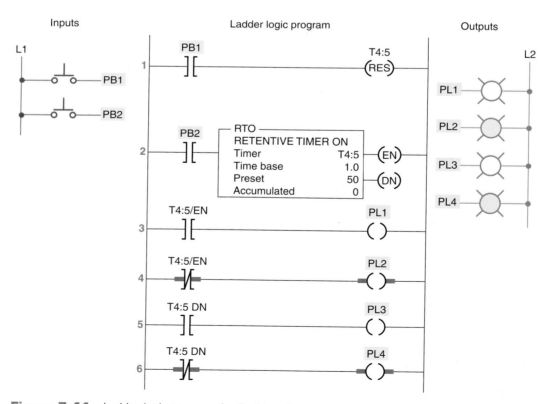

Figure 7-41 Ladder logic program for Problem 5.

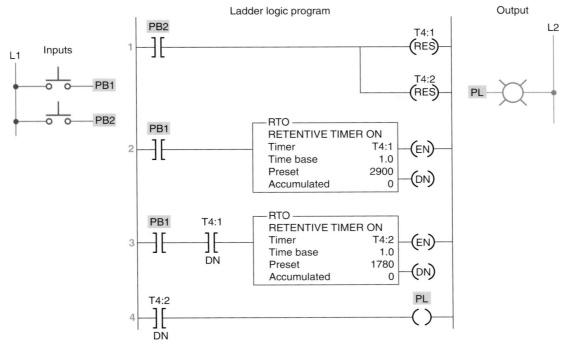

Figure 7-42 Ladder logic program for Problem 6.

i. What happens if inputs PB1 and PB2 are both true at the same time?

6. Study the ladder logic program in Figure 7-42 and answer the questions that follow:
   a. What is the purpose of interconnecting the two timers?
   b. How much time must elapse before output PL is energized?
   c. What two conditions must be satisfied for timer T4:2 to start timing?
   d. Assume that output PL is on and power to the system is lost. When power is restored, what will the status of this output be?
   e. When input PB2 is on, what will happen?
   f. When input PB1 is on, how much accumulated time must elapse before rung 3 will be true?

7. You have a machine that cycles on and off during its operation. You need to keep a record of its total run time for maintenance purposes. Which timer would accomplish this?

8. Write a ladder logic program that will turn on a light, PL, 15 s after switch S1 has been turned on.

9. Study the on-delay timer ladder logic program in Figure 7-43, and from each of the conditions stated, determine whether the timer is reset, timing, or timed out or if the conditions stated are not possible.
   a. The input is true, and EN is 1, TT is 1, and DN is 0.

b. The input is true, and EN is 1, TT is 1, and DN is 1.
c. The input is false, and EN is 0, TT is 0, and DN is 0.
d. The input is true, and EN is 1, TT is 0, and DN is 1.

10. Study the off-delay timer ladder logic program in Figure 7-44, and from each of the conditions stated, determine whether the timer is reset, timing, or timed out or if the conditions stated are not possible.
   a. The input is true, and EN is 0, TT is 0, and DN is 1.

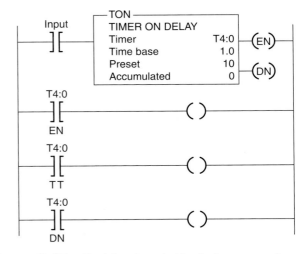

Figure 7-43 On-delay timer ladder logic program for Problem 9.

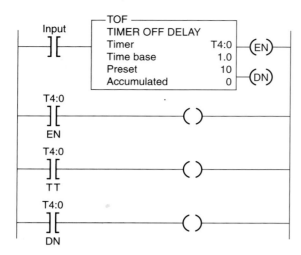

Figure 7-44 Off-delay timer ladder logic program for Problem 10.

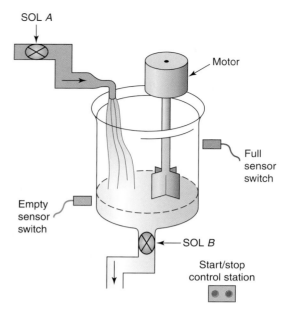

Figure 7-45 Process for Problem 13.

b. The input is true, and EN is 1, TT is 1, and DN is 1.

c. The input is true, and EN is 1, TT is 0, and DN is 1.

d. The input is false, and EN is 0, TT is 1, and DN is 1.

e. The input is false, and EN is 0, TT is 0, and DN is 0.

11. Write a program for an "anti–tie down circuit" that will disallow a punch press solenoid from operating unless both hands are on the two palm start buttons. Both buttons must be pressed at the same time within 0.5 s. The circuit also will not allow the operator to tie down one of the buttons and operate the press with just one button. (Hint: Once either of the buttons is pressed, begin timing 0.5 s. Then, if both buttons are not pressed, prevent the press solenoid from operating.)

12. Modify the program for the control of traffic lights in two directions so that there is a 3-s period when both directions will have their red lights illuminated.

13. Write a program to implement the process illustrated in Figure 7-45. The sequence of operation is to be as follows:
- Normally open start and normally closed stop pushbuttons are used to start and stop the process.
- When the start button is pressed, solenoid A energizes to start filling the tank.
- As the tank fills, the empty level sensor switch closes.
- When the tank is full, the full level sensor switch closes.

- Solenoid A is de-energized.
- The agitate motor starts automatically and runs for 3 min to mix the liquid.
- When the agitate motor stops, solenoid B is energized to empty the tank.
- When the tank is completely empty, the empty sensor switch opens to de-energize solenoid B.
- The start button is pressed to repeat the sequence.

14. When the lights are turned off in a building, an exit door light is to remain on for an additional 2 min, and the parking lot lights are to remain on for an additional 3 min after the door light goes out. Write a program to implement this process.

15. Write a program to simulate the operation of a sequential taillight system. The light system consists of three separate lights on each side of the car. Each set of lights will be activated separately, by either the left or right turn signal switch. There is to be a 1 s delay between the activation of each light, and a 1-s period when all the lights are off. Ensure that when both switches are on, the system will not operate. Use the least number of timers possible. The sequence of operation should be as follows:
- The switch is operated.
- Light 1 is illuminated.
- Light 2 is illuminated 1 s later.
- Light 3 is illuminated 1 s later.
- Light 3 is illuminated for 1 s.
- All lights are off for 1 s.
- The system repeats while the switch is on.

8

Programming Counters

Image Used with Permission of Rockwell Automation, Inc.

Chapter Objectives

After completing this chapter, you will be able to:

8.1 List and describe the functions of PLC counter instructions

8.2 Describe the operating principle of a transitional, or one-shot, contact

8.3 Analyze and interpret typical PLC counter ladder logic programs

8.4 Apply the PLC counter function and associated circuitry to control systems

8.5 Apply combinations of counters and timers to control systems

All PLCs include both up-counters and down-counters. Counter instructions and their function in ladder logic are explained in this chapter. Typical examples of PLC counters include the following: straight counting in a process, two counters used to give the sum of two counts, and two counters used to give the difference between two counts.

8.1 Counter Instructions

Programmed counters can serve the same function as mechanical counters. Figure 8-1 shows the construction of a simple mechanical counter. Every time the actuating lever is moved over, the counter adds one number; the actuating lever then returns automatically to its original position. Resetting to zero is done with a pushbutton located on the side of the unit.

Electronic counters, such as those shown in Figure 8-2, can count up, count down, or be combined to count up and down. Although the majority of counters used in industry are up-counters, numerous applications require the implementation of down-counters or of combination up/down-counters.

All PLC manufacturers offer some form of counter instruction as part of their instruction set. One common counter application is keeping track of the number of items moving past a given point as illustrated in Figure 8-3.

Counters are similar to timers except that they do not operate on an internal clock but are dependent on external

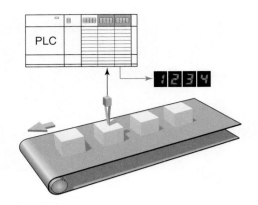

Figure 8-3 Counter application.

or program sources for counting. The two methods used to represent a counter within a PLC's ladder logic program are the coil format and the block format. Figure 8-4 shows a typical coil-formatted up-counter instruction. The up-counter increments its accumulated value by 1 each time the counter rung makes a false-to-true transition. When the accumulated count equals the preset count the counter output is energized or set to 1. Shown as part of the instruction are the:

Counter type
Counter address
Counter preset value
Accumulated count

The counter reset instruction must be used in conjunction with the counter instruction. Up-counters are always reset to zero. Down-counters may be reset to zero or to some preset value. Some manufacturers include the reset function as a part of the general counter instruction, whereas others dedicate a separate instruction for resetting the counter. Figure 8-5 shows a coil-formatted counter instruction with a separate instruction for resetting the counter. When programmed, the counter reset coil (CTR) is given the same reference address as the

Figure 8-1 Mechanical counter.

Figure 8-2 Electronic counters.
Source: Photo courtesy Omron Industrial Automation, **www.ia.omron.com**.

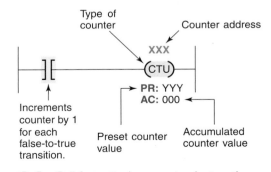

Figure 8-4 Coil-formatted up-counter instruction.

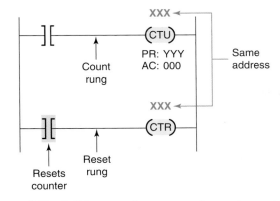

Figure 8-5 Coil-formatted counter and reset instructions.

counter (CTU) that it is to reset. In this example the reset instruction is activated whenever the CTR rung condition is true.

Figure 8-6 shows a *block-formatted* counter. The instruction block indicates the type of counter (up or down), along with the counter's preset value and accumulated or current value. The counter has two input conditions associated with it, namely, the count and reset. All PLC counters operate, or count, on the leading edge

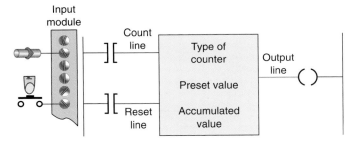

Figure 8-6 Block-formatted counter instruction.

of the input signal. The counter will either increment or decrement whenever the count input transfers from an off state to an on state. The counter will *not* operate on the trailing edge, or on-to-off transition, of the input condition.

Some manufacturers require the reset rung or line to be true to reset the counter, whereas others require it to be false to reset the counter. For this reason, it is wise to consult the PLC's operations manual before attempting any programming of counter circuits.

PLC counters are normally retentive; that is, whatever count was contained in the counter at the time of a processor shutdown will be restored to the counter on power-up. The counter may be reset, however, if the reset condition is activated at the time of power restoration.

PLC counters can be designed to count up to a preset value or to count down to a preset value. The *up-counter* is incremented by 1 each time the rung containing the counter is energized. The *down-counter* decrements by 1 each time the rung containing the counter is energized. These rung transitions can result from events occurring in the program, such as parts traveling past a sensor or actuating a limit switch. The preset value of a programmable controller counter can be set by the operator or can be loaded into a memory location as a result of a program decision.

Figure 8-7 illustrates the counting sequence of an up-counter and a down-counter. The value indicated by the counter is termed the *accumulated value*. The counter will increment or decrement, depending on the type of counter, until the accumulated value of the counter is equal to or greater than the preset value, at which time an output will be produced. A counter reset is always provided to cause the counter accumulated value to be reset to a predetermined value.

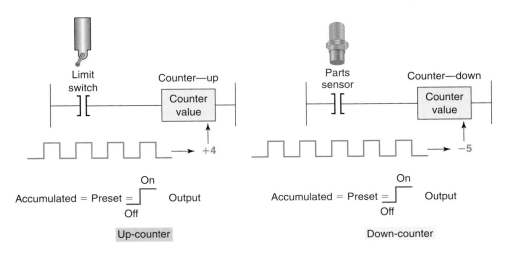

Figure 8-7 Counter counting sequence.

8.2 Up-Counter

The up-counter is an output instruction whose function is to increment its accumulated value on false-to-true transitions of its instruction. It thus can be used to count false-to-true transitions of an input instruction and then trigger an event after a required number of counts or transitions. The up-counter output instruction will increment by 1 each time the counted event occurs.

Figure 8-8 shows the program and timing diagram for an SLC 500 Count-Up Counter. This control application is designed to turn the red pilot light on and the green pilot light off after an accumulated count of 7. The operation of the program can be summarized as follows:

- Operating pushbutton PB1 provides the off-to-on transition pulses that are counted by the counter.
- The preset value of the counter is set for 7.

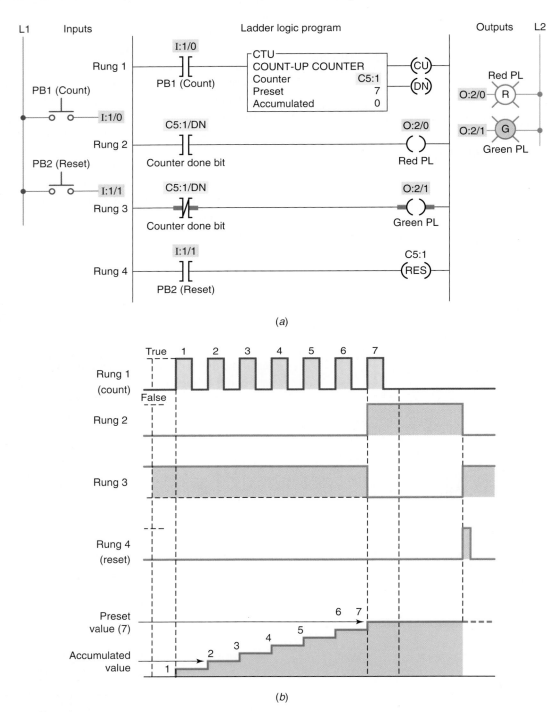

(a)

(b)

Figure 8-8 Simple up-counter program. (a) Program. (b) Timing diagram.

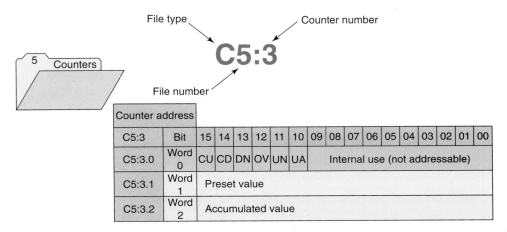

Figure 8-9 SLC 500 counter file.

- Each false-to-true transition of rung 1 increases the counter's accumulated value by 1.
- After 7 pulses, or counts, when the preset counter value equals the accumulated counter value, output DN is energized.
- As a result, rung 2 becomes true and energizes output O:2/0 to switch the red pilot light on.
- At the same time, rung 3 becomes false and de-energizes output O:2/1 to switch the green pilot light off.
- The counter is reset by closing pushbutton PB2, which makes rung 4 true and resets the accumulated count to zero.
- Counting can resume when rung 4 goes false again.

The Allen-Bradley SLC 500 counter file is file 5 (Figure 8-9). Each counter is composed of three 16-bit words, collectively called a counter element. These three data words are the control word, preset word, and accumulated word. Each of the three data words shares the same base address, which is the address of the counter itself. There can be up to 256 counter elements. Addresses for counter file 5, counter element 3 (C5:3), are listed below.

C5 = counter file 5

:3 = counter element 3 (0–255 counter elements per file)

C5:3/DN is the address for the done bit of the counter.

C5:3/CU is the address for the count-up enable bit of the counter.

C5:3/CD is the address for the count-down enable bit of the counter.

C5:3/OV is the address for the overflow bit of the counter.

C5:3/UN is the address for the underflow bit of the counter.

C5:3/UA is the address for the update accumulator bit of the counter.

Figure 8-10 shows the counter table for the Allen-Bradley SLC 500 controller. The *control word* uses status control bits consisting of the following:

Count-Up (CU) Enable Bit—The count-up enable bit is used with the count-up counter and is true whenever the count-up counter instruction is true. If the count-up counter instruction is false, the CU bit is false.

Count-Down (CD) Enable Bit—The count-down enable bit is used with the count-down counter and is true whenever the count-down counter instruction is true. If the count-down counter instruction is false, the CD bit is false.

Done (DN) Bit—The done bit is true whenever the accumulated value is equal to or greater than the preset value of the counter, for either the count-up or the count-down counter.

Overflow (OV) Bit—The overflow bit is true whenever the counter counts past its maximum value, which is 32,767. On the next count, the counter will wrap around to 32,768 and will continue counting

Counter Table

	/CU	/CD	/DN	/OV	/UN	/UA	.PRE	.ACC
C5:0	0	0	0	0	0	0	0	0
C5:1	0	0	0	0	0	0	0	0
C5:2	0	0	0	0	0	0	0	0
C5:3	0	0	0	0	0	0	50	0
C5:4	0	0	0	0	0	0	0	0
C5:5	0	0	0	0	0	0	0	0

Address C5:3   Table: C5: Counter ▼

Figure 8-10 SLC 500 counter table.

from there toward 0 on successive false-to-true transitions of the count-up counter.

Underflow (UN) Bit—The underflow bit will go true when the counter counts below 32,768. The counter will wrap around to +32,767 and continue counting down toward 0 on successive false-to-true rung transitions of the count-down counter.

Update Accumulator (UA) Bit—The update accumulator bit is used only in conjunction with an external HSC (high-speed counter).

The *preset value (PRE) word* specifies the value that the counter must count to before it changes the state of the done bit. The preset value is the set point of the counter and ranges from −32,768 through +32,767. The number is stored in binary form, with any negative numbers being stored in 2's complement binary.

The *accumulated value (ACC) word* is the current count based on the number of times the rung goes from false to true. The accumulated value either increments with a false-to-true transition of the count-up counter instruction or decrements with a false-to-true transition of the count-down counter instruction. It has the same range as the preset: −32,768 through +32,767. The accumulated value will continue to count past the preset value instead of stopping at the preset like a timer does.

Figure 8-11 shows an example of the count-up counter and its status bits used in the SLC 500 controller

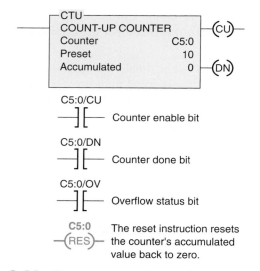

C5:0/CU
┤├── Counter enable bit

C5:0/DN
┤├── Counter done bit

C5:0/OV
┤├── Overflow status bit

C5:0
─(RES)─ The reset instruction resets the counter's accumulated value back to zero.

Figure 8-11 Count-up counter instruction.

instruction set. The address for counters begins at C5:0 and continues through C5:255. The information to be entered includes:

Counter Number—This number must come from the counter file. In the example shown, the counter number is C5:0, which represents counter file 5, counter 0 in that file. The address for this counter should not be used for any other count-up counter.

Preset Value—The preset value can range from −32,768 to +32,767. In the example shown, the preset value is 10.

Accumulated Value—The accumulated value can also range from −32,768 through +32,767. Typically, as in this example, the value entered in the accumulated word is 0. Regardless of what value is entered, the reset instruction will reset the accumulated value to 0.

Figure 8-12 shows the timer/counter menu tab from the RSLogix toolbar. Several timer and counter instructions appear when this tab is selected. The first three are timer instructions that are covered in Chapter 7. The next two instructions from the left are the up-counter (CTU) and down-counter (CTD) instructions. To the right of the CTU and CTD instructions is the reset (RES) instruction, which is used by both counters and timers. The counter commands can be summarized as follows:

CTU (Count-Up)—Increments the accumulated value at each false-to-true transition and retains the accumulated value when an off/on power cycle occurs.

CTD (Count-Down)—Decrements the accumulated value at each false-to-true transition and retains the accumulated value when an on/off power cycle occurs.

HSC (High-Speed Counter)—Counts high-speed pulses from a high-speed input.

Figure 8-13 shows a PLC counter program used to stop a motor from running after 10 operations. The operation of the program can be summarized as follows:

• Up-counter C5:0 counts the number of off/on operations of the motor.

• The preset value of the counter is set to 10.

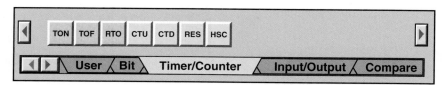

Figure 8-12 Counter selection toolbar.

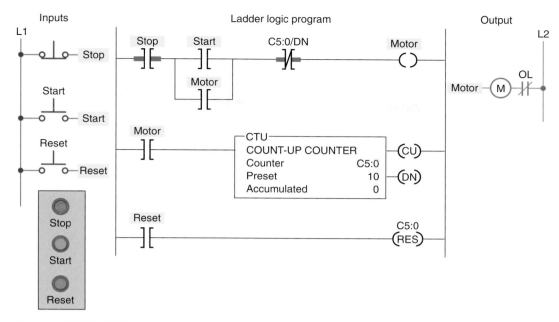

Figure 8-13 PLC counter program used to stop a motor from running after 10 operations.

- A counter done bit examine-off instruction is programmed in series with the motor output instruction.
- A motor output examine-on instruction is used to increment the accumulated value of the counter for each off/on operation.
- After the count of 10 is reached the counter done bit examine-off instruction goes false preventing the motor from being started.
- Closure of the reset pushbutton resets the accumulated count to zero.

Figure 8-14 shows a PLC can-counting program that uses three up-counters. The operation of the program can be summarized as follows:

- Counter C5:2 counts the total number of cans coming off an assembly line for final packaging.
- Each package must contain 10 parts.
- When 10 cans are detected, counter C5:1 sets bit B3/1 to initiate the box closing sequence.
- Counter C5:3 counts the total number of packages filled in a day. (The maximum number of packages per day is 300.)
- A pushbutton is used to restart the total part and package count from zero daily.

One-Shot Instruction

Figure 8-15 shows the program for a *one-shot,* or *transitional, contact circuit* that is often used to automatically clear or reset a counter. The program is designed to generate an output pulse that, when triggered, goes on for the duration of one program scan and then goes off. The one-shot can be triggered from a momentary signal or from a signal that comes on and stays on for some time. Whichever signal is used, the one-shot is triggered by the leading-edge (off-to-on) transition of the input signal. It stays on for one scan and goes off. It stays off until the trigger goes off, and then comes on again. The one-shot is perfect for resetting both counters and timers since it stays on for one scan only.

Some PLCs provide transitional contacts or one-shot instructions in addition to the standard NO and NC contact instructions. The *off-to-on* transitional contact instruction, shown in Figure 8-16a, is programmed to provide a one-shot pulse when the referenced trigger signal makes a positive (off-to-on) transition. This contact will close for exactly one program scan whenever the trigger signal goes from off to on. The contact will allow logic continuity for one scan and then open, even though the triggering signal may stay on. The *on-to-off* transitional contact, shown in Figure 8-16b, provides the same operation as the off-to-on transitional contact instruction, except that it allows logic continuity for a single scan whenever the trigger signal goes from an on to an off state.

The conveyor motor PLC program of Figure 8-17 illustrates the application of an up-counter along with a programmed one-shot (OSR) transitional contact instruction. The counter counts the number of cases coming off the conveyor. When the total number of cases reaches 50, the conveyor motor stops automatically. The trucks being loaded will take a total of only 50 cases of this particular

Ladder logic program

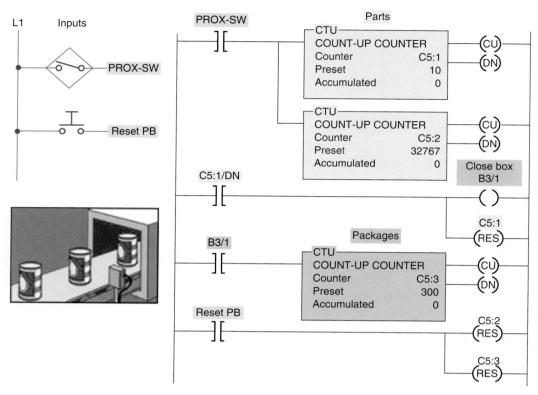

Figure 8-14 Can-counting program.

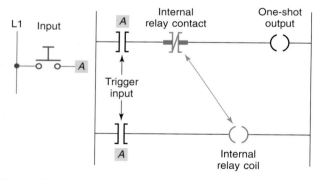

Figure 8-15 One-shot, or transitional, contact program.

product; however, the count can be changed for different product lines. The operation of the program can be summarized as follows:

- The momentary start button is pressed to start the conveyor motor M1.
- The passage of cases is sensed by the proximity switch.
- Cases move past the proximity switch and increment the counter's accumulated value with each false-to-true transition of the switch.
- After a count of 50, the done bit of the counter changes state to stop the conveyor motor automatically and reset the counter's accumulated value to zero.

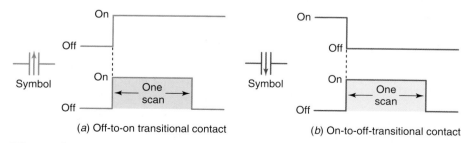

(a) Off-to-on transitional contact

(b) On-to-off-transitional contact

Figure 8-16 Transitional contact instructions.

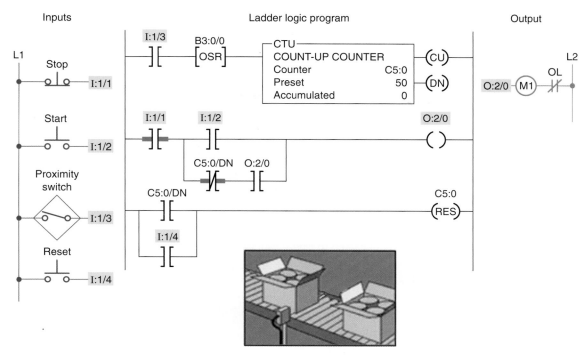

Figure 8-17 Case-counting program.

- The conveyor motor can be stopped and started manually at any time without loss of the accumulated count.
- The accumulated count of the counter can be reset manually at any time by means of the count reset button.

The Allen-Bradley SLC 500 one-shot rising (OSR) instruction is an input instruction that triggers an event to occur one time. The OSR instruction is placed in the ladder logic before the output instruction. When the rung conditions preceding the OSR instructions go from false-to-true, the OSR instruction goes true also but for only one scan. Figure 8-18 illustrates the operation of an OSR rung which can be summarized as follows:

- The OSR, one-shot rising instruction is used to make the counter reset instruction (RES) true for one scan when limit switch input LS1 goes from false to true.
- The OSR is assigned a Boolean bit (B3:0/0) that is not used anywhere else in the program.
- The OSR instruction must immediately precede the output instruction.
- When the limit switch closes the LS1 and OSR, input instructions go from false to true. The OSR instruction conditions the rung so that the counter C5:1 reset output instruction goes true for one program scan.
- The output reset instruction goes false and remains false for successive scans until the input makes another false-to-true transition.
- The OSR bit is set to 1 as long as the limit switch remains closed.
- The OSR bit is reset to 0 when the limit switch is opened.

Applications for the OSR instruction include freezing rapidly displayed LED values. Figure 8-19 shows a one-shot instruction used to send data to an output LED display. The one-shot allows the rapidly changing accumulated time from the timer to be frozen to ensure a readable, stable display. The operation of the program is summarized as follows:

- The accumulated value of timer T4:1 is converted to Binary Coded Decimal (BCD) and moved to output word O:6 where an LED display is connected.

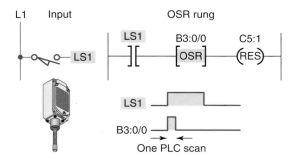

Figure 8-18 One-shot rising (OSR) instruction.

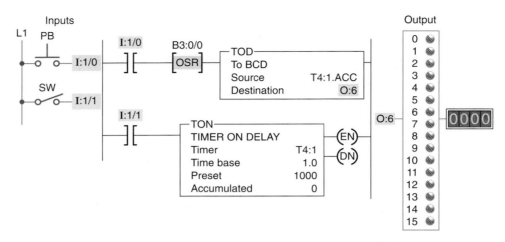

Figure 8-19 OSR instruction used to freeze rapidly displayed LED values.

- When the timer is running, SW (I:1/1) closed, the accumulated value changes rapidly.
- Closing the momentary pushbutton PB (I:1/0) will freeze and display the value at that point in time.

The alarm monitor PLC program of Figure 8-20 illustrates the application of an up-counter used in conjunction with the programmed timed oscillator circuit studied in Chapter 7. The operation of the program can be summarized as follows:

- The alarm is triggered by the closing of float switch FS.
- The light will flash whenever the alarm condition is triggered and has not been acknowledged,

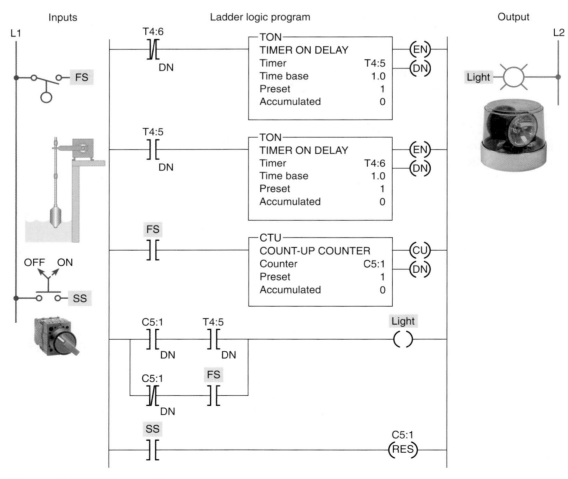

Figure 8-20 Alarm monitor program.

even if the alarm condition clears in the meantime.
- The alarm is acknowledged by closing selector switch SS.
- The light will operate in the steady on mode when the alarm trigger condition still exists but has been acknowledged.

8.3 Down-Counter

The down-counter instruction will count down or decrement by 1 each time the counted event occurs. Each time the down-count event occurs, the accumulated value is decremented. Normally the down-counter is used in conjunction with the up-counter to form an up/down-counter.

Figure 8-21 shows the program and timing diagram for a generic, block-formatted up/down-counter. The operation of the program can be summarized as follows:

- Separate count-up and count-down inputs are provided.
- Assuming the preset value of the counter is 3 and the accumulated count is 0, pulsing the count-up input (PB1) three times will switch the output light from off to on.
- This particular PLC counter keeps track of the number of counts received above the preset value. As a result, three additional pulses of the count-up input (PB1) produce an accumulated value of 6 but no change in the output.
- If the count-down input (PB2) is now pulsed four times, the accumulated count is reduced to 2 (6 − 4). As a result, the accumulated count drops below the preset count and the output light switches from on to off.
- Pulsing the reset input (PB3) at any time will reset the accumulated count to 0 and turn the output light off.

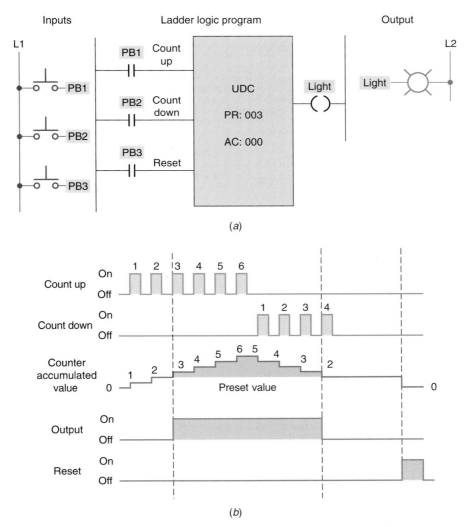

(a)

(b)

Figure 8-21 Generic up/down-counter program. (a) Program. (b) Counting diagram.

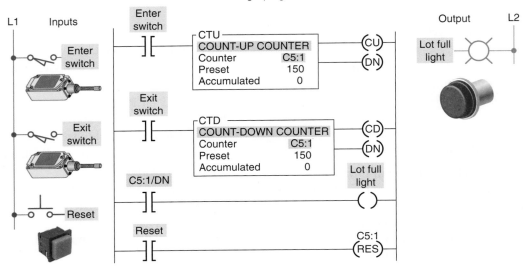

Figure 8-22   Parking garage counter.

Not all counter instructions count in the same manner. Some up-counters count only to their preset values, and additional counts are ignored. Other up-counters keep track of the number of counts received above the counter's preset value. Conversely, some down-counters will simply count down to zero and no further. Other down-counters may count below zero and begin counting down from the largest preset value that can be set for the PLC's counter instruction. For example, a PLC up/down-counter that has a maximum counter preset limit of 999 may count up as follows: 997, 998, 999, 000, 001, 002, and so on. The same counter would count down in the following manner: 002, 001, 000, 999, 998, 997, and so on.

One application for an up/down-counter is to keep count of the cars that enter and leave a parking garage. Figure 8-22 shows a typical PLC program that could be used to implement this. The operation of the program can be summarized as follows:

- As a car enters, the enter switch triggers the up-counter output instruction and increments the accumulated count by 1.
- As a car leaves, the exit switch triggers the down-counter output instruction and decrements the accumulated count by 1.
- Because both the up- and down-counters have the same address, C5:1, the accumulated value will be the same in both instructions as well as the preset.
- Whenever the accumulated value of 150 equals the preset value of 150, the counter output is energized by the done bit to light up the Lot Full sign.

- A reset button has been provided to reset the accumulated count.

Figure 8-23 shows an example of the count-down counter instruction used as part of the Allen-Bradley SLC 500 controller instruction set. The information to be entered into the instruction is the same as for the count-up counter instruction.

The CTD instruction decrements its accumulated value by 1 every time it is transitioned. It sets its done bit when the accumulated value is equal to or greater than the preset value. The CTD instruction requires the RES instruction to reset its accumulated value and status bits. Because it resets its accumulated value to 0, the CTD instruction then

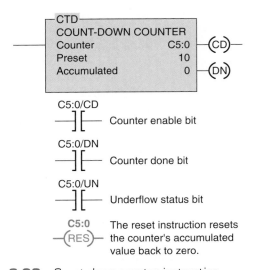

Figure 8-23   Count-down counter instruction.

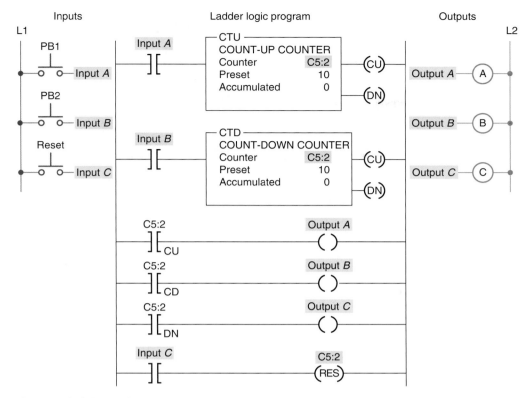

Figure 8-24 Up/down-counter program.

counts negative when it transitions. If the CTD instruction were used by itself with a positive preset value, its done bit would be reset when the accumulated value reached 0. Then, counting in a negative direction, the accumulated value would never reach its preset value and set the done bit. However, the preset can be entered with a negative value; then the done bit is cleared when the accumulated value becomes less than the preset value.

Figure 8-24 shows an up/down-counter program that will increase the counter's accumulated value when pushbutton PB1 is pressed and will decrease the counter's accumulated value when pushbutton PB2 is pressed. Note that the same address is given to the *up-counter* instruction, the down-counter instruction, and the reset instruction. All three instructions will be looking at the *same address* in the counter file. When input A goes from false to true, one count is added to the accumulated value. When input B goes from false to true, one count is subtracted from the accumulated value. The operation of the program can be summarized as follows:

- When the CTU instruction is true, C5:2/CU will be true, causing output A to be true.
- When the CTD instruction is true, C5:2/CD will be true, causing output B to be true.

- When the accumulated value is greater than or equal to the preset value, C5:2/DN will be true, causing output C to be true.
- Input C going true will cause both counter instructions to reset. When *reset* by the RES instruction, the accumulated value will be reset to 0 and the done bit will be reset.

Figure 8-25 illustrates the operation of the up/down-counter program used to provide continuous monitoring of items in process. An in-feed photoelectric sensor counts raw parts going into the system, and an out-feed photoelectric sensor counts finished parts leaving the machine. The number of parts between the in-feed and out-feed is indicated by the accumulated count of the counter. Counts applied to the up-input are added, and counts applied to the down-input are subtracted. The operation of the program can be summarized as follows:

- Before start-up, the system is completely empty of parts, and the counter is reset manually to 0.
- When the operation begins, raw parts move through the in-feed sensor, with each part generating an up count.

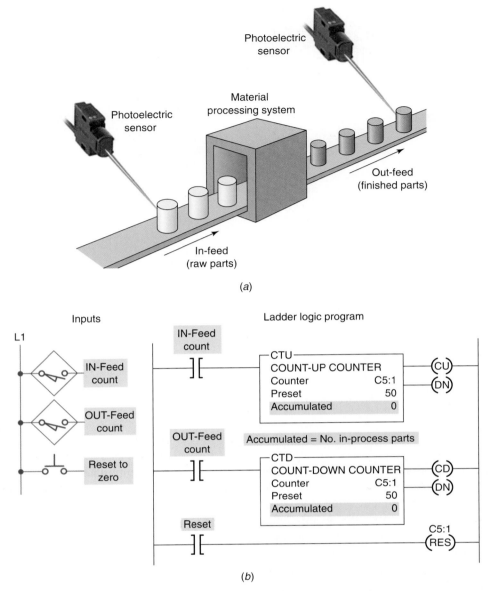

Figure 8-25 In-process monitoring program. (a) Process. (b) Program.

- After processing, finished parts appearing at the out-feed sensor generate down counts, so the accumulated count of the counter continuously indicates the number of in-process parts.
- The counter preset value is irrelevant in this application. It does not matter whether the counter outputs are on or off. The output on-off logic is not used. We have arbitrarily set the counter's preset values to 50.

The maximum speed of transitions that you can count is determined by your program's scan time. For a reliable count, your counter input signal must be fixed for one scan time. If the input changes faster than one scan period, the count value will become unreliable because counts will be missed. When this situation occurs, you need to use a high-speed counter input or a separate counter I/O module designed for high-speed applications.

8.4 Cascading Counters

Depending on the application, it may be necessary to count events that exceed the maximum number allowable per counter instruction. One way of accomplishing this count is by interconnecting, or cascading, two counters. The program of Figure 8-26 illustrates the application of

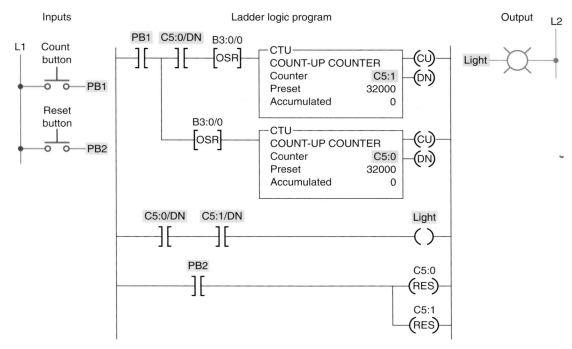

Figure 8-26 Counting beyond the maximum count.

the technique. The operation of the program can be summarized as follows:

- The output of the first counter is programmed into the input of the second counter.
- The status bits of both counters are programmed in series to produce an output.
- These two counters allow twice as many counts to be measured.

Another method of cascading counters is sometimes used when an extremely large number of counts must be stored. For example, if you require a counter to count up to 250,000, it is possible to achieve this by using only two counters. Figure 8-27 shows how the two counters would be programmed for this purpose. The operation of the program can be summarized as follows:

- Counter C5:1 has a preset value of 500 and counter C5:2 has a preset value of 500.
- Whenever counter C5:1 reaches 500, its done bit resets counter C5:1 and increments counter C5:2 by 1.
- When the done bit of counter C5:1 has turned on and off 500 times, the output light becomes energized. Therefore, the output light turns on after 500 × 500, or 250,000, transitions of the count input.

Some PLCs include a real-time clock as part of their instruction set. A real-time clock allows you to display the time of day or to log data pertaining to the operation of the process. The logic used to implement a clock as part of a PLC's program is straightforward and simple to accomplish. A single timer instruction and counter instructions are all you need.

Figure 8-28 illustrates a timer-counter program that produces a time-of-day clock measuring time in hours and minutes. The operation of the program can be summarized as follows:

- An RTO timer instruction (T4:0) is programmed first with a preset value of 60 seconds.
- The T4:0 timer times for a 60-s period, after which its done bit is set.
- This, in turn, causes the up-counter (C5:0) of rung 001 to increment 1 count.
- On the next processor scan, the timer is reset and begins timing again.
- The C5:0 counter is preset to 60 counts, and each time the timer completes its time-delay period, its count is incremented.
- When the C5:0 counter reaches its preset value of 60, its done bit is set.
- This, in turn, causes the up-counter (C5:1) of rung 002, which is preset for 24 counts, to increment 1 count.
- Whenever the C5:1 counter reaches its preset value of 24, its done bit is set to reset itself.

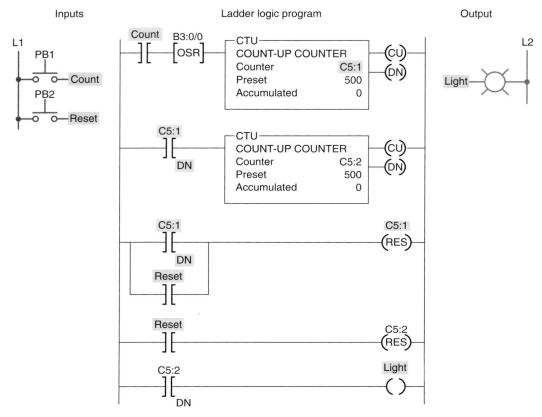

Figure 8-27 Cascading counters for extremely large counts.

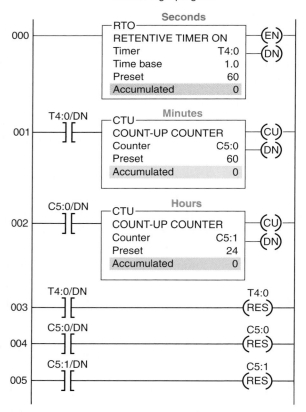

Figure 8-28 24-hour clock program.

- The time of day is generated by examining the current, or accumulated, count or time for each counter and the timer.
- Counter C5:1 indicates the hour of the day in 24-h military format, while the current minutes are represented by the accumulated count value of counter C5:0.
- The timer displays the seconds of a minute as its current, or accumulated, time value.

The 24-hour clock can be used to record the time of an event. Figure 8-29 illustrates the principle of this technique. In this application the time of the opening of a pressure switch is to be recorded. The operation of the program can be summarized as follows:

- The circuit is set into operation by pressing the reset button and setting the clock for the time of day.
- This starts the 24-hour clock and switches the set indicating light on.
- Should the pressure switch open at any time, the clock will automatically stop and the trip indicating light will switch on.
- The clock can then be read to determine the time of opening of the pressure switch.

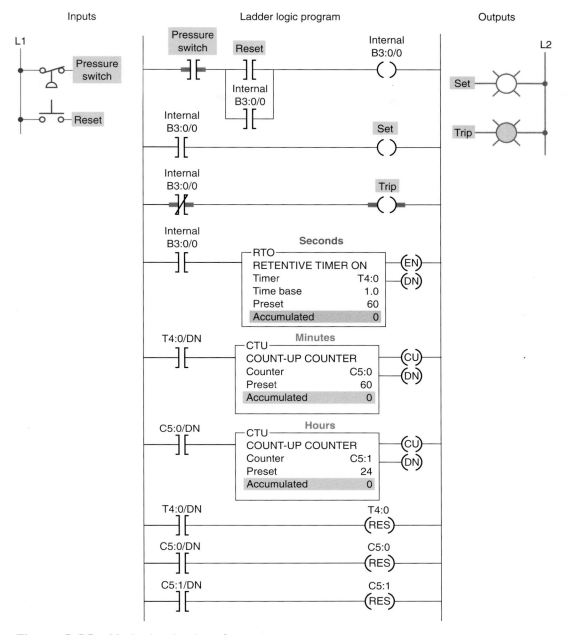

Inputs | Ladder logic program | Outputs

Figure 8-29 Monitoring the time of an event.

8.5 Incremental Encoder-Counter Applications

The incremental optical encoder shown in Figure 8-30 creates a series of square waves as its shaft is rotated. The encoder disk interrupts the light as the encoder shaft is rotated to produce the square wave output waveform.

The number of square waves obtained from the output of the encoder can be made to correspond to the mechanical movement required. For example, to divide a shaft revolution into 100 parts, an encoder could be selected to supply 100 square wave-cycles per revolution. By using a counter to count those cycles, we could tell how far the shaft had rotated.

Figure 8-31 illustrates an example of cutting objects to a specified length. The object is advanced for a specified distance and measured by encoder pulses to determine the correct length for cutting.

Figure 8-32 shows a counter program used for length measurement. This system accumulates the total length of random pieces of bar stock moved on a conveyor. The operation of the program can be summarized as follows:

- Count input pulses are generated by the magnetic sensor, which detects passing teeth on a conveyor drive sprocket.

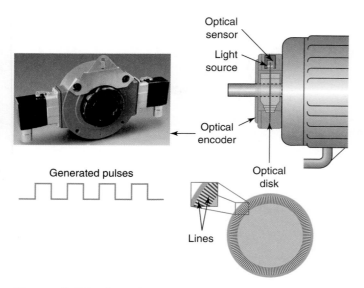

Figure 8-30 Optical incremental encoder.
Source: Photo courtesy Avtron, **www.avtron.com**.

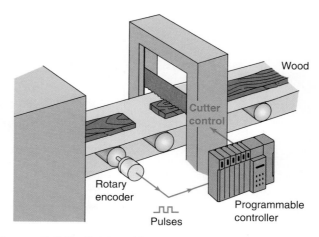

Figure 8-31 Cutting objects to a specified length.

- If 10 teeth per foot of conveyor motion pass the sensor, the accumulated count of the counter would indicate feet in tenths.
- The photoelectric sensor monitors a reference point on the conveyor. When activated, it prevents the unit from counting, thus permitting the

counter to accumulate counts only when bar stock is moving.
- The counter is reset by closing the reset button.

8.6 Combining Counter and Timer Functions

Many PLC applications use both the counter function and the timer function. Figure 8-33 illustrates an automatic stacking program that requires both a timer and counter.

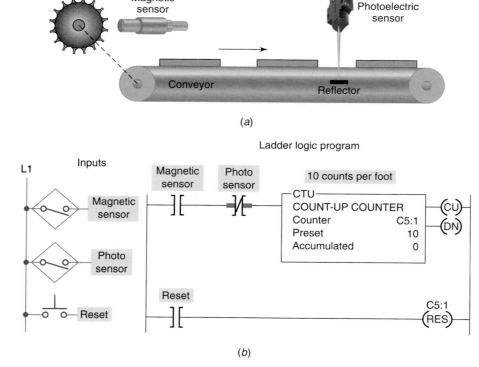

Figure 8-32 Counter used for length measurement. (a) Process. (b) Program.

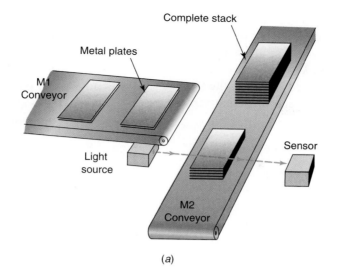

(a)

Ladder logic program

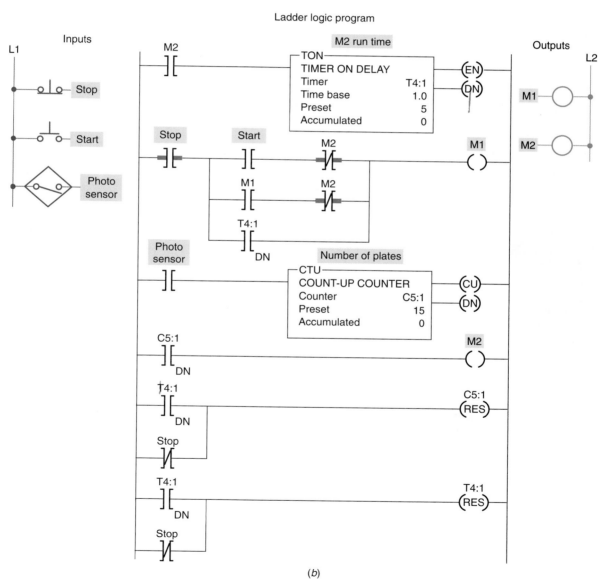

(b)

Figure 8-33 Automatic stacking program. (a) Process. (b) Program.

In this process, conveyor M1 is used to stack metal plates onto conveyor M2. The photoelectric sensor provides an input pulse to the PLC counter each time a metal plate drops from conveyor M1 to M2. When 15 plates have been stacked, conveyor M2 is activated for 5 s by the PLC timer. The operation of the program can be summarized as follows:

- When the start button is pressed, conveyor M1 begins running.

- After 15 plates have been stacked, conveyor M1 stops and conveyor M2 begins running.
- After conveyor M2 has been operated for 5 s, it stops and the sequence is repeated automatically.
- The done bit of the timer resets the timer and the counter and provides a momentary pulse to automatically restart conveyor M1.

Figure 8-34 shows a motor lock-out program. This program is designed to prevent a machine operator from

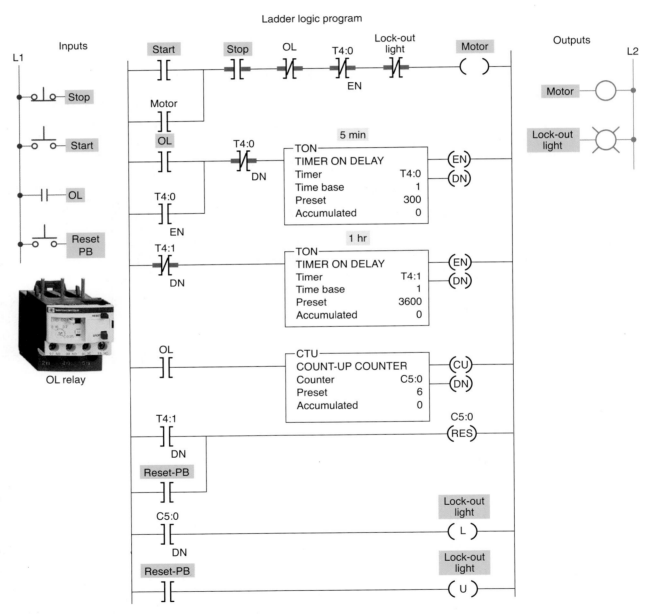

Figure 8-34 Motor lock-out program.
Source: This material and associated copyrights are proprietary to, and used with the permission of Schneider Electric.

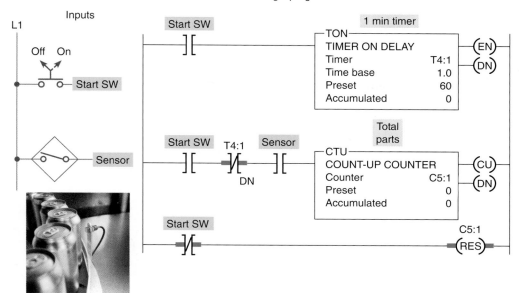

Figure 8-35 Product flow rate program.
Source: Photo courtesy Omron Industrial Automation, **www.ia.omron.com**.

starting a motor that has tripped off more than 5 times in an hour. The operation of the program can be summarized as follows:

- The normally open overload (OL) relay contact momentarily closes each time an overload current is sensed.
- Every time the motor stops due to an overload condition, the motor start circuit is locked out for 5 min.
- If the motor trips off more than 5 times in an hour, the motor start circuit is permanently locked out and cannot be started until the reset button is actuated.
- The lock-out pilot light is switched on whenever a permanent lock-out condition exists.

Figure 8-35 shows a product part flow rate program. This program is designed to indicate how many parts pass a given process point per minute. The operation of the program can be summarized as follows:

- When the start switch is closed, both the timer and counter are enabled.
- The counter is pulsed for each part that passes the parts sensor.
- The counting begins and the timer starts timing through its 1-minute time interval.
- At the end of 1 minute, the timer done bit causes the counter rung to go false.

- Sensor pulses continue but do not affect the PLC counter.
- The number of parts for the past minute is represented by the accumulated value of the counter.
- The sequence is reset by momentarily opening and closing the start switch.

A timer is sometimes used to drive a counter when an extremely long time-delay period is required. For example, if you require a timer to time to 1,000,000 s, you can achieve this by using a single timer and counter. Figure 8-36 shows how the timer and counter would be programmed for such a purpose. The operation of the program can be summarized as follows:

- Timer T4:0 has a preset value of 10,000, and counter C5:0 has a preset value of 100.
- Each time the timer T4:0 input contact closes for 10,000 s, its done bit resets timer T4:0 and increments counter C5:0 by 1.
- When the done bit of timer T4:0 has turned on and off 100 times, the output light becomes energized.
- Therefore, the output light turns on after 10,000 × 100, or 1,000,000, seconds after the timer input contact closes.

Ladder logic program

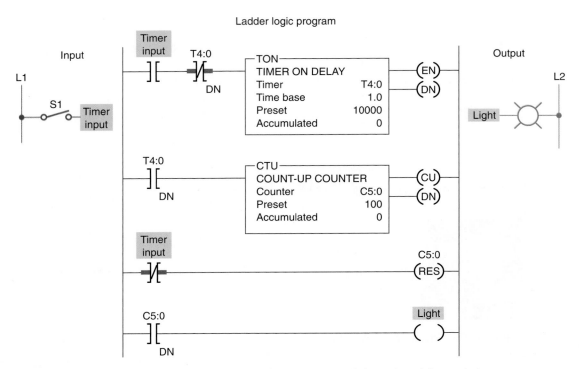

Figure 8-36 Timer driving a counter to produce an extremely long time-delay period.

1. Name the three forms of PLC counter instructions, and explain the basic operation of each.
2. State four pieces of information usually associated with a PLC counter instruction.
3. In a PLC counter instruction, what rule applies to the addressing of the counter and reset instructions?
4. When is the output of a PLC counter energized?
5. When does the PLC counter instruction increment or decrement its current count?
6. The counter instructions of PLCs are normally retentive. Explain what this means.
7. **a.** Compare the operation of a standard Examine-on contact instruction with that of an off-to-on transitional contact.
   b. What is the normal function of a transitional contact used in conjunction with a counter?
8. Explain how an OSR (one-shot rising) instruction can be used to freeze rapidly changing data.

9. Identify the type of counter you would choose for each of the following situations:
   a. Count the total number of parts made during each shift.
   b. Keep track of the current number of parts in a stage of a process as they enter and exit.
   c. There are 10 parts in a full hopper. As parts leave, keep track of the number of parts remaining in the hopper
10. Describe the basic programming process involved in the cascading of two counters.
11. **a.** When is the overflow bit of an up-counter set?
    b. When is the underflow bit of a down-counter set?
12. Describe two common applications for counters.
13. What determines the maximum speed of transitions that a PLC counter can count? Why?

1. Study the ladder logic program in Figure 8-37, and answer the questions that follow:
   a. What type of counter has been programmed?
   b. When would output O:2/0 be energized?
   c. When would output O:2/1 be energized?

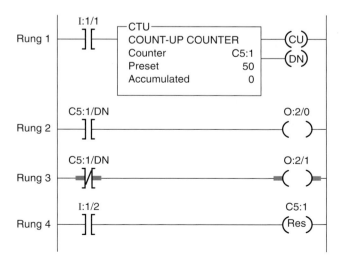

Ladder logic program

Figure 8-37 Program for Problem 1.

d. Suppose your accumulated value is 24 and you lose ac line power to the controller. When power is restored to your controller, what will your accumulated value be?
e. Rung 4 goes true and while it is true, rung 1 goes through five false-to-true transitions of rung conditions. What is the accumulated value of the counter after this sequence of events?
f. When will the count be incremented?
g. When will the count be reset?

2. Study the ladder logic program in Figure 8-38, and answer the questions that follow:
   a. Suppose the input pushbutton is actuated from off to on and remains held on. How will the status of output B3:0/9 be affected?
   b. Suppose the input pushbutton is now released to the normally off position and remains off. How will the status of output B3:0/9 be affected?
3. Study the ladder logic program in Figure 8-39, and answer the questions that follow:
   a. What type of counter has been programmed?
   b. What input address will cause the counter to increment?

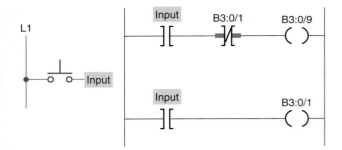

Figure 8-38 Program for Problem 2.

Ladder logic program

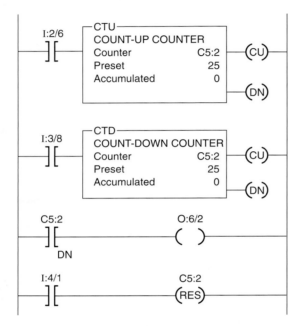

Figure 8-39 Program for Problem 3.

c. What input address will cause the counter to decrement?

d. What input address will reset the counter to a count of zero?

e. When would output O:6/2 be energized?

f. Suppose the counter is first reset, and then input I:2/6 is actuated 15 times and input I:3/8 is actuated 5 times. What is the accumulated count value?

4. Design a PLC program and prepare a typical I/O connection diagram and ladder logic program for the following counter specifications:
   • Counts the number of times a pushbutton is closed.
   • Decrements the accumulated value of the counter each time a second pushbutton is closed.

• Turns on a light anytime the accumulated value of the counter is less than 20.
• Turns on a second light when the accumulated value of the counter is equal to or greater than 20.
• Resets the counter to 0 when a selector switch is closed.

5. Design a PLC program and prepare a typical I/O connection diagram and ladder logic program that will execute the following control circuit correctly:
   • Turns on a nonretentive timer when a switch is closed (preset value of timer is 10 s).
   • Resets timer automatically through a programmed transitional contact when it times out.
   • Counts the number of times the timer goes to 10 s.
   • Resets counter automatically through a second programmed transitional contact at a count of 5.
   • Latches on a light at the count of 5.
   • Resets light to off and counter to 0 when a selector switch is closed.

6. Design a PLC program and prepare a typical I/O connection diagram and ladder logic program that will correctly execute the industrial control process in Figure 8-40. The sequence of operation is as follows:
   • Product in position (limit switch LS1 contacts close).
   • The start button is pressed and the conveyor motor starts to move the product forward toward position *A* (limit switch LS1 contacts open when the actuating arm returns to its normal position).
   • The conveyor moves the product forward to position *A* and stops (position detected by 8 off-to-on output pulses from the encoder, which are counted by an up-counter).
   • A time delay of 10 s occurs, after which the conveyor starts to move the product to limit switch LS2 and stops (LS2 contacts close when the actuating arm is hit by the product).

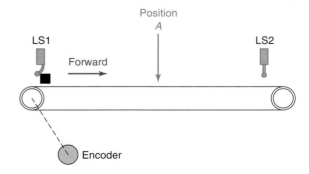

Figure 8-40 Control process for Problem 6.

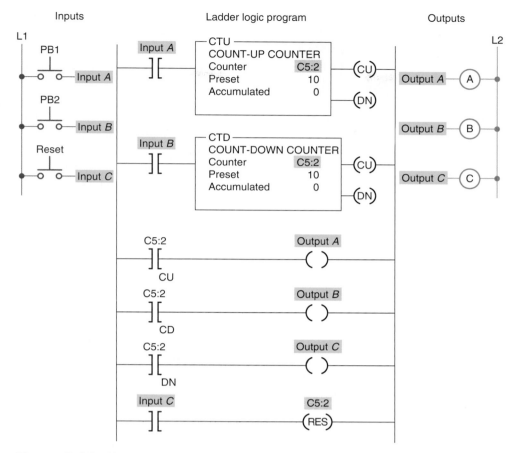

Figure 8-41 Program for Problem 7.

- An emergency stop button is used to stop the process at any time.
- If the sequence is interrupted by an emergency stop, counter and timer are reset automatically.

7. Answer the following questions with reference to the up/down-counter program shown in Figure 8-41. Assume that the following sequence of events occurs:
   - Input *C* is momentarily closed.
   - 20 on/off transitions of input *A* occur.
   - 5 on/off transitions of input *B* occur.

As a result:
   a. What is the accumulated count of counter CTU?
   b. What is the accumulated count of counter CTD?
   c. What is the state of output *A*?
   d. What is the state of output *B*?
   e. What is the state of output *C*?

8. Write a program to implement the process illustrated in Figure 8-42. An up-counter must be programmed as part of a batch-counting operation to sort parts automatically for quality control. The counter is installed to divert 1 part out of every

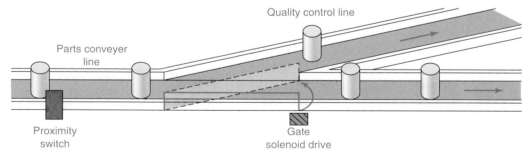

Figure 8-42 Control process for Problem 8.

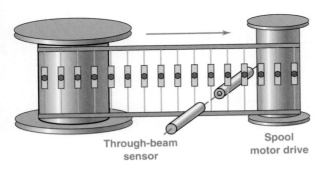

Through-beam sensor

Spool motor drive

Figure 8-43 Control process for Problem 10.

1000 for quality control or inspection purposes. The circuit operates as follows:

- A start/stop pushbutton station is used to turn the conveyor motor on and off.
- A proximity sensor counts the parts as they pass by on the conveyor.
- When a count of 1000 is reached, the counter's output activates the gate solenoid, diverting the part to the inspection line.
- The gate solenoid is energized for 2 s, which allows enough time for the part to continue to the quality control line.
- The gate returns to its normal position when the 2-s time period ends.
- The counter resets to 0 and continues to accumulate counts.
- A reset pushbutton is provided to reset the counter manually.

9. Write a program that will increment a counter's accumulated value 1 count every 60 s. A second counter's accumulated value will increment 1 count every time the first counter's accumulated value reaches 60. The first counter will reset when its accumulated value reaches 60, and the second counter will reset when its accumulated value reaches 12.

10. Write a program to implement the process illustrated in Figure 8-43. A company that makes electronic assembly kits needs a counter to count and control the number of resistors placed into each

kit. The controller must stop the take-up spool at a predetermined amount of resistors (100). A worker on the floor will then cut the resistor strip and place it in the kit. The circuit operates as follows:

- A start/stop pushbutton station is used to turn the spool motor drive on and off manually.
- A through-beam sensor counts the resistors as they pass by.
- A counter preset for 100 (the amount of resistors in each kit) will automatically stop the take-up spool when the accumulated count reaches 100.
- A second counter is provided to count the grand total used.
- Manual reset buttons are provided for each counter.

11. Write a program that will latch on a light 20 s after an input switch has been turned on. The timer will continue to cycle up to 20 s and reset itself until the input switch has been turned off. After the third time the timer has timed to 20 s, the light will be unlatched.

12. Write a program that will turn a light on when a count reaches 20. The light is then to go off when a count of 30 is reached.

13. Write a program to implement the box-stacking process illustrated in Figure 8-44. This application requires the control of a conveyor belt that feeds a mechanical stacker. The stacker can stack various numbers of cartons of ceiling tile onto each pallet (depending on the pallet size and the preset value of the counter). When the required number of cartons has been stacked, the conveyor is stopped until the loaded pallet is removed and an empty pallet is placed onto the loading area. A photoelectric sensor will be used to provide count pulses to the counter after each carton passes by. In addition to a conveyor motor start/stop station, a remote reset button is provided to allow the operator to reset the system from the forklift after an empty pallet is placed onto the loading area.

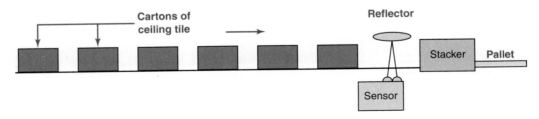

Cartons of ceiling tile

Reflector

Stacker **Pallet**

Sensor

Figure 8-44 Control process for Problem 13.

The operation of this system can be summarized as follows:

- The conveyor is started by pressing the start button.
- As each box passes the photoelectric sensor, a count is registered.
- When the preset value is reached (in this case 12), the conveyor belt turns off.
- The forklift operator removes the loaded pallet.
- After the empty pallet is in position, the forklift operator presses the remote reset button, which then starts the whole cycle over again.

14. Write a program to operate a light according to the following sequence:
- A momentary pushbutton is pressed to start the sequence.
- The light is switched on and remains on for 2 s.
- The light is then switched off and remains off for 2 s.
- A counter is incremented by 1 after this sequence.
- The sequence then repeats for a total of 4 counts.
- After the fourth count, the sequence will stop and the counter will be reset to zero.

Program Control Instructions

9

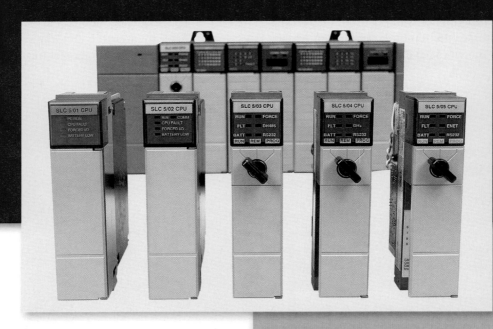

Image Used with Permission of Rockwell Automation, Inc.

The program control instructions covered in this chapter are used to alter the program scan from its normal sequence. The use of program control instructions can shorten the time required to complete a program scan. Portions of the program not being utilized at any particular time can be jumped over, and outputs in specific zones in the program can be left in their desired states. Typical industrial program control applications are explained.

Chapter Objectives

After completing this chapter, you will be able to:

9.1 State the purpose of program control instructions

9.2 Describe the operation of the master control reset instruction and develop an elementary program illustrating its use

9.3 Describe the operation of the jump instruction and the label instruction

9.4 Explain the function of subroutines

9.5 Describe the immediate input and output instructions function

9.6 Describe the forcing capability of the PLC

9.7 Describe safety considerations built into PLCs and programmed into a PLC installation

9.8 Explain the differences between standard and safety PLCs

9.9 Describe the function of the selectable timed interrupt and fault routine files

9.10 Explain how the temporary end instruction can be used to troubleshoot a program

9.1 Master Control Reset Instruction

Several output-type instructions, which are often referred to as *override* instructions, provide a means of executing sections of the control logic if certain conditions are met. These program control instructions allow for greater program flexibility and greater efficiency in the program scan. Portions of the program not being utilized at any particular time can be jumped over, and outputs in specific zones in the program can be left in their desired states.

Program control instructions are used to enable or disable a block of logic program or to move execution of a program from one place to another place. Figure 9-1 shows the *Program Control* menu tab for the Allen-Bradley SLC 500 PLC and its associated RSLogix software. The program control commands can be summarized as follows:

JMP (Jump to Label)—Jump forward/backward to a corresponding label instruction.

LBL (Label)—Specifies label location.

JSR (Jump to Subroutine)—Jump to a designated subroutine instruction.

RET (Return from Subroutine)—Exits current subroutine and returns to previous condition.

SBR (Subroutine)—Identifies the subroutine program.

TND (Temporary End)—Makes a temporary end that halts program execution.

MCR (Master Control Reset)—Clears all set nonretentive output rungs between the paired MCR instructions.

SUS (Suspend)—Identifies conditions for debugging and system troubleshooting.

Hardwired master control relays are used in relay control circuitry to provide input/output power shutdown of an entire circuit. Figure 9-2 shows a typical hardwired master control relay circuit. In this circuit, unless the master control relay coil is energized, there is no power flow to the load side of the MCR contacts.

PLC manufacturers offer a form of a master control relay as part of their instruction set. These instructions function in a similar manner to the hardwired master control relay; that is, when the instruction is true, the circuit functions normally, and when the instruction is false, nonretentive outputs are switched off. Because these instructions are not hardwired but programmed, for safety reasons they should *not* be used as a substitute for a hardwired master control relay, which provides *emergency* I/O power shutdown.

A *Master Control Reset (MCR)* instruction is an output coil instruction that functions like a master control

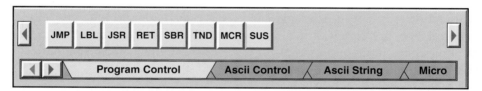

Figure 9-1 Program Control menu tab.

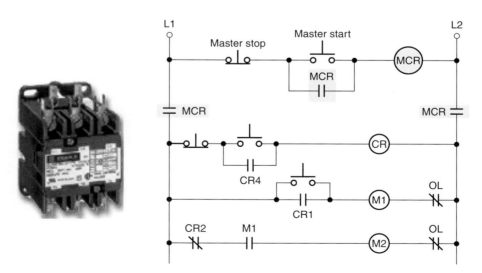

Figure 9-2 Hardwired master control relay.
Source: This material and associated copyrights are proprietary to, and used with the permission of Schneider Electric.

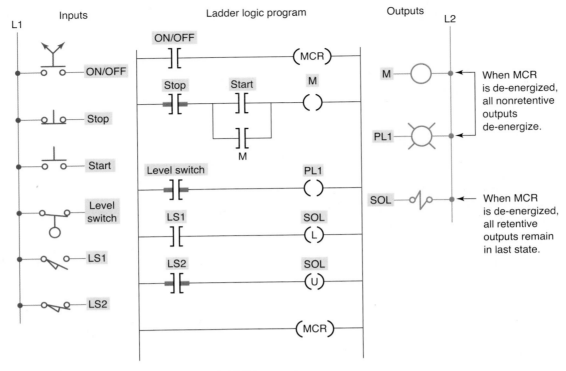

Figure 9-3 Master Control Reset (MCR) instruction.

relay. MCR coil instructions are used in pairs and can be programmed to control an entire circuit or to control only selected rungs of a circuit. In the program of Figure 9-3, the MCR is programmed to control an entire circuit. The operation of the program can be summarized as follows:

- When the MCR instruction is false, or de-energized, all nonretentive (nonlatched) rungs below the MCR will be de-energized even if the programmed logic for each rung is true.
- All *retentive* rungs will remain in their *last state.*
- The MCR instruction establishes a zone in the user program in which all nonretentive outputs can be turned off simultaneously.
- *Retentive* instructions should not normally be placed within an MCR zone because the MCR zone maintains retentive instructions in the last active state when the instruction goes false.
- An off-delay timer will start timing when in a de-energized MCR zone.

Allen-Bradley SLC 500 controllers use the master control reset instruction to set up single or multiple zones within a program. The MCR instruction is used in pairs to disable or enable a zone within a ladder program, and

it has *no address.* Figure 9-4 shows the programming of an MCR fenced zone with the zone true. The operation of the program can be summarized as follows:

- The MCR zone is enclosed by a *start fence,* which is a rung with a conditional MCR, and an *end fence,* which is a rung with an unconditional MCR.
- Input *A* of the start rung is true so all outputs act according to their rung logic as if the zone did not exist.

Figure 9-5 shows the programmed MCR fenced zone with the zone false. The operation of the program can be summarized as follows:

- When the MCR in the start fence is false, all rungs within the zone are treated as false. The scan ignores the inputs and de-energizes all nonretentive outputs (that is, the output energize instruction, the on-delay timer, and the off-delay timer).
- All retentive devices, such as latches, retentive timers, and counters, remain in their last state.
- Input *A* of the start rung is false so output *A* and T4:1 will be false and output *B* will remain in its last state.
- The input conditions in each rung will have no effect on the output conditions.

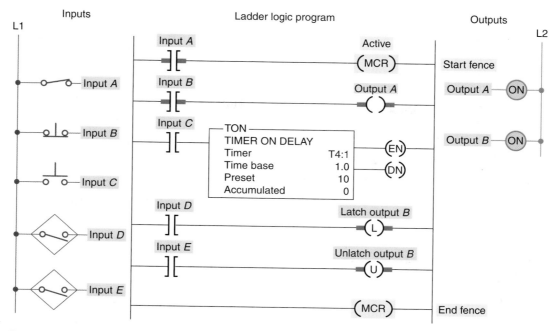

Figure 9-4 MCR fenced zone with the zone true.

A common application of an MCR zone control involves examining one or more fault bits as part of the start fence and enclosing the portion of the program you want de-energized in case of a fault in the MCR zone. In case of a detected fault condition, the outputs in that zone would be de-energized automatically.

If you start instructions such as timers or counters in an MCR zone, instruction operation ceases when the zone is disabled. The TOF timer will activate when placed inside a false MCR zone. When troubleshooting a program that contains an MCR zone, you need to be aware of which rungs are within zones in order to correctly edit the circuit.

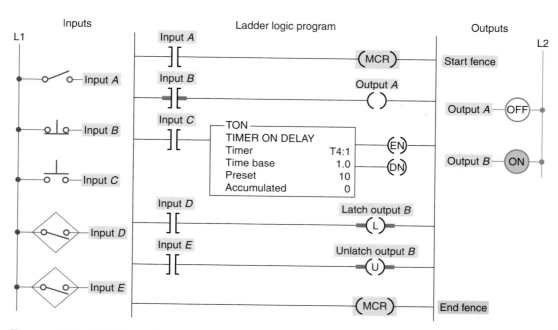

Figure 9-5 MCR fenced zone with the zone false.

MCR-controlled areas must contain only two MCR instructions—one to define the start and one to define the end. Never overlap or nest MCR zones. Any additional MCR instructions, or a jump instruction programmed to jump to an MCR zone, could produce unexpected and damaging results to your program and to machine operation.

9.2 Jump Instruction

In PLC programming it is sometimes desirable to be able to jump over certain program instructions when certain conditions exist. The *jump (JMP)* instruction is an output instruction used for this purpose. When the jump instruction is used, the PLC will not execute the instructions of a rung that is jumped. The jump instruction is often used to jump over instructions not pertinent to the machine's operation at that instant. In addition, sections of a program may be programmed to be jumped should a production fault occur.

Some manufacturers provide a skip instruction, which is essentially the same as the jump instruction.

The program of Figure 9-6 illustrates the use of a jump instruction in conjunction with Allen-Bradley SLC 500 programmable controllers. Addresses Q2:0 through Q2:255 are the addresses used for the *jump (JMP)* instructions. The *label (LBL)* instruction is a target for the jump instruction. In addition, the jump instruction with its associated label must have the same address. The area of the program that the processor jumps over is defined by the locations of the jump and label instructions in the program. If the jump coil is energized, all logic between the jump and label instructions is bypassed and the processor continues scanning after the LBL instruction.

The operation of the program can be summarized as follows:

- When the switch is open the jump instruction is not activated.
- With the switch open, closing PB turns on all three pilot lights.
- When the switch is closed the jump (JMP) instruction will activate.
- With the switch closed, pressing PB turns on pilot lights PL1 and PL3 only.
- Rung 3 is skipped over during the PLC program scan so PL2 is not turned on.

Figure 9-7 illustrates the effect on input and output instructions of jumped rungs in a program. The label instruction is used to identify the ladder rung that is the target destination but does not contribute to logic continuity. For practical purposes the label instruction is always considered to be logically true. The operation of the program can be summarized as follows:

- When rung 4 has logic continuity, the processor is instructed to jump to rung 8 and continue to execute the main program from that point.
- Jumped rungs 5, 6, and 7 are not scanned by the processor.
- Input conditions for the jumped rungs are not examined and outputs controlled by these rungs remain in their last state.
- Any timers or counters programmed within the jump area cease to function and will not update themselves during this period. For this reason they should be programmed outside the jumped section in the main program zone.

You can jump to the same label from multiple jump locations, as illustrated in the program of Figure 9-8. In this example, there are two jump instructions addressed Q2:20. There is a single label instruction addressed Q2:20. The scan can then jump from either jump instruction to label Q2:20, depending on whether input *A* or input *D* is true.

It is possible to jump backward in the program, but this should not be done an excessive number of times. Care must be taken that the scan does not remain in a loop too long. The processor has a watchdog timer that sets the maximum allowable time for a total program scan. If this time is exceeded, the processor will indicate a fault and shut down.

The forward jump is similar to an MCR instruction in that both permit an input logic condition to skip over a block of PLC ladder logic. The main difference between

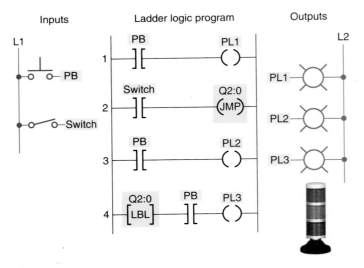

Figure 9-6 Jump (JMP) operation.

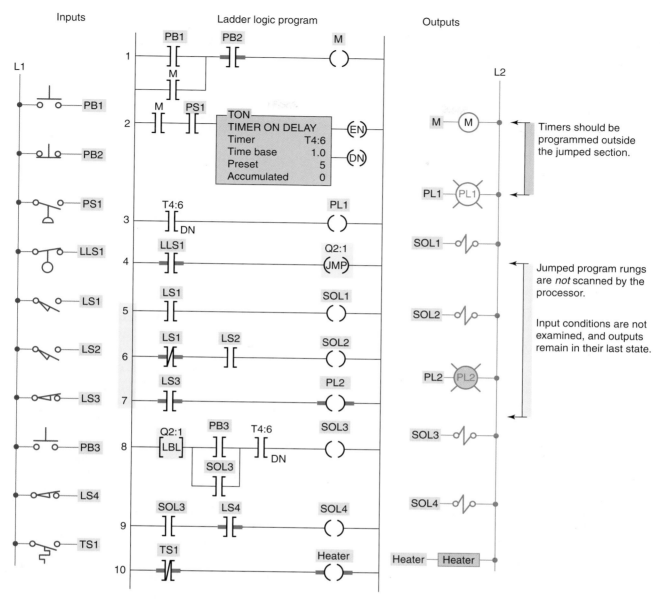

Figure 9-7 Effect on input and output instructions of jumped rungs.

the two is in how the outputs are handled when the instructions are executed. The MCR instruction sets all nonretentive outputs to the false state and keeps the retentive outputs in their last state. The JMP instruction leaves all outputs in their last state. You should never jump into a Master Control Reset zone. If you do, instructions that are programmed within the MCR zone starting at the LBL instruction and ending at the end MCR instruction will always be evaluated as though the MCR zone is true, without consideration to the state of the start MCR instruction.

9.3 Subroutine Functions

In addition to the main ladder logic program, PLC programs may also contain additional program files known as *subroutines*. A subroutine is a short program that is used by the main program to perform a specific function. Large programs are often broken into subroutine program files, which are called and executed from the main program. In the SLC 500 series PLCs, the main ladder logic program is in program file two (shown as LAD 2). Ladder logic programs for subroutines can be placed in file number three (LAD 3) through file number 255 (LAD 255).

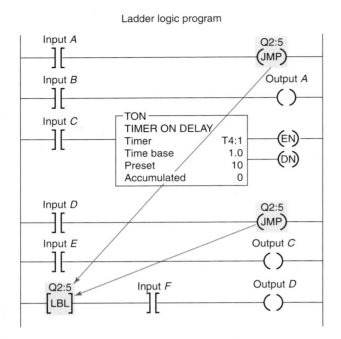

Ladder logic program

Figure 9-8 Jump-to-label from two locations.

Use of subroutines is a valuable tool in PLC programming. At times it is better to construct programs that consist of several subroutines than a lengthy single program. When programs are written with subroutines, each subroutine can be tested individually for functionality. These subroutines can then be called from the main program as illustrated in Figure 9-9.

When a subroutine is called from the main program, the program is able to escape from the main program and *go to* a program *subroutine* to perform certain functions and *then return* to the main program. In situations in which a machine has a portion of its cycle that must be repeated several times during one machine cycle, the subroutine can save a great deal of duplicate programming.

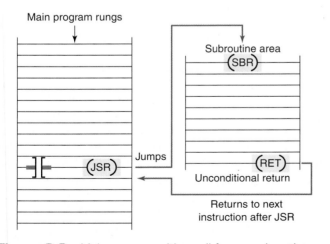

Figure 9-9 Main program with a call from a subroutine.

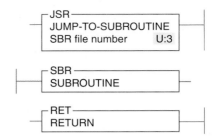

Figure 9-10 Allen-Bradley subroutine-related instructions.

The sequence of rungs could be programmed one time into a subroutine and just called when needed.

The subroutine concept is the same for all programmable controllers, but the method used to call and return from a subroutine uses different commands, depending on the PLC manufacturer. The subroutine-related instructions used in the Allen-Bradley PLCs shown in Figure 9-10 are the jump to subroutine (JSR) output instruction, the subroutine (SBR) input instruction, and the return (RET) output instruction.

The subroutine instructions can be summarized as follows:

Jump to Subroutine (JSR)—The JSR instruction is an output instruction that causes the scan to jump to the program file designated in the instruction. It is the only parameter entered in the instruction. When rung conditions are true for this output instruction, it causes the processor to jump to the targeted subroutine file. Each subroutine must have a unique file number (decimal 3–255).

Subroutine (SBR)—The SBR instruction is the first input instruction on the first rung in the subroutine file. It serves as an identifier that the program file is a subroutine. This file number is used in the JSR instruction to identify the target to which the program should jump. It is always true, and although its use is optional, it is still recommended.

Return (RET)—The RET instruction is an output instruction that marks the end of the subroutine file. It causes the scan to return to the main program at the instruction following the JSR instruction where it exited the program. The scan returns from the end of the file if there is no RET instruction. The rung containing the RET instruction may be conditional if this rung precedes the end of the subroutine. In this way, the processor omits the balance of a subroutine only if its rung condition is true.

The jump to subroutine (JSR), subroutine (SBR), and return (RET) instructions are used to direct the controller to execute a subroutine file. Figure 9-11 shows a materials

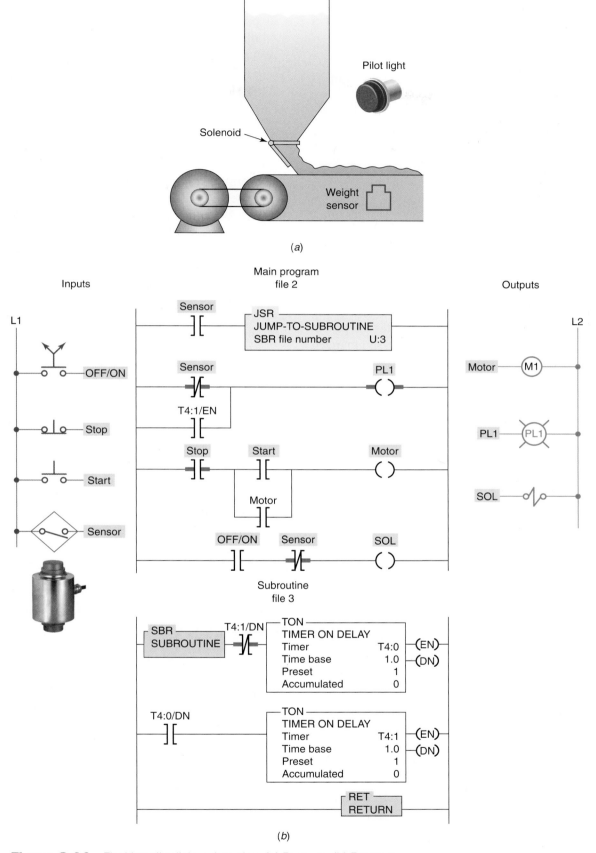

Figure 9-11 Flashing pilot light subroutine. (a) Process. (b) Program.

conveyor system with a flashing pilot light as a subroutine. The operation of the program can be summarized as follows:

- If the weight on the conveyor exceeds a preset value, the solenoid is de-energized and pilot light PL1 will begin flashing.
- When the weight sensor switch closes, the JSR is activated and directs the processor scan to jump to the subroutine U:3.
- The subroutine program is scanned and pilot light PL1 begins flashing.
- When the weight sensor switch opens, the processor will no longer scan the subroutine area and pilot light PL1 will return to its normal on state.

The Allen-Bradley SLC 500 controller main program is located in program file 2 whereas subroutines are assigned to program file numbers 3 to 255. Each subroutine must be programmed in its own program file by assigning it a unique file number. Figure 9-12 illustrates the procedure for setting up a subroutine and can be summarized as follows:

- Note each ladder location where a subroutine should be called.
- Create a subroutine file for each location. Each subroutine file should begin with an SBR instruction.

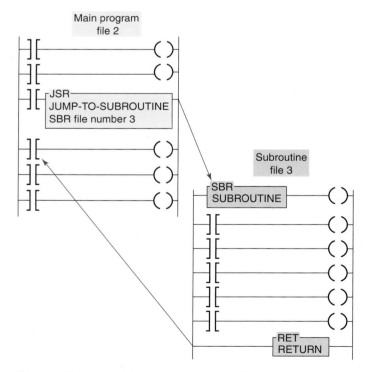

Figure 9-12 Setting up a subroutine file.

- At each ladder location where a subroutine is called, program a JSR instruction specifying the subroutine file number.
- The RET instruction is optional.
  - The end of a subroutine program will cause a return to the main program.
  - If you want to end a subroutine program before it executes to the end of program file, a conditional return (RET) instruction may be used.

An optional SBR instruction is the header instruction that stores incoming parameters. This feature lets you *pass* selected values to a subroutine before execution so the subroutine can perform mathematical or logical operations on the data and return the results to the main program. For example, the program shown in Figure 9-13 will cause the scan to jump from the main program file to program file 4 when input *A* is true. When the scan jumps to program file 4, data will also be passed from N7:30 to N7:40. When the scan returns to the main program from program file 4, data will be passed from N7:50 to N7:60.

Nesting subroutines allows you to direct program flow from the main program to a subroutine and then to another subroutine, as illustrated in Figure 9-14. Nested subroutines make complex programming easier and program operation faster because the programmer does not have to continually return from one subroutine to enter another. Programming nested subroutines may cause scan time problems because while the subroutine is being scanned, the main program is not. Excessive delays in scanning the main program may cause the outputs to operate later than required. This situation may be avoided by updating critical I/O using immediate input and/or immediate output instructions.

9.4 Immediate Input and Immediate Output Instructions

The immediate input and immediate output instructions interrupt the normal program scan to update the input image table file with current input data or to update an output module group with the current output image table file data. These instructions are intended to be used only for time-critical I/O data.

The *immediate input (IIN)* Allen-Bradley PLC-5 instruction is used to read an input condition before the I/O update is performed. This operation interrupts the program scan when it is executed. After the immediate input instruction is executed, normal program scan resumes. This instruction is used with critical input devices that require updating in advance of the I/O scan.

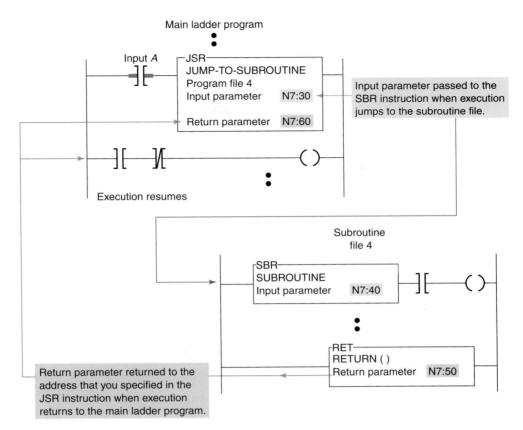

Figure 9-13   Passing subroutine parameters.

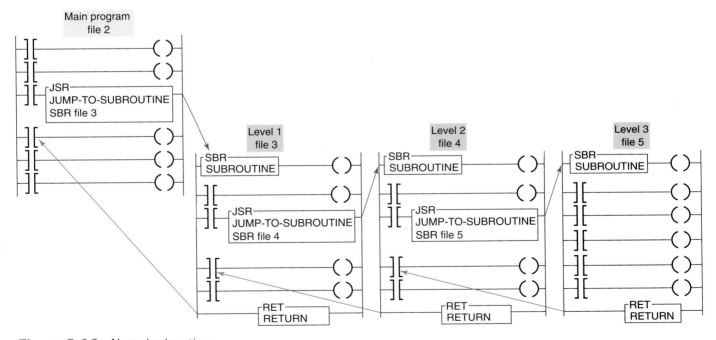

Figure 9-14   Nested subroutines.

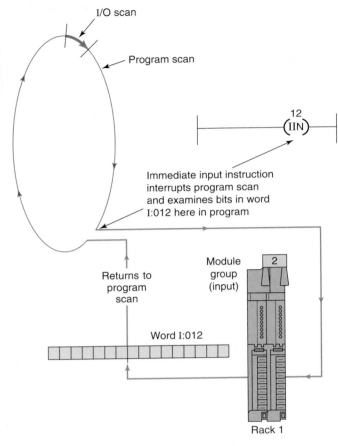

Figure 9-15 Immediate input instruction.

The *immediate output (IOT)* Allen-Bradley PLC-5 instruction is a special version of the output energize instruction used to update the status of an output device before the I/O update is performed. The immediate output is used with critical output devices that require updating in advance of the I/O scan. When the program scan reaches the immediate output instruction, the scan is interrupted and the bits of the addressed word are updated. The operation of the immediate output instruction is illustrated in Figure 9-16 and can be summarized as follows:

- When the program scan reaches a true IOT instruction, the scan is interrupted and the data in the output image table at the word address on the instruction are transferred to the real-world outputs.
- In this example, the IOT instruction follows the output energize instruction.
- Thus, the output image table word is updated first, and then the data are transferred to the real-world outputs.

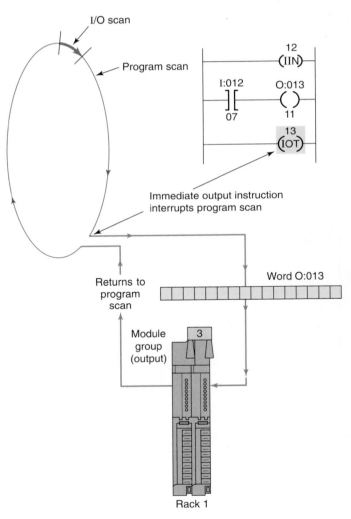

Figure 9-16 Immediate output instruction.

The operation of the immediate input instruction is illustrated in Figure 9-15. When the program scan reaches the immediate input instruction, the scan is interrupted and the bits of the addressed word are updated. The immediate input is most useful if the instruction associated with the critical input device is at the middle or toward the end of the program. The immediate input is not needed near the beginning of the program since the I/O scan has just occurred at that time. Although the immediate input instruction speeds the updating of bits, its scan-time interruption increases the total scan time of the program. The operation of the program can be summarized as follows:

- When the scan reaches a true IIN instruction, the scan is interrupted.
- The processor updates 16 bits in the input image table at the location indicated on the IIN instruction.
- The two-digit address on the IIN instruction is composed of the rack number (first digit) and the I/O group number (second digit) containing the input or inputs and needs immediate updating.

The Allen-Bradley SLC 500 PLC's immediate I/O instructions contain a few improvements over those of the PLC-5. The SLC 500's instructions, which are called *immediate input with mask (IIM)* and *immediate output with mask (IOM),* allow the programmer to specify which of the 16 bits are to be copied from an input module to the input image data table (or from the output image table to an output module). The other bits in the input image table or output module are not affected by these instructions. In addition, the SLC 500 instructions allow you to input or output a series of data words from a single input module or output a series of data words to an output module.

The immediate input with mask (IIM) instruction is shown in Figure 9-17. The IIM instruction operates on the inputs assigned to a particular word of a slot. When the IIM rung is true, the program scan is interrupted, and data from a specific input slot are transferred through the mask to the input data file. These data are then available to the commands in the ladder following the IIM instruction. The following parameters are entered in the instruction:

Slot Specifies the slot and word that contain the data to be updated. For example, I:3.0 means the input of slot 3, word 0.

Mask Specifies either a hex constant or a register address. For the mask, a 1 in the bit position passes data from the source to the destination. A 0 inhibits or blocks bits from passing from the source to the destination.

Length Used to transfer more than one word per slot.

The program operation of the instruction is summarized as follows:

- The IIM instruction retrieves data from I:1.0 and passes it through the mask.
- The mask permits only the four least significant bits to be moved to the input register I:1.0.

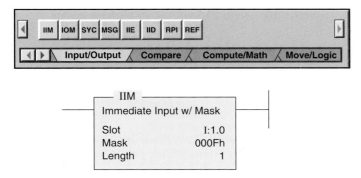

Figure 9-17 Immediate input with mask (IIM) instruction.

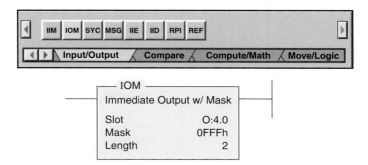

Figure 9-18 Immediate output with mask (IOM) instruction.

- This allows the programmer to update only sections of the inputs to be used throughout the rest of the program.

The immediate output with mask (IOM) instruction is shown in Figure 9-18. The IOM operates on the physical outputs assigned to a particular word of a slot. When the IOM rung is true, the program scan is interrupted to update output data to the module located in the slot specified in the instruction. These data are then available to the commands in the ladder following the IOM instruction. The parameters entered are basically the same as those entered for the IIM instruction.

Processor communication with the local chassis is many times faster than communication with the remote chassis. This is due to the fact that local I/O scan is synchronous with the program scan and communication is in *parallel* with the processor, whereas the remote I/O scan is asynchronous with the program scan and communication with remote I/O is *serial.* For this reason, fast-acting devices should be wired into the local chassis.

9.5 Forcing External I/O Addresses

The force function is essentially a manual override control function. Forcing allows the PLC user to turn an external input or output on or off from the keyboard of the programming device. This is accomplished regardless of the actual state of the field device. The forcing capability allows a machine or process to continue operation until a faulty field device can be repaired. It is also valuable during start-up and troubleshooting of a machine or process to simulate the action of portions of the program that have not yet been implemented.

Forcing inputs manipulates the input image table file bits and thus affects *all* areas of the program that use those bits. The forcing of inputs is done just after the input scan. When we force an input address, we are forcing the status bit of the instruction at the I/O address to an on or

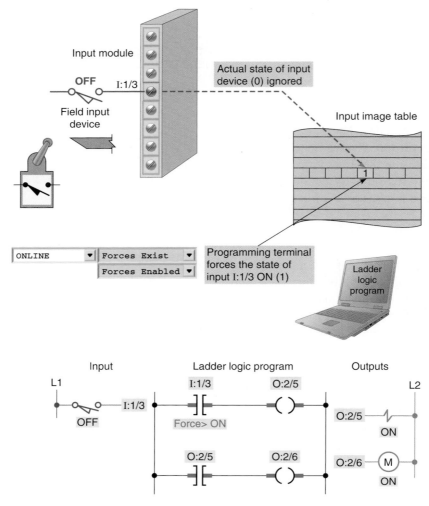

Figure 9-19 Forcing an input on.

off state. Figure 9-19 illustrates how an input is forced on. The operation of the program can be summarized as follows:

- The processor ignores the actual state of input limit switch I:1/3.
- Although limit switch I:1/3 is off (0 or false) the processor considers it as being in the on (1 or true) state.
- The program scan records this, and the program is executed with this forced status.
- In other words, the program is executed as if the limit switch were actually closed.

Forcing outputs affects *only* the addressed output terminal. Therefore, since the output image table file bits are unaffected, your program will be unaffected. The forcing of outputs is done just before the output image table file is updated. When we force an output address, we are forcing only the output terminal to an on or off state. The status

bit of the output instruction at the address is usually not affected. Figure 9-20 illustrates how an output is forced on. The operation of the program can be summarized as follows:

- The processor ignores the actual state of solenoid output O:2/5.
- The programming device sets the force state in the output force data file and the PLC implements the force to turn solenoid output O:2/5 on even though the output image table file indicates that the user logic is setting the point to off.
- M output O:2/6 remains off because the status bit of output O:2/5 is not affected by the force instruction.
- Not all brands of PLCs operate this way. For example, forcing an output with a GE Fanuc controller will cause the contacts that have the same address as the output to also change to the appropriate state.

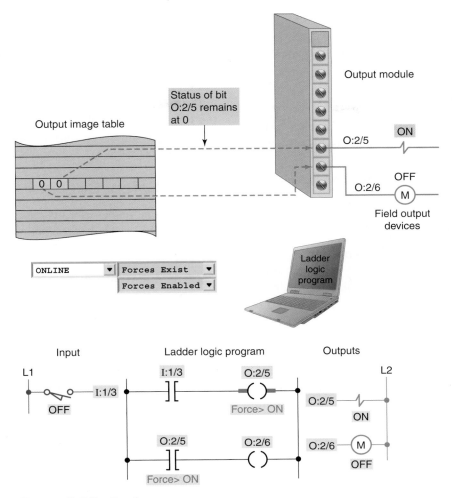

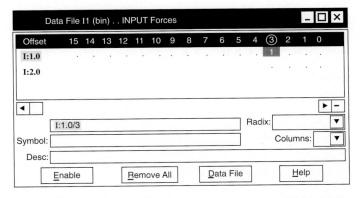

Figure 9-20 Forcing an output on.

Overriding of physical inputs on conventional relay control systems can be accomplished by installing hardwire jumpers. With PLC control hardwire jumpers are not necessary because the input data table values can be forced to an on or off state. The force function allows you to override the actual status of external input circuits by forcing external data bits on or off. Similarly, you can override the processor logic and status of output data file bits by forcing output bits on or off. By forcing outputs off, you can prevent the controller from energizing those outputs even though the ladder logic, which normally controls them, may be true. In other instances, outputs may be forced on even though logic for the rungs controlling those outputs may be false.

Figure 9-21 shows the forces version of the data table with bit I:1/3 forced on. You can enter and enable or disable forces while you are monitoring your file offline, or in any processor mode while monitoring your

Figure 9-21 Forces version of the data table with bit I:1/3 forced on.

file online. With RSLogix 500 software, the steps are as follows:

1. Open the program file in which you want to force the logic on or off.

2. With the right mouse button, click the I/O bit you want to force.

3. From the menu that appears, select Go to Data Table or select Force On or Force Off.

4. From the associated data table that appears, click on the Forces button.

5. The Forces version of the data table appears with the selected bit highlighted. Click on this bit with the right mouse button.

6. From the menu that appears, you can force the selected bit on or off.

Exercise care when you use forcing functions. **If used incorrectly, force functions can cause injuries to persons working around a system, and/or equipment damage.** For this reason, forcing functions should be used only by personnel who completely understand the circuit and the process machinery or driven equipment (Figure 9-22). You must understand the potential effect that forcing given inputs or outputs will have on machine operation in order to avoid possible personal injury and equipment damage. Before using a force function, check whether the force acts on the I/O point only or whether it acts on the user logic as well as on the I/O point. Most programming terminals and PLC CPUs provide some visible means of alerting the user that a force is in effect.

In situations in which rotating equipment is involved, the force instruction can be extremely dangerous. For

Figure 9-22 Exercise care when you use forcing functions.
Source: Courtesy Givens Engineering Inc.

example, if maintenance personnel are performing routine maintenance on a de-energized motor, the machine may suddenly become energized by someone forcing the motor to turn on. This is why a hardwired master control circuit is required for the I/O rack. The hardwired circuit will provide a method of physically removing power to the I/O system, thereby ensuring that it is impossible to energize any inputs or outputs when the master control is off.

9.6 Safety Circuitry

Sufficient emergency circuits must be provided to stop either partially or totally the operation of the controller or the controlled machine or process. These circuits should be hardwired outside the controller so that in the event of total controller failure, independent and rapid shutdown is available.

Figure 9-23 shows typical safety wiring requirements for a PLC installation. The safety requirements of this installation can be summarized as follows:

• A main disconnect switch is installed on the incoming power lines as a means of removing power from the entire programmable controller system.

• The main power disconnect switch should be located where operators and maintenance personnel have quick and easy access to it. Ideally, the disconnect switch is mounted on the outside of the PLC enclosure so that it can be accessed without opening the enclosure.

• In addition to disconnecting electrical power, you should de-energize, lock out, and tag all other sources of power (pneumatic and hydraulic) before you work on a machine or process controlled by the controller.

• An isolation transformer is used to isolate the controller from the main power distribution system and step the voltage down to 120 VAC.

• A hardwired master control relay is included to provide a convenient means for emergency controller shutdown. Because the master control relay allows the placement of several emergency-stop switches in different locations, its installation is important from a safety standpoint.

• Overtravel limit switches or mushroom head emergency stop pushbuttons are wired in series so that when one of them opens, the master control is de-energized.

• This removes power to input and output device circuits. Power continues to be supplied to the

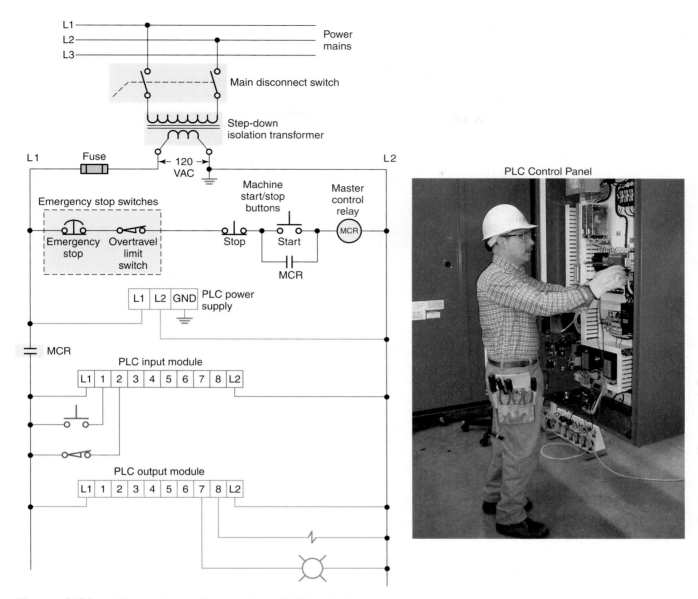

Figure 9-23 Safety wiring requirements for a PLC installation.
Source: Courtesy Minarik Automation & Control.

controller power supply so that any diagnostic indicators on the processor module can still be observed.

- Note that the master control relay is not a substitute for a disconnect switch. When you are replacing any module, replacing output fuses, or working on equipment, the main disconnect switch should be pulled and locked out.

The **master control relay** must be able to inhibit all machine motion by removing power to the machine I/O devices when the relay is de-energized. This hardwired electromechanical component must not be dependent on electronic components (hardware or software). Any part can fail, including the switches in a master control relay circuit. The failure of one of these switches would most likely cause an open circuit, which would be a safe power-off failure. However, if one of these switches shorts out, it no longer provides any safety protection. These switches should be tested periodically to ensure that they will stop machine motion when needed. Never alter these circuits to defeat their function. Serious injury or machine damage could result.

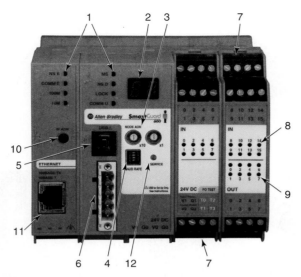

Number	Feature
1	Module status indicators
2	Alphanumeric display
3	Node address switches
4	Baud rate switches
5	USB port
6	DeviceNet communication connector
7	Terminal connectors
8	Input status indicators
9	Output status indicators
10	IP address desplay switch
11	Ethernet connector
12	Service switch

Figure 9-24 Safety PLC.

Source: Image Used with Permission of Rockwell Automation, Inc.

Safety PLCs, such as the one shown in Figure 9-24, are now available for applications that require more advanced safety functionality. A safety PLC is typically certified by third parties to meet rigid safety and reliability requirements of international standards. Both standard and safety PLCs have the ability to perform control functions but a standard PLC was not initially designed to be fault tolerant and fail-safe. That is the fundamental difference.

Some of the differences between standard and safety PLCs include the following:

- A standard PLC has one microprocessor that executes the program, Flash memory area that stores the program, RAM for making calculations, ports for communications, and I/O for detection and control of the machine. In contrast, a safety PLC has redundant microprocessors, Flash and RAM that are continuously monitored by a watchdog circuit, and a synchronous detection circuit. *Redundancy* is duplication. The probability of hazards arising from one malfunction in an electrical circuit can be minimized by creating partial or complete redundancy (duplication).

- Standard PLC inputs provide no internal means for testing the functionality of the input circuitry. By contrast, safety PLCs have an internal output circuit associated with each input for the purpose of testing the input circuitry. Inputs are driven both high and low for very short cycles during runtime to verify their functionality.

- Safety PLCs use power supplies designed specifically for use in safety control systems and redundant backplane circuitry between the controller and I/O modules.

Safety considerations should be developed as part of the PLC program. A PLC program for any application will be only as safe as the time and thought spent on both personnel and hardware considerations make it. One such consideration involves the use of a motor starter *auxiliary seal-in contact,* shown in Figure 9-25, in place of the programmed contact referenced to the output coil instruction. The use of the field-generated starter auxiliary contact status in the program is more costly in terms of field wiring and hardware, but it is *safer* because it provides positive feedback to the processor about the exact status of the motor. Assume, for example, that the OL contact of the starter opens under an overload condition. The motor, of course, would stop operating because power would be lost to the starter coil. If the program was written using an examine-on contact instruction referenced to the output coil instruction as the seal-in for the circuit, the processor would never know that power had been lost to the motor. When the OL was reset, the motor would restart instantly, creating a potentially unsafe operating condition.

Another safety consideration concerns the *wiring of stop buttons.* A stop button is generally considered a safety function as well as an operating function. As such, *all stop buttons should be wired using a normally closed contact programmed to examine for an on*

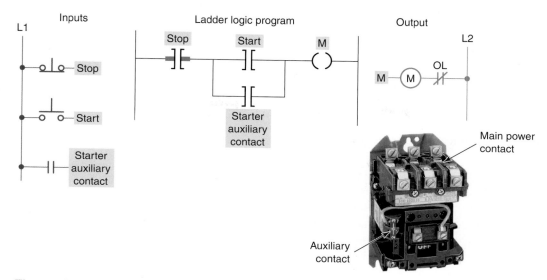

Figure 9-25 Motor starter programmed using the starter auxiliary seal-in contact.
Source: Image Used with Permission of Rockwell Automation, Inc.

condition (Figure 9-26). Using a normally open contact programmed to examine for an off condition will produce the same logic but is not considered to be as safe. Assume that the latter configuration is used. If, by some chain of events, the circuit between the button and the input point were to be broken, the stop button could be depressed forever, but the PLC logic could never react to the stop command because the input would never be true. The same holds true if power were lost to the stop button control circuit. If the normally closed wiring configuration is used, the input point receives power continuously unless the stop function is desired. Any faults occurring with the stop circuit wiring, or a loss of circuit power, would effectively be equivalent to an intentional stop.

9.7 Selectable Timed Interrupt

The *selectable timed interrupt (STI)* instruction is used to interrupt the scan of the main program file automatically, on a time basis, to scan a specified subroutine file. For Allen-Bradley SLC 500 controllers, the time base at which the program file is executed and the program file assigned as the selectable timed interrupt file are determined by the values stored in words S:30 and S:31 of the status section of the data files. The value in S:30 stores the time base, which may be from 1 through 32,767, at 10 millisecond increments. Word S:31 stores the program file assigned as the selectable interrupt file, which may be any program file from 3 through 999. Entering a 0 in the time-base word disables the selectable timed interrupt.

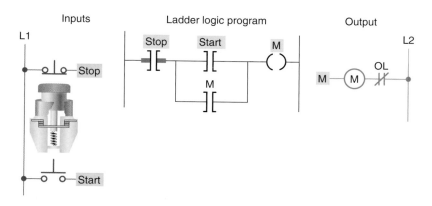

Figure 9-26 Wiring of stop buttons.

Programming the selectable timed interrupt is done when a section of program needs to be executed on a *time basis* rather than on an *event basis.* For example, a program may require certain calculations to be executed at a repeatable time interval for accuracy. These calculations can be accomplished by placing this programming in the selectable timed-interrupt file. This instruction can also be used for process applications that require periodic lubrication.

The immediate input and immediate output instructions are often located in a selectable timed interrupt file, so that a particular section of program is updated on a timed basis. This process could be done on a high-speed line, when items on the line are being examined and the rate at which they pass the sensor is faster than the scan time of the program. In this way, the item can be scanned multiple times during the program scan, and the appropriate action may be taken before the end of the scan.

The *selectable timed disable (STD)* instruction is generally paired with the *selectable timed enable (STE)* instruction to create zones in which STI interrupts cannot occur. Figure 9-27 illustrates the use of the STD and STE instructions and can be summarized as follows:

- In this program, the STI instruction is assumed to be in effect.
- The STD and STE instructions in rungs 6 and 12 are included in the ladder program to avoid having STI subroutine execution at any point in rungs 7 through 11.
- The STD instruction (rung 6) resets the STI enable bit, and the STE instruction (rung 12) sets the enable bit again.
- The first pass bit S:1/15 and the STE instruction in rung 0 are included to ensure that the STI function is initialized after a power cycle.

9.8 Fault Routine

Allen-Bradley SLC 500 controllers allow you to designate a subroutine file as a fault routine. If used, it determines how the processor responds to a programming error. The program file assigned as the fault routine is determined by the value stored in word S:29 of the status file. Entering a 0 in word S:29 disables the fault routine.

There are two kinds of major faults that result in a processor fault: recoverable and nonrecoverable faults. When the processor detects a major fault, it looks for a fault routine. If a fault routine exists, it is executed; if one

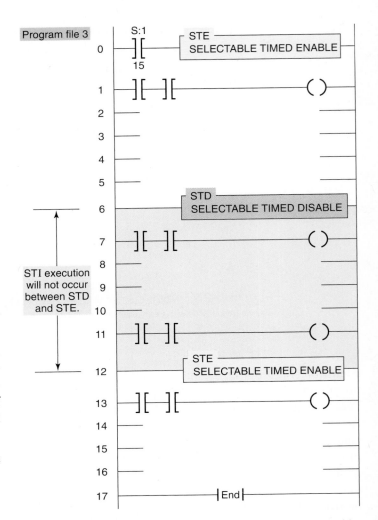

Figure 9-27 Selectable timed disable (STD) and selectable timed enable (STE) instructions.

does not exist, the processor shuts down. When there is a fault routine, and the fault is *recoverable,* the fault routine is executed. If the fault is *nonrecoverable,* the fault routine is scanned once and shuts down. Either way, the fault routine allows for an orderly shutdown.

9.9 Temporary End Instruction

The *temporary end (TND)* instruction is an output instruction used to progressively debug a program or conditionally omit the balance of your current program file or subroutines. When rung conditions are true, this instruction stops the program scan, updates the I/O, and resumes scanning at rung 0 of the main program file.

Figure 9-28 illustrates the use of the TND instruction in troubleshooting a program. The TND instruction lets your program run only up to this instruction. You can move it progressively through your program as you debug each new section. You can program the TND instruction

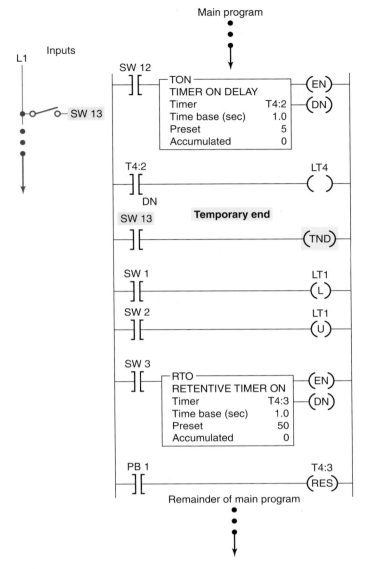

Figure 9-28 Temporary end (TND) instruction.

unconditionally, or you can condition its rung according to your debugging needs.

9.10 Suspend Instruction

The *suspend (SUS)* instruction is used to trap and identify specific conditions during system troubleshooting and program debugging. Figure 9-29 shows a suspend instruction in a ladder logic rung. The execution of the instruction can be summarized as follows:

- When you program the SUS instruction, you must enter a suspend ID number (number 100 is used in this example).
- When the rung is true, the SUS output instruction places the controller in the suspend mode and the PLC immediately terminates scan cycling.
- All ladder logic outputs are de-energized, but other status files have the data present when the suspend instruction is executed.
- The SUS instruction writes the suspend ID number (100) to S:7 as it executes.
- You can include several SUS instructions in a program, each with a different suspend ID and read S:7 to determine which SUS instruction caused the PLC to halt.
- Status file S:8 will contain the number of the program file that was executing when the SUS instruction executed.

Figure 9-29 Suspend (SUS) instruction.

1. **a.** Two MCR output instructions are to be programmed to control a section of a program. Explain the programming procedure to be followed.
   b. State how the status of the output devices within the fenced zone will be affected when the MCR instruction makes a false-to-true transition.
   c. State how the status of the output devices within the fenced zone will be affected when the MCR instruction makes a true-to-false transition.

2. What is the main advantage of the jump instruction?

3. What types of instructions are not normally included inside the jumped section of a program? Why?

4. **a.** What is the purpose of the label instruction in the jump-to-label instruction pair?
   b. When the jump-to-label instruction is executed, in what way are the jumped rungs affected?

5. **a.** Explain what the jump-to-subroutine instruction allows the program to do.
   b. In what type of machine operation can this instruction save a great deal of duplicate programming?

6. What advantage is there to the nesting of subroutines?

7. **a.** When are the immediate input and immediate output instructions used?
   b. Why is it of little benefit to program an immediate input or immediate output instruction near the beginning of a program?

8. **a.** What does the forcing capability of a PLC allow the user to do?
   b. Outline two practical uses for forcing functions.
   c. Why should extreme care be exercised when using forcing functions?

9. Why should emergency stop circuits be hardwired instead of programmed?

10. State the function of each of the following in the basic safety wiring for a PLC installation:
    a. Main disconnect switch
    b. Isolation transformer
    c. Emergency stops
    d. Master control relay

11. Compare standard and safety PLCs with regard to:
    a. Processors
    b. Input circuitry
    c. Output circuitry
    d. Power supplies

12. When programming a motor starter circuit, why is it safer to use the starter seal-in auxiliary contact in place of a programmed contact referenced to the output coil instruction?

13. When programming stop buttons, why is it safer to use an NC pushbutton programmed to examine for an on condition than an NO pushbutton programmed to examine for an off condition?

14. Explain the selectable timed interrupt function.

15. Explain the function of the fault routine file.

16. How is the temporary end instruction used to troubleshoot a program?

CHAPTER 9 PROBLEMS

1. Answer the questions, in sequence, for the MCR program in Figure 9-30, assuming the program has just been entered and the PLC is placed in the RUN mode with all switches turned off.
   a. Switches S2 and S3 are turned on. Will outputs PL1 and PL2 come on? Why?
   b. With switches S2 and S3 still on, switch S1 is turned on. Will output PL1 or PL2 or both come on? Why?
   c. With switches S2 and S3 still on, switch S1 is turned off. Will both outputs PL1 and PL2 de-energize? Why?
   d. With all other switches off, switch S6 is turned on. Will the timer time? Why?
   e. With switch S6 still on, switch S5 is turned on. Will the timer time? Why?
   f. With switch S6 still on, switch S5 is turned off. What happens to the timer? If the timer was an

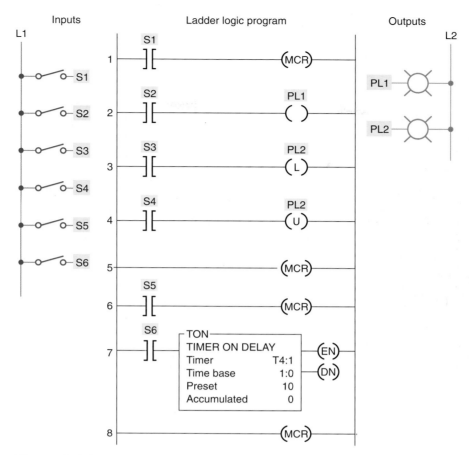

Figure 9-30 Program for Problem 1.

RTO type instead of a TON, what would happen to the accumulated value?

2. Answer the questions, in sequence, for the jump-to-label program in Figure 9-31. Assume all switches are turned *off after each operation.*

   a. Switch S3 is turned on. Will output PL1 be energized? Why?

   b. Switch S2 is turned on *first,* then switch S5 is turned on. Will output PL4 be energized? Why?

   c. Switch S3 is turned on and output PL1 is energized. Next, switch S2 is turned on. Will output PL1 be energized or de-energized after turning on switch S2? Why?

   d. All switches are turned on in order according to the following sequence: S1, S2, S3, S5, S4. Which pilot lights will turn on?

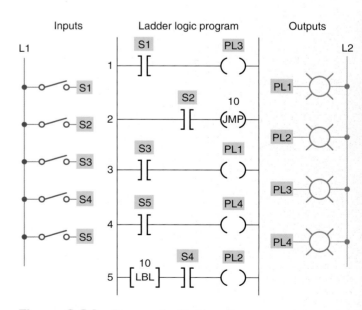

Figure 9-31 Program for Problem 2.

3. Answer the questions, in sequence, for the jump-to-subroutine and return program in Figure 9-32. Assume all switches are *turned off after each operation.*

  a. Switches S1, S3, S4, and S5 are all turned on. Which pilot light will *not* be turned on? Why?

  b. Switch S2 is turned on and then switch S4 is turned on. Will output PL3 be energized? Why?

  c. To what rung does the RET instruction return the program scan?

4. Answer the questions, in sequence, for Figure 9-33. Assume all switches are *turned off after each operation.*

  a. Switches S2, S12, and S5 are turned on in order. Will output PL5 be energized? Why?

  b. All switches except S7 are turned off. Will RTO start timing? Why?

  c. Switches S3 and S8 are turned on in order. Will pilot light PL2 come on? Why?

  d. When will timer TON function?

  e. Assume all switches are turned on. In what order will the rungs be scanned?

  f. Assume all switches are turned off. In what order will the rungs be scanned?

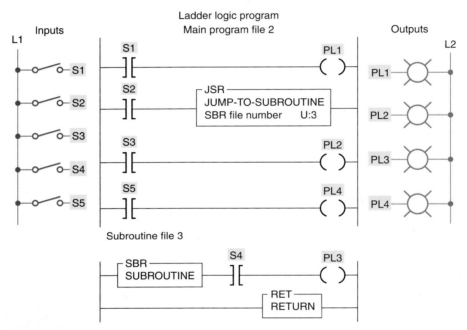

Figure 9-32 Program for Problem 3.

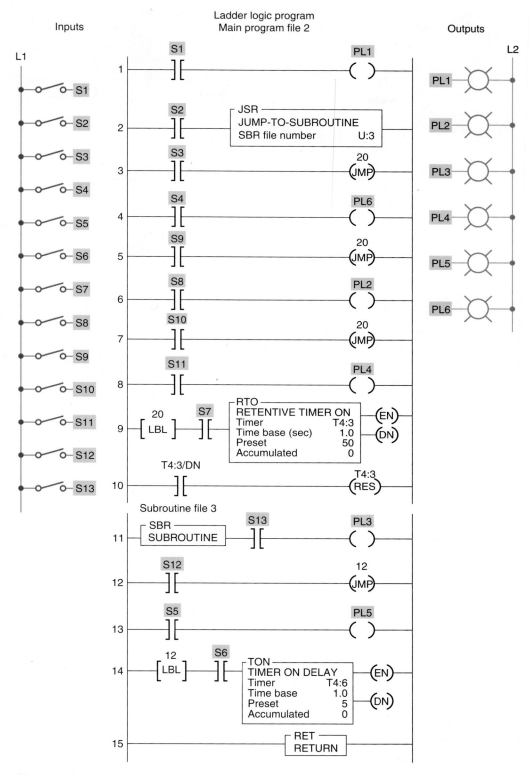

Figure 9-33 Program for Problem 4.

10

Data Manipulation Instructions

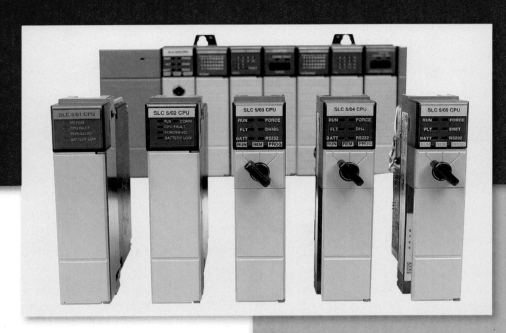

Image Used with Permission of Rockwell Automation, Inc.

Data manipulation involves transferring data and operating on data with math functions, data conversions, data comparison, and logical operations. This chapter covers both data manipulation instructions that operate on word data and those that operate on file data, which involve multiple words. Data manipulations are performed internally in a manner similar to that used in microcomputers. Examples of processes that need these operations on a fast and continuous basis are studied.

Chapter Objectives

After completing this chapter, you will be able to:

10.1 Execute data transfer of word and file level instructions from one memory location to another

10.2 Interpret data transfer and data compare instructions as they apply to a PLC program

10.3 Compare the operation of discrete I/Os with that of multibit and analog types

10.4 Understand the basic operation of PLC closed-loop control systems

10.1 Data Manipulation

Data manipulation instructions allow numerical data stored in the controller's memory to be operated on within the control program. This category of word operation instructions allows the user to truly exploit the computer capabilities of the PLC.

The use of data manipulation extends a controller's capability from that of simple on/off control based on binary logic, to quantitative decision making involving data comparisons, arithmetic, and conversions—which in turn can be applied to analog and positioning control.

There are two basic classes of instructions to accomplish data manipulation: instructions that operate on word data and those that operate on file, or block, data, which involve multiple words.

Each data manipulation instruction requires two or more words of data memory for operation. The words of data memory in singular form may be referred to either as *registers* or as *words,* depending on the manufacturer. The terms *table* or *file* are generally used when a *consecutive* group of related data memory words is referenced. Figure 10-1 illustrates the difference between a word and a file. The data contained in files and words will be in the form of binary *bits* represented as series of 1s and 0s. A group of consecutive elements or words in an Allen-Bradley SLC 500 are referred to as a file.

The data manipulation instructions allow the movement, manipulation, or storage of data in either single- or multiple-word groups from one data memory area of the PLC to another. Use of these PLC instructions in applications that require the generation and manipulation of large quantities of data greatly reduces the complexity and quantity of the programming required. Data manipulation

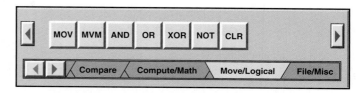

Figure 10-2 Move/Logical menu tab.

can be placed in two broad categories: *data transfer* and *data comparison.*

The manipulation of entire words is an important feature of a programmable controller. This feature enables PLCs to handle inputs and outputs containing multiple bit configurations such as analog inputs and outputs. Arithmetic functions also require data within the programmable controller to be handled in word or register format. To simplify the explanation of the various data manipulation instructions available, the instruction protocol for the Allen-Bradley SLC 500 families of PLCs will be used. Again, even though the format and instructions vary with each manufacturer, the concepts of data manipulation remain the same.

Figure 10-2 shows the **Move/Logical** menu tab for the SLC 500 PLC and its associated RSLogix software. The commands can be summarized as follows:

MOV (Move)—Moves the source value to the destination.

MVM (Masked Move)—Moves data from a source location to a selected portion of the destination.

AND (And)—Performs a bitwise AND operation.

OR (Or)—Performs a bitwise OR operation.

XOR (Exclusive Or)—Performs a bitwise XOR operation.

NOT (Not)—Performs a bitwise NOT operation.

CLR (Clear)—Sets all bits of a word to zero.

10.2 Data Transfer Operations

Data transfer instructions simply involve the transfer of the contents from one word or register to another. Figure 10-3a and b illustrate the concept of moving numerical binary data from one memory location to another. Figure 10-3a shows the original data are in register N7:30 and N7:20. Figure 10-3b shows that after the data transfer has occurred register N7:20 now holds a duplicate of the information that is in register N7:30. The previously existing data stored in register N7:20 have been replaced with those of N7:30. This process is referred to as *writing over the existing data.*

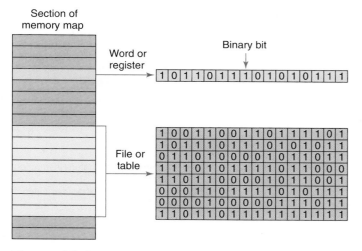

Figure 10-1 Data files, words, and bits.

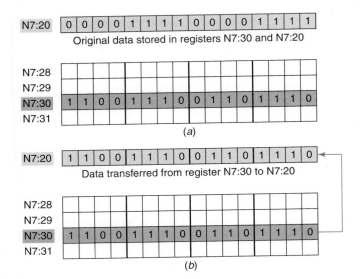

N7:20 `0 0 0 0 1 1 1 1 0 0 0 0 1 1 1 1`
Original data stored in registers N7:30 and N7:20

N7:28
N7:29
N7:30 `1 1 0 0 1 1 1 0 0 1 1 0 1 1 1 0`
N7:31

(a)

N7:20 `1 1 0 0 1 1 1 0 0 1 1 0 1 1 1 0`
Data transferred from register N7:30 to N7:20

N7:28
N7:29
N7:30 `1 1 0 0 1 1 1 0 0 1 1 0 1 1 1 0`
N7:31

(b)

Figure 10-3 Data transfer concept.

Data transfer instructions can address almost any location in the memory. Prestored values can be automatically retrieved and placed in any new location. That location may be the preset register for a timer or counter or even an output register that controls a seven-segment display.

SLC 500 controllers use a block-formatted *move (MOV)* instruction to accomplish data moves. The MOV instruction is used to copy the value in one register or word to another. This instruction copies data from a *source* register to a *destination* register. Figure 10-4 shows an example of the MOV instruction. The operation of the program can be summarized as follows:

- When the rung is true, input switch *A* closed, the value stored at the source address, N7:30, is copied into the destination address, N7:20.
- When the rung goes false, input switch *A* opened, the destination address will retain the value unless it is changed elsewhere in the program.
- The source value remains unchanged and no data conversion occurs.

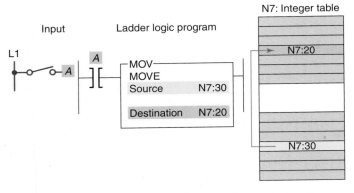

Figure 10-4 SLC 500 block-formatted move instruction.

- The instruction may be programmed with input conditions preceding it, or it may be programmed unconditionally.

The *move with mask (MVM)* instruction differs slightly from the MOV instruction because a *mask* word is involved in the move. The data being moved must pass through the mask to get to their destination address. Masking refers to the action of hiding a portion of a binary word before transferring it to the destination address. The operation of a mask word can be summarized as follows:

- The pattern of characters in the mask determines which source bits will be passed through to the destination address.
- The bits in the mask that are set to zero (0) do not pass data.
- Only the bits in the mask that are set to one (1) will pass the source data through to the destination.
- Bits in the destination are not affected when the corresponding bits in the mask are zero.
- The MVM instruction is used to copy the desired part of a 16-bit word by masking the rest of the value.

Figure 10-5 shows an example of a mask move (MVM) instruction. This instruction transfers data through the mask from the source address, B3:0, to the destination address, B3:4. The operation of the program can be summarized as follows:

- The mask may be entered as an address or in hexadecimal format, and its value will be displayed in hexadecimal.

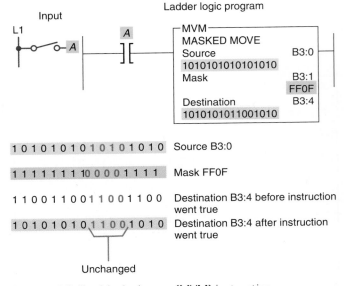

Figure 10-5 Masked move (MVM) instruction.

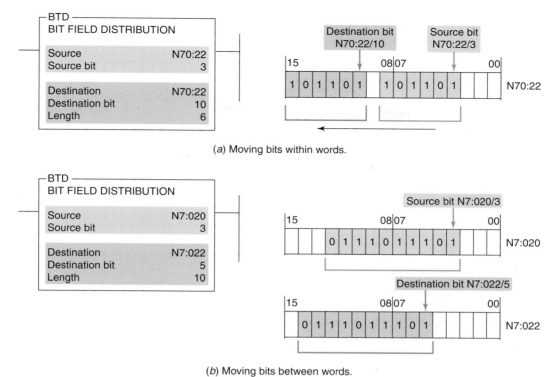

(a) Moving bits within words.

(b) Moving bits between words.

Figure 10-6 Bit distribute (BTD) instruction.

- Where there is a 1 in the mask, data will pass from the source to the destination.
- Where there is a 0 in the mask, data in the destination will remain in their last state.
- Status in bits 4–7 are unchanged due to zeroes in the mask (remained in their last state).
- Status in bits 0–3 and 8–15 were copied from the source to destination when the MVM instruction went true.
- The mask must be the same word size as the source and destination.

The *bit distribute (BTD)* instruction is used to move bits within a word or between words, as illustrated in Figure 10-6. On each scan, when the rung that contains the BTD instruction is true, the processor moves the bit field from the source word to the destination word. Bits are lost if they extend beyond the destination word; the bits are not wrapped to the next higher word. To move data within a word, enter the same address for the source and destination. The source data will remain unchanged but the instruction writes over the destination with the specified bits.

The program of Figure 10-7 illustrates how the move (MOV) instruction can be used to create variable preset

timer values. A two-position selector switch is operated to select one of two preset timer values. Operation of the program can be summarized as follows:

- When the selector switch is in the open 10 s position, rung 2 has logic continuity and rung 3 does not.
- As a result, the value 10 stored at the source address, N7:1, is copied into the destination address, T4:1.PRE.
- Therefore, the preset value of timer T4:1 will change from 0 to 10.
- When pushbutton PB1 is closed, there will be a 10 s delay period before the pilot light is energized.
- When the selector switch is in the closed 5 s position, rung 3 has logic continuity and rung 2 does not.
- As a result, the value 5 stored at the source address, N7:2, is copied into the destination address, T4:1. PRE.
- Closing pushbutton PB1 will now result in a 5 s time-delay period before the pilot light is energized.

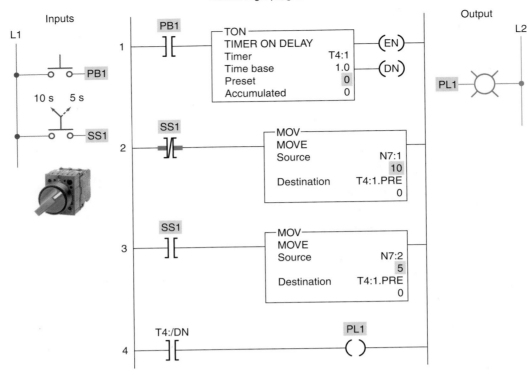

Figure 10-7 Move instruction used to change the preset time of a timer.

The program of Figure 10-8 illustrates how the move (MOV) instruction can be used to create variable preset counter values. The operation of the program can be summarized as follows:

- Limit switch LSI is programmed to the input of up-counter C5:1 and counts the number of parts coming off a conveyor line onto a storage rack.
- Three different types of products are run on this line.
- The storage rack has room for only 300 boxes of product *A* or 175 boxes of product *B* or 50 boxes of product *C*.
- Three momentary switches are used to select the desired preset counter value depending on the product line (*A*, *B*, or *C*) being manufactured.
- A reset button is provided to reset the accumulated count to 0.
- A pilot lamp is switched on to indicate when the storage rack is full.
- The program has been constructed so that normally only one of the three switches will be closed at any one time. If more than one of the preset counter switches is closed, the *last* value is selected.

A *file* is a group of related consecutive words in the data table that have a defined start and end and are used to store information. For example, a batch process program may contain several separate recipes in different files that can be selected by an operator.

In some instances it may be necessary to shift complete files from one location to another within the programmable controller memory. Such data shifts are termed *file-to-file shifts*. File-to-file shifts are used when the data in one file represent a set of conditions that must interact with the programmable controller program several times and, therefore, must remain *intact* after each operation. Because the data within this file must also be changed by the program action, a second file is used to handle the data changes, and the information within that file is allowed to be altered by the program. The data in the first file, however, remain constant and therefore can be used many times. Other types of data manipulation used with file instructions include word-to-file and file-to-word moves, as illustrated in Figure 10-9.

Files allow large amounts of data to be scanned quickly and are useful in programs requiring the transfer, comparison, or conversion of data. Most PLC manufacturers display file instructions in block format on the programming terminal screen. Figure 10-10 compares the SLC

Ladder logic program

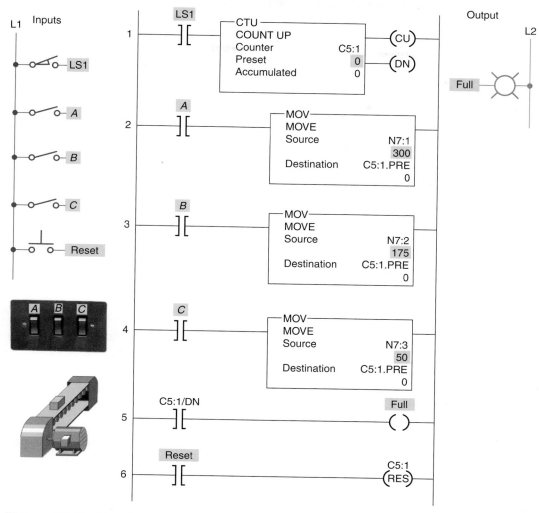

Figure 10-8 Move instruction used to change the preset count of a counter.

500 controller word and file addressing. The addressing formats can be summarized as follows:

- The address that defines the beginning of a file or group of words starts with the pound sign #.
- The # prefix is omitted in a single word or element address.

- Address N7:30 is a word address that represents a single word: word number 30 in integer file 7.
- Address #N7:30 represents the starting address of a group of consecutive words in integer file 7. The length is eight words, which is determined by the instruction where the file address is used.

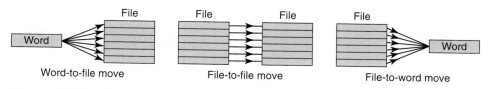

Figure 10-9 Moving data using file instructions.

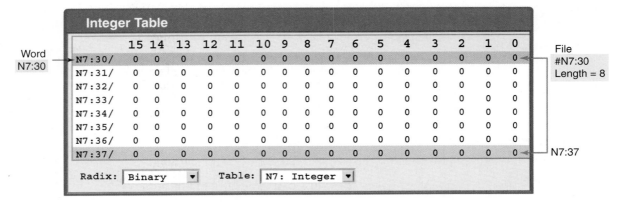

Figure 10-10 SLC 500 word and file address.

The *file arithmetic and logic (FAL)* instruction is used to copy data from one file to another and to do file math and file logic. This instruction is available only on Allen-Bradley PLC-5 and ControlLogix platforms. An example of the FAL instruction is shown in Figure 10-11.

The basic operation of the FAL instruction is similar in all functions and requires the following parameters and PLC-5 addresses to be entered in the instruction:

Control

- Is the first entry and the address of the control structure in the control area (R) of processor memory.
- The processor uses this information to run the instruction.
- The default file for the control file is data file 6.
- The control element for the FAL instruction must be unique for that instruction and may not be used to control any other instruction.
- The control element is made up of three words.
- The control word uses four control bits: bit 15 (enable bit), bit 13 (done bit), bit 11 (error bit), and bit 10 (unload bit).

Length

- Is the second entry and represents the file length.
- This entry will be in words, except for the floating-point file, for which the length is in

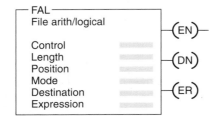

Figure 10-11 File arithmetic/logic (FAL) instruction.

elements. (A floating-point element consists of two words.)

- The maximum length possible is 1000 elements. Enter any decimal number from 1 to 1000.

Position

- Is the third entry and represents the current location in the data block that the processor is accessing.
- It points to the word being operated on.
- The position starts with 0 and indexes to 1 less than the file length.
- You generally enter a 0 to start at the beginning of a file. You may also enter another position at which you want the FAL to start its operation.
- When the instruction resets, however, it will reset the position to 0.
- You can manipulate the position from the program.

Mode

- Is the fourth entry and represents the number of file elements operated on per program scan. There are three choices: all mode, numeric mode, and incremental mode.

All Mode

- For this mode you enter the letter *A*.
- In the all mode, the instruction will transfer the complete file of data in *one* scan.
- The enable (EN) bit will go true when the instruction goes true and will follow the rung condition.
- When all of the data have been transferred, the done (DN) bit will go true. This change will occur on the same scan during which the instruction goes true.
- If the instruction does not go to completion due to an error in the transfer of data (such as trying to store too large or too small a number for the data-table

type), the instruction will stop at that point and set the error (ER) bit. The scan will continue, but the instruction will not continue until the error bit is reset.

- If the instruction goes to completion, the enable bit and the done bit will remain set until the instruction goes false, at which point the position, the enable bit, and the done bit will all be reset to 0.

Numeric Mode

- For this mode you enter a decimal number (1–1000).
- In the numeric mode, the file operation is distributed over a number of program scans.
- The value you enter sets the number of elements to be transferred per scan.
- The numeric mode can decrease the time it takes to complete a program scan. Instead of waiting for the total file length to be transferred in one scan, the numeric mode breaks up the transfer of the file data into multiple scans, thereby cutting down on the instruction execution time per scan.

Incremental mode

- For this mode you enter the letter I.
- In the incremental mode, one element of data is operated on for every false-to-true transition of the instruction.
- The first time the instruction sees a false-to-true transition and the position is at 0, the data in the first element of the file are operated on. The position will remain at 0 and the UL bit will be set. The EN bit will follow the instruction's condition.
- On the second false-to-true transition, the position will index to 1, and data in the second word of the file will be operated on.
- The UL bit controls whether the instruction will operate just on data in the current position, or whether it will index the position and then transfer data. If the UL bit is reset, the instruction—on a false-to-true transition of the instruction—will operate on the data in the current position and set the UL bit. If the UL bit is set, the instruction—on a false-to-true transition of the instruction—will index the position by 1 and operate on the data in their new position.

Destination

- Is the fifth entry and is the address at which the processor stores the result of the operation.
- The instruction converts to the data type specified by the destination address.
- It may be either a file address or an element address.

Expression

- Is the last entry and contains addresses, program constants, and operators that specify the source of data and the operations to be performed.
- The expression entered determines the function of the FAL instruction.
- The expression may consist of file addresses, element addresses, or a constant and may contain only one function because the FAL instruction may perform only one function.

Figure 10-12 shows an example of a file-to-file copy function using the FAL instruction. The operation of the program can be summarized as follows:

- When input A goes true, data from the expression file #N7:20 will be copied into the destination file #N7:50.
- The length of the two files is set by the value entered in the control element word R6:1.LEN.
- In this instruction, we have also used the ALL mode, which means all of the data will be transferred in the first scan in which the FAL instruction sees a false-to-true transition.
- The DN bit will also come on in that scan unless an error occurs in the transfer of data, in which case the ER bit will be set, the instruction will stop operation at that position, and then the scan will continue at the next instruction.

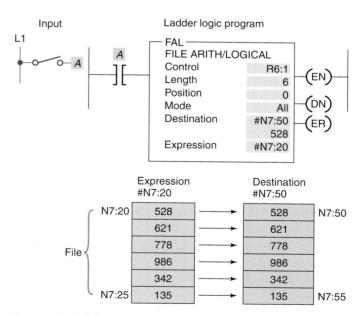

Figure 10-12   File-to-file copy function using the FAL instruction.

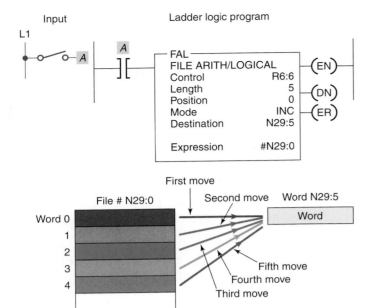

Figure 10-13 File-to-word copy function using the FAL instruction.

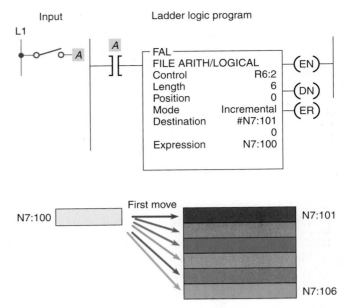

Figure 10-14 Word-to-file copy function using the FAL instruction.

Figure 10-13 shows an example of a file-to-word copy function using the FAL instruction. The operation of the program can be summarized as follows:

- With each false-to-true rung transition of input *A,* the processor reads one word of integer file N29.
- The processor starts reading at word 0, and writes the image into word 5 of integer file N29.
- The instruction writes over any data in the destination.

Figure 10-14 shows an example of a word-to-file copy function using the FAL instruction. It is similar to the file-to-word copy function except that the instruction copies data from a word address into a file. The operation of the program can be summarized as follows:

- The expression is a word address (N7:100) and the destination is a file address (#N7:101).
- If we start with position 0, the data from N7:100 will be copied into N7:101 on the first false-to-true transition of input *A.*
- The second false-to-true transition of input *A* will copy the data from N7:100 into N7:102.
- On successive false-to-true transitions of the instruction, the data will be copied into the next position in the file until the end of the file, N7:106, is reached.

The exceptions to the rule that file addresses must take consecutive words in the data table are in the *timer, counter,* and *control data* files for the FAL instruction. In these three

data files, if you designate a file address, the FAL instruction will take every third word in that file and make a file of preset, accumulated, length, or position data within the corresponding file type. This might be done, for example, so that recipes storing values for timer presets can be moved into the timer presets, as illustrated in Figure 10-15.

The *file copy (COP)* instruction and the *fill file (FLL)* instruction are high-speed instructions that operate more quickly than the same operation with the FAL instruction. Unlike the FAL instruction, there is no control element to monitor or manipulate. Data conversion does not

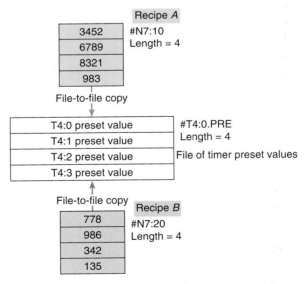

Figure 10-15 Copying recipes and storing values for timer presets.

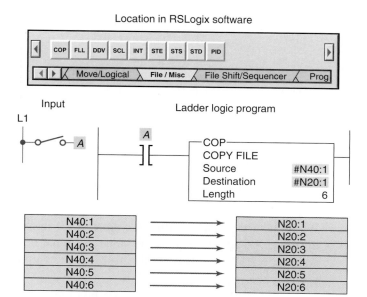

Figure 10-16   File copy (COP) instruction.

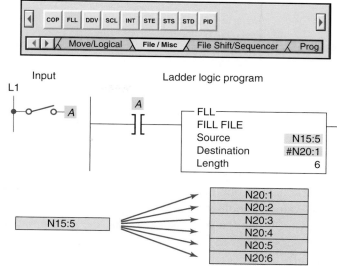

Figure 10-17   Fill file (FLL) instruction.

take place, so the source and destination should be the same file types. An example of the file COP instruction is shown in Figure 10-16. The operation of the program can be summarized as follows:

- Both the source and destination are file addresses.
- When input *A* goes true, the values in file N40 are copied to file N20.
- The instruction copies the entire file length for each scan during which the instruction is true.

An example of the fill file (FLL) instruction is shown in Figure 10-17. It operates in a manner similar to the FAL instruction that performs the word-to-file copy in the ALL mode. The operation of the program can be summarized as follows:

- When input *A* goes true, the value in N15:5 is copied into N20:1 through N20:6.
- Because the instruction transfers to the end of the file, the file will be filled with the same data value in each word.

The FLL instruction is frequently used to zero all of the data in a file, as illustrated in the program of Figure 10-18. The operation of the program can be summarized as follows:

- Momentarily pressing pushbutton PB1 copies the contents of file #N10:0 into file #N12:0.
- Momentarily pressing pushbutton PB2 then clears file #N12:0.
- Note that 0 is entered for the source value.

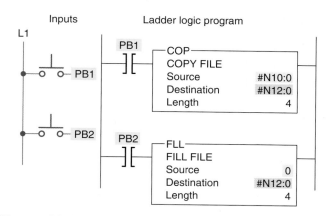

Figure 10-18   Using the FLL instruction to change all the data in a file to zero.

10.3   Data Compare Instructions

Data transfer operations are all output instructions, whereas *data compare* instructions are *input* instructions. Data compare instructions are used to compare numerical values. These instructions compare the data stored in two or more words (or registers) and make decisions based on the program instructions. Numeric values in two words of memory can be compared for each of the basic data compare instructions shown in Figure 10-19, depending on the PLC.

Data comparison concepts have already been used with the timer and counter instructions. In both these instructions, an output was turned on or off when the accumulated value of the timer or counter equaled its preset

Name	Symbol
Equal to	(=)
Not equal to	(≠)
Less than	(<)
Greater than	(>)
Less than or equal to	(≤)
Greater than or equal to	(≥)

Figure 10-19 Basic PLC data compare instructions.

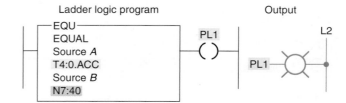

Figure 10-21 EQU logic rung.

value. What actually occurred was that the accumulated numeric data in one memory word was compared to the preset value of another memory word on each scan of the processor. When the processor saw that the accumulated value was equal to the preset value, it switched the output on or off.

Comparison instructions are used to test pairs of values to determine if a rung is true. Figure 10-20 shows the Compare menu tab for the Allen-Bradley SLC 500 PLC and its associated RSLogix software. The compare instructions can be summarized as follows:

LIM (Limit test)—Tests whether one value is within the limit range of two other values.

MEQ (Masked Comparison for Equal)—Tests portions of two values to see whether they are equal. Compares 16-bit data of a source address to 16-bit data at a reference address through a mask.

EQU (Equal)—Tests whether two values are equal.

NEQ (Not Equal)—Tests whether one value is not equal to a second value.

LES (Less Than)—Tests whether one value is less than a second value.

GRT (Greater Than)—Tests whether one value is greater than a second value.

LEQ (Less Than or Equal)—Tests whether one value is less than or equal to a second value.

GEQ (Greater Than or Equal)—Tests whether one value is greater than or equal to a second value.

The *equal (EQU)* instruction is an input instruction that compares source A to source B: when source A is equal to source B, the instruction is logically true; otherwise it is logically false. Figure 10-21 shows an example of an

EQU logic rung. The operation of the rung can be summarized as follows:

- When the accumulated value of counter T4:0 stored in source A's address equals the value in source B's address, N7:40, the instruction is true and the output is energized.
- Source A may be a word address or a floating-point address.
- Source B may be a word address, a floating-point address, or a constant value.
- With the equal instruction, the floating-point data is not recommended because of the exactness required. One of the other comparison instructions, such as the limit test, is preferred.

The *not equal (NEQ)* instruction is an input instruction that compares source A to source B: when source A is not equal to source B, the instruction is logically true; otherwise it is logically false. Figure 10-22 shows an example of an NEQ logic rung. The operation of the rung can be summarized as follows:

- When the value stored at source A's address, N7:5, is not equal to 25, the output will be true; otherwise, the output will be false.
- The value stored at Source A is 30.
- The value stored at Source B is 25.
- Since the two values are not the same the output will be true or on.
- In all input-comparison instructions, Source A must be an address and Source B can be an address or a constant.

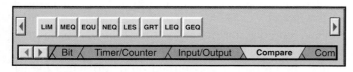

Figure 10-20 Compare menu tab.

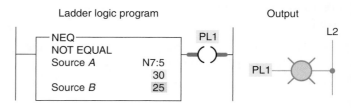

Figure 10-22 NEQ logic rung.

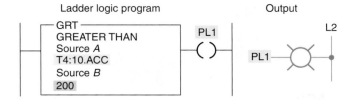

Figure 10-23 GRT logic rung.

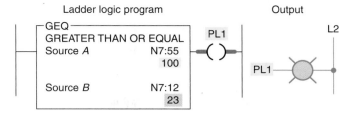

Figure 10-25 GEQ logic rung.

The *greater than (GRT)* instruction is an input instruction that compares source A to source B: when source A is greater than source B, the instruction is logically true; otherwise it is logically false. Figure 10-23 shows an example of a GRT logic rung. The operation of the rung can be summarized as follows:

- The instruction is either true or false, depending on the values being compared.
- When the accumulated value of timer T4:10, stored at the address of source A, is greater than the constant 200 of source B, the output will be on; otherwise the output will be off.

The *less than (LES)* instruction is an input instruction that compares source A to source B: when source A is less than source B, the instruction is logically true; otherwise it is logically false. Figure 10-24 shows an example of an LES logic rung. The operation of the rung can be summarized as follows:

- The instruction is either true or false, depending on the values being compared.
- When the accumulated value of counter C5:10, stored at the address of source A, is less than the constant 350 of source B, the output will be on; otherwise, it will be off.

The *greater than or equal (GEQ)* instruction is an input instruction that compares source A to source B: when source A is greater than or equal to source B, the instruction is logically true; otherwise it is logically false. Figure 10-25 shows an example of a GEQ logic rung. The operation of the rung can be summarized as follows:

- When the value stored at the address of source A, N7:55, is greater than or equal to the value stored at

the address of source B, N7:12, the output will be true; otherwise, it will be false.

- The value stored at source A is 100.
- The value stored at source B is 23.
- Therefore the output will be true or on.

The *less than or equal (LEQ)* instruction is an input instruction that compares source A to source B: when source A is less than or equal to source B, the instruction is logically true; otherwise it is logically false. Figure 10-26 shows an example of an LEQ logic rung. The operation of the rung can be summarized as follows:

- When the accumulated count of counter C5:1 is less than or equal to 457, the pilot light will turn on.
- The accumulated value of the counter is less than 457.
- Therefore the output will be false or off.

The *limit test (LIM)* instruction is used to test whether values are within or outside the specified range. Applications in which the limit test instruction is used include allowing a process to operate as long as the temperature is within or outside a specified range.

Programming the LIM instruction consists of entering three parameters: low limit, test, and high limit. The limit test instruction functions in the following two ways:

- ***The instruction is true if***—The lower limit is equal to or less than the higher limit, and the test parameter value is equal to or inside the limits. Otherwise the instruction is false.
- ***The instruction is true if***—The lower limit has a value greater than the higher limit, and the instruction is equal to or outside the limits. Otherwise the instruction is false.

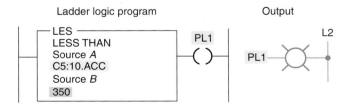

Figure 10-24 LES logic rung.

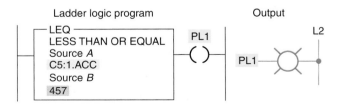

Figure 10-26 LEQ logic rung.

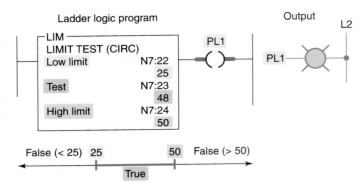

Figure 10-27 LIM instruction where the low limit value is less than the high limit value.

The limit test instruction is said to be circular because it can function in either of two ways. Figure 10-27 shows an example of an LIM instruction where the low limit value is less than the high limit value. The operation of the logic rung can be summarized as follows:

- The high limit has a value of 50, and the low limit has a value of 25.
- Instruction is true for values of the test from 25 through 50.
- Instruction is false for test values less than 25 or greater than 50.
- Instruction is true because the test value is 48.

Figure 10-28 shows an example of an LIM instruction where the low limit value is greater than the high limit value. The operation of the logic rung can be summarized as follows:

- The high limit has a value of 50, and the low limit has a value of 100.
- Instruction is true for test values of 50 and less than 50 and for test values of 100 and greater than 100.

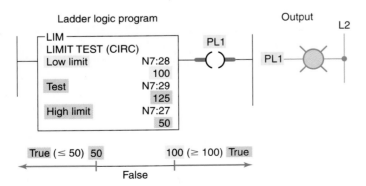

Figure 10-28 LIM instruction where the low limit value is greater than the high limit value.

Figure 10-29 MEQ instruction can be used to monitor the state of limit switches.
Source: Courtesy Jayashree Electrodevices.

- Instruction is false for test values greater than 50 and less than 100.
- Instruction is true because the test value is 125.

The *masked comparison for equal (MEQ)* instruction compares a value from a source address with data at a compare address and allows portions of the data to be masked. One application for the MEQ instruction is to compare the correct position of up to 16 limit switches when the source contains the limit switch address and the compare stores their desired states. The mask can block out the switches you don't want to compare (Figure 10-29).

Figure 10-30 shows an example of an MEQ instruction. The operation of the logic rung can be summarized as follows:

- When the data at the source address match the data at the compare address bit-by-bit (less masked bits), the instruction is true.
- The instruction goes false as soon as it detects a mismatch.
- A mask passes data when the mask bits are set (1); a mask blocks data when the mask bits are reset (0).

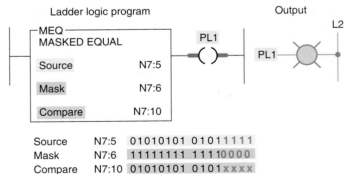

Source	N7:5	01010101 01011111
Mask	N7:6	11111111 11110000
Compare	N7:10	01010101 0101xxxx

Figure 10-30 Masked comparison for equal (MEQ) logic rung.

- The mask must be the same element size (16 bits) as the source and compare addresses.
- You must set mask bits to 1 to compare data. Bits in the compare address that correspond to 0s in the mask are not compared.
- If you want the ladder program to change mask value, store the mask at a data address. Otherwise, enter a hexadecimal value for a constant mask value.
- The instruction is true because reference bits XXXX are not compared.

10.4 Data Manipulation Programs

Data manipulation instructions give new dimension and flexibility to the programming of control circuits. For example, consider the hardwired relay-operated, time-delay circuit in Figure 10-31. This circuit uses three electromechanical time-delay relays to control four solenoid valves. The operation of the hardwired circuit can be summarized as follows:

- When the momentary start pushbutton is pressed solenoid A is energized immediately.
- Solenoid B is energized 5 s later than solenoid A.
- Solenoid C is energized 10 s later than solenoid A.
- Solenoid D is energized 15 s later than solenoid A.

The hardwired time-delay circuit could be implemented using a conventional PLC program and three internal timers. However, the same circuit can be programmed using only *one* internal timer along with data compare instructions. Figure 10-32 shows the program required to implement the circuit using only one internal timer. The operation of the program can be summarized as follows:

- The momentary stop button is closed.
- When the momentary start button is pressed, SOL A output energizes immediately to switch on solenoid A.
- SOL A examine-on contact becomes true to seal in output SOL A and to start on-delay timer T4:1 timing.
- The timer preset time is set to 15 seconds.
- Output SOL D will energize (through the timer done bit T4:1/DN) after a total time delay of 15 seconds to energize solenoid D.
- Output SOL B will energize after a total time delay of 5 seconds, when the accumulated time becomes equal to and then greater than 5 seconds. This, in turn, will energize solenoid B.
- Output SOL C will energize after a total time delay of 10 seconds, when the accumulated time becomes equal to and then greater than 10 seconds. This, in turn, will energize solenoid C.

Figure 10-33 shows an application of an on-delay timer program implemented using the EQU instruction. The operation of the program can be summarized as follows:

- When the switch (S1) is closed, timer T4:1 will begin timing.
- Both EQU instructions' source As are addressed to get the accumulated value from the timer while it is running.
- The EQU instruction of rung 2 has the value of 5 stored in source B.
- When the accumulated value of the timer reaches 5, the EQU instruction of rung 2 will become logic true for 1 second.
- As a result, the latch output will energize to switch the pilot light PL1 on.
- When the accumulated value of the timer reaches 15, the EQU instruction of rung 3 will be true for 1 second.
- As a result, the unlatch output will energize to switch the pilot light PL1 off.
- Therefore, when the switch is closed, the pilot light will come on after 5 seconds, stay on for 10 seconds, and then turn off.

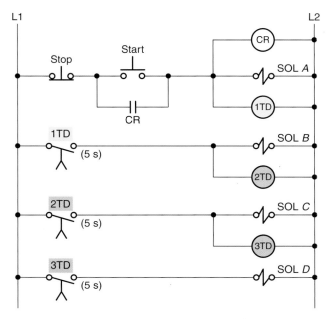

Figure 10-31 Three electromechanical time-delay relays used to control four solenoid valves.

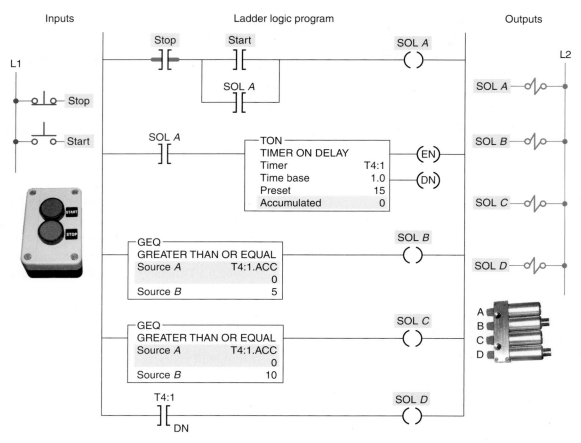

Figure 10-32 Controlling multiple loads using one timer and the GEQ instruction.

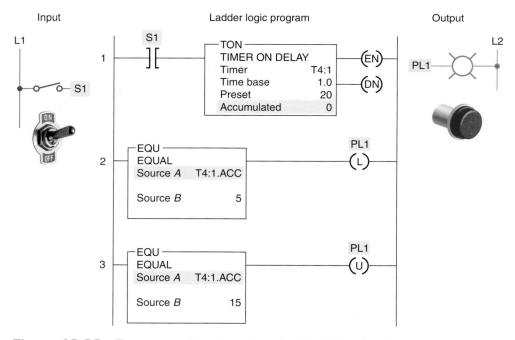

Figure 10-33 Timer program implemented using the EQU instruction.

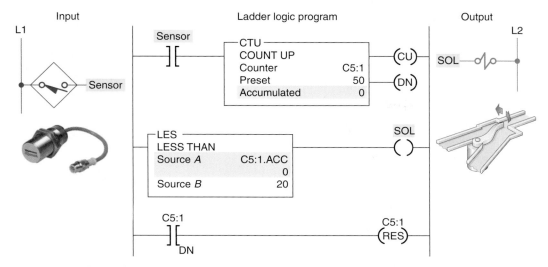

Figure 10-34 Counter program implemented using the LES instruction.
Source: Photo courtesy Turck, Inc., **www.turck.com**.

Figure 10-34 shows an application of an up-counter program implemented using the LES instruction. The operation of the program can be summarized as follows:

- Up-counter C5:1 will increment by 1 for every false-to-true transition of the proximity sensor switch.

- Source *A* of the LES instruction is addressed to the accumulated value of the counter and source *B* has a constant value of 20.

- The LES instruction will be true as the long as the value contained in source *A* is *less than* that of source *B*.

- Therefore, output solenoid SOL will be energized when the accumulated value of the counter is between 0 and 19.

- When the counter's accumulated value reaches 20, the LES instruction will go false, de-energizing output solenoid SOL.

- When the counter's accumulated value reaches its preset value of 50, the counter reset will be energized through the counter done bit (C5:1/DN) to reset the accumulated count to 0.

The use of comparison instructions is generally straightforward. However, one precaution involves the use of these instructions in PLC programs used to control the flow in vessel filling operations (Figure 10-35). This control scenario can be summarized as follows:

- The receiving vessel has its weight monitored continuously by the PLC program as it fills.

- When the weight reaches a preset value, the flow is cut off.

Figure 10-35 Vessel filling operation.
Source: Courtesy Feige Filling.

- While the vessel fills, the PLC performs a comparison between the vessel's current weight and a predetermined constant programmed in the processor.

- If the programmer uses only the equal instruction, problems may result.

- As the vessel fills, the comparison for equality will be false. At the instant the vessel weight reaches the desired preset value of the equal instruction, the instruction becomes true and the flow is stopped.

- However, should the supply system leak additional material into the vessel, the total weight of the material could rise *above* the preset value, causing the instruction to go false and the vessel to overfill.

- The simplest solution to this problem is to program the comparison instruction as a greater than or equal to instruction. This way, any excess material entering the vessel will not affect the filling operation.

- It may be necessary, however, to include additional programming to indicate a serious overfill condition.

10.5 Numerical Data I/O Interfaces

The expanding data manipulation processing capabilities of PLCs led to the development of I/O interfaces known as numerical data I/O interfaces. In general, numerical data I/O interfaces can be divided into two groups: those that provide interface to *multibit digital* devices and those that provide interface to *analog* devices.

The multibit digital devices are like the discrete I/O because processed signals are discrete (on/off). The difference is that, with the discrete I/O, only a *single* bit is required to read an input or control an output. Multibit interfaces allow a group of bits to be input or output *as a unit.* They can be used to accommodate devices that require BCD inputs or outputs.

The *thumbwheel switches (TWS),* shown in Figure 10-36, are typical BCD input devices. Each one of the four switches provides four binary digits at its output that correspond to the decimal number selected on the switch. The conversion from a single decimal digit to four binary digits is performed by the TWS device. The BCD input module allows the processor to accept the 4-bit digital codes and input their data into specific register or word locations in memory to be used by the control program. Data manipulation instructions can be used to access the data from the input module allowing a person to change set points, timer, or counter presets *externally* without modifying the control program.

The *seven-segment LED* display board, shown in Figure 10-37, is a typical Binary Coded Decimal (BCD) output device. It displays a decimal number that corresponds to the BCD value it receives at its input. Conversion of the four binary bits to a single decimal digit on the display is

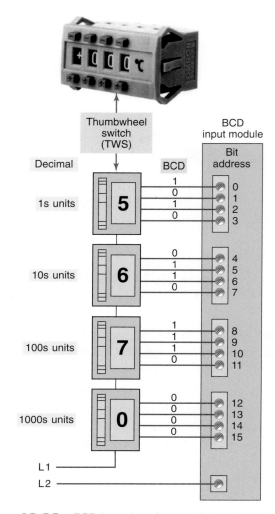

Figure 10-36 BCD input interface module connected to a thumbwheel switch.
Source: Photo courtesy Omron Industrial Automation, **www.ia.omron.com**.

performed by the LED display device. The BCD output module is used to output data from a specific register or word location in memory. This type of output module enables a PLC to operate devices that require BCD coded signals.

Figure 10-38 shows a PLC program that uses a BCD input interface module connected to a thumbwheel switch and a BCD output interface module connected to an LED display board. The program is designed so that the LEDs display the setting of the thumbwheel switch. Both the MOV and EQU instructions form part of the program. The operation of the program can be summarized as follows:

- The LED display board monitors the decimal setting of the thumbwheel switch.

- The MOV instruction is used to move the data from the thumbwheel switch input to the LED display output.

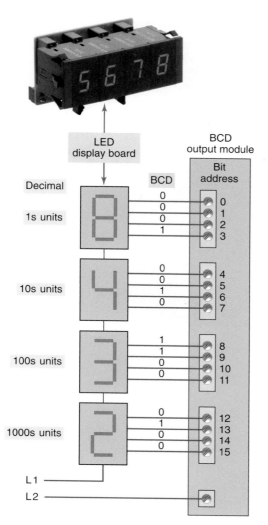

Figure 10-37 BCD output interface module connected to a seven-segment LED display board.

Source: Photo courtesy Omron Industrial Automation, **www.ia.omron.com**.

- Setting of the thumbwheel switch is compared to the reference number 1208 stored in source *B* by the EQU instruction.

- Pilot light output PL is energized whenever the input switch S1 is true (closed) and the value of the thumbwheel switch is equal to 1208.

Input and output modules can be addressed either at the bit level or at the word level. Analog modules convert analog signals to 16-bit digital signals (input) or 16-bit digital signals to analog values (output). An analog I/O will allow monitoring and control of analog voltages and currents. Figure 10-39 illustrates how an analog input interface operates. The operation of this input module can be summarized as follows:

- The analog input module contains the circuitry necessary to accept analog voltage or current signals from field devices.

- The input signal is converted from an analog to a digital value by an analog-to-digital (A/D) converter circuit.

- The conversion value, which is proportional to the analog signal, is passed through the controller's data bus and stored in a specific register or word location in memory for later use by the control program.

An analog output interface module receives numerical data from the processor; these data are then translated into a proportional voltage or current to control an analog field device. Figure 10-40 illustrates how an analog output interface operates. The operation of this output module can be summarized as follows:

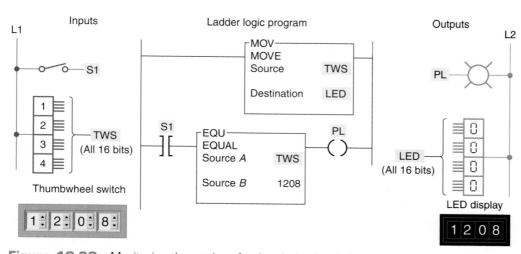

Figure 10-38 Monitoring the setting of a thumbwheel switch.

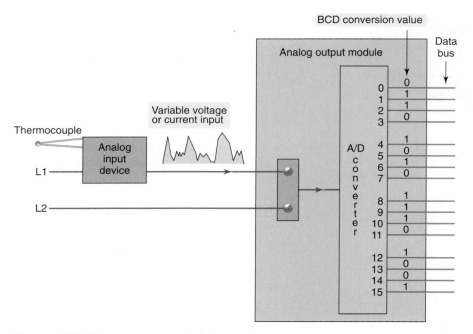

Figure 10-39 Analog input interface module.

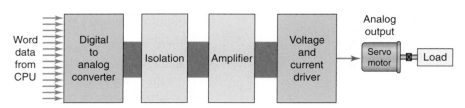

Figure 10-40 Analog output interface module.

- The function of the analog output module is to accept a range of numeric values output from the PLC program and to produce a varying current or voltage signal required to control a connected analog output device.

- Data from a specific register or word location in the CPU memory are passed through the controller's data bus to the digital-to-analog (D/A) converter.

- The analog output from the D/A converter is then used to control the analog output device.

- The level of the analog signal output is based on the digital value of the data word supplied by the CPU and manipulated by the control program.

- These output interfaces normally require an external power supply that meets certain current and voltage requirements.

10.6 Closed-Loop Control

In open-loop control, no feedback loop is employed and system variations which cause the output to deviate from the desired value are not detected or corrected. A closed-loop system utilizes feedback to measure the actual system operating parameter being controlled such as temperature, pressure, flow, level, or speed. This feedback signal is sent back to the PLC where it is compared with the desired system set-point. The controller develops an error signal that initiates corrective action and drives the final output device to the desired value.

PLC *set-point control* in its simplest form compares an input value, such as analog or thumbwheel inputs, to a set-point value. A discrete output signal is provided if the input value is less than, equal to, or greater than the set-point value. The temperature control program of

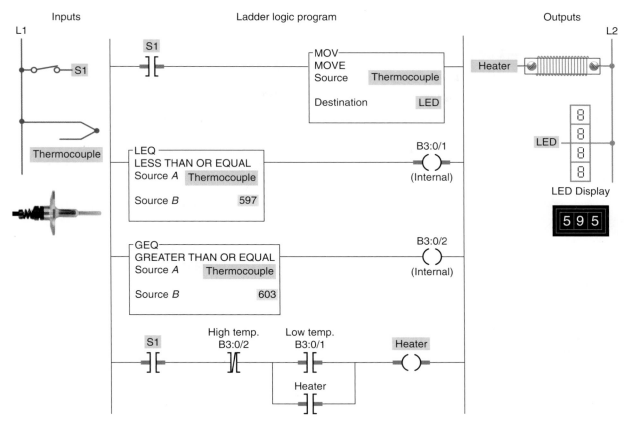

Figure 10-41 Set-point control program.

Figure 10-41 is one example of set-point control. In this application, a PLC is to provide for simple off/on control of the electric heating elements of an oven. The operation of the program can be summarized as follows:

- Oven is to maintain an average set-point temperature of 600°F with a variation of about 1 percent between the off and on cycles.
- The electric heaters are turned on when the temperature of the oven is 597°F or less and will stay on until the temperature rises to 603°F or more.
- The electric heaters stay off until the temperature drops to 597°F, at which time the cycle repeats itself.
- Whenever the less than or equal (LEQ) instruction is true, a low-temperature condition exists and the program switches on the heater.
- Whenever the greater than or equal (GEQ) instruction is true, a high-temperature condition exists and the program switches off the heater.
- For the program as shown the temperature is 595°F so the LEQ instruction and B3:0/1 will both be true and the heater output will be switched

on and sealed-in through the heater examine-on instruction.

- Once the temperature increases to 598°F the LEQ instruction goes false but the heater output remains on until the temperature rises to 603°F.
- At the 603°F point the GEQ instruction and B3:0/2 will both be true and the heater will be switched off.

Several set-point control schemes can be performed by different PLC models. These include on/off control, proportional (P) control, proportional-integral (PI) control, and proportional-integral-derivative (PID) control. Each involves the use of some form of closed-loop control to maintain a process characteristic such as a temperature, pressure, flow, or level at a desired value. When a control system is designed such that it receives operating information from the machine and makes adjustments to the machine based on this operating information, the system is said to be a closed-loop system.

The block diagram of a closed-loop control system is shown in Figure 10-42. A measurement is made of the variable to be controlled. This measurement is then compared to a reference point, or set-point. If a difference

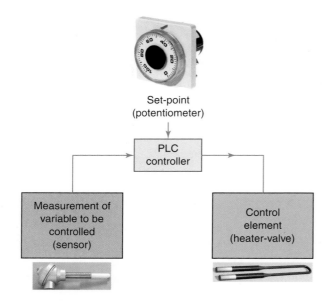

Figure 10-42 Closed-loop control system.

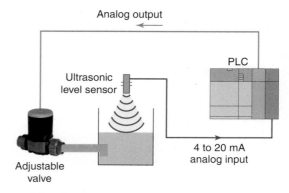

Figure 10-43 Proportional control process.

(error) exists between the actual and desired levels, the PLC control program will take the necessary corrective action. Adjustments are made continuously by the PLC until the difference between the desired and actual output is as small as is practical.

With on/off PLC control (also known as *two-position* and *bang-bang control*), the output or final control element is either on or off—one for the occasion when the value of the measured variable is above the set-point and the other for the occasion when the value is below the set-point. The controller will never keep the final control element in an intermediate position. Most residential thermostats are on/off type controllers.

On/off control is inexpensive but not accurate enough for most process and machine control applications. On/off control almost always means overshoot and resultant system cycling. For this reason a *deadband* usually exists around the set-point. The deadband or hysteresis of the control loop is the difference between the on and off operating points.

Proportional controls are designed to eliminate the hunting or cycling associated with on/off control. They allow the final control element to take intermediate positions between on and off. This permits *analog control* of the final control element to vary the amount of energy to the process, depending on how much the value of the measured variable has shifted from the desired value.

The process illustrated in Figure 10-43 is an example of a proportional control process. The PLC analog output module controls the amount of fluid placed in the holding tank by adjusting the percentage of valve opening. The valve is initially open 100 percent. As the fluid level in the tank approaches the preset point, the processor modifies

the output to degrade closing the valve by different percentages, adjusting the valve to maintain a set-point.

Proportional-integral-derivative (PID) control is the most sophisticated and widely used type of process control. PID operations are more complex and are mathematically based. PID controllers produce outputs that depend on the *magnitude, duration,* and *rate of change* of the system error signal. Sudden system disturbances are met with an aggressive attempt to correct the condition. A PID controller can reduce the system error to 0 faster than any other controller.

A typical PID control loop is illustrated in Figure 10-44. The loop measures the process, compares it to a set-point, and then manipulates the output in the direction which should move the process toward the set-point. The terminology used in conjunction with a PID loop can be summarized as follows:

- Operating information that the controller receives from the machine is called the *process variable (PV)* or *feedback.*
- Input from the operator that tells the controller the desired operating point is called the *set-point (SP).*
- When operating, the controller determines whether the machine needs adjustment by comparing (by subtraction) the set-point and the process variable

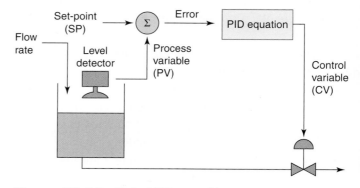

Figure 10-44 Typical PID control loop.

to produce a difference (the difference is called the *error*).

- Output from the loop is called the *control variable (CV)*, which is connected to the controlling part of the process.
- The PID loop takes appropriate action to modify the process operating point until the control variable and the set-point are very nearly equal.

Programmable controllers are either equipped with PID I/O modules that produce PID control or have sufficient mathematical functions of their own to allow PID control to be carried out. Figure 10-45 shows an SLC 500 PID instruction with typical addresses for the parameters entered. The PID instruction normally controls a closed loop using inputs from an analog input module and provides an output to an analog output module. Explanation

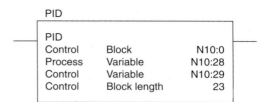

Figure 10-45   SLC 500 PID instruction.

of the PID instruction parameters can be summarized as follows:

- Control Block is the file that stores the data required to operate the instruction.
- Process Variable (PV) is an element address that stores the process input value.
- Control Variable (CV) is an element address that stores the output of the PID instruction.

1. In general, what do data manipulation instructions allow the PLC to do?

2. Explain the difference between a register or word and a table or file.

3. Into what two broad categories can data manipulation instructions be placed?

4. What takes place with regard to a data transfer instruction?

5. The MOV instruction is to be used to copy the information stored in word N7:20 to N7:35. What address is entered into the source and the destination?

6. What is the purpose of the mask word in the MVM instruction?

7. What is the purpose of the bit distribute instruction?

8. List three types of data shifts used with file instructions.

9. List the six parameters and addresses that must be entered into the file arithmetic and logic (FAL) instruction.

10. Assume the ALL mode has been entered as part of a FAL instruction. How will this affect the transfer of data?

11. What is the advantage of using the file copy (COP) or fill file (FLL) instruction rather than the FAL instruction for the transfer of data?

12. What are data compare instructions used for?

13. Name and draw the symbols for the six different types of data compare instructions.

14. Explain what each of the logic rungs in Figure 10-46 is instructing the processor to do.

15. What does the limit test (LIM) instruction test values for?

16. How are multibit I/O interfaces different from the discrete type?

17. Assume that a thumbwheel switch is set for the decimal number 3286.
    a. What is the equivalent BCD value for this setting?
    b. What is the equivalent binary value for this setting?

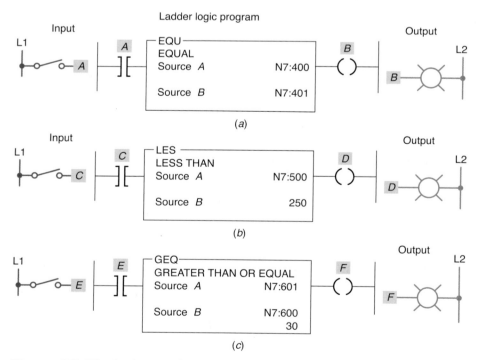

Figure 10-46 Logic rungs for Question 14.

18. Assume that a thermocouple is connected to an analog input module. Explain how the temperature of the thermocouple is communicated to the processor.

19. Outline the process by which an analog output interface module operates the field device connected to it.

20. Compare the operation of open-loop and closed-loop PLC systems.

21. Outline the control process involved with simple PLC set-point control.

22. Compare the operation of the final control element in on/off and proportional control systems.

23. Explain the meaning of the following terms as they apply to a PID control:
    a. Process variable
    b. Set-point
    c. Error
    d. Control variable

CHAPTER 10 PROBLEMS

1. Study the data transfer program of Figure 10-47 and answer the following questions:
   a. When S1 is open, what decimal number will be stored in integer word address N7:13 of the MOV instruction?
   b. When S1 is on, what decimal number will be stored in integer word address N7:112 of the MOV instruction?
   c. When S1 is on, what decimal number will appear in the LED display?
   d. What is required for the decimal number 216 to appear in the LED display?

2. Study the data transfer counter program of Figure 10-48 and answer the following questions:
   a. What determines the preset value of the counter?
   b. Outline the steps to follow to operate the program so that the PL1 output is energized after 25 off-to-on transitions of the count PB input.

3. Construct a nonretentive timer program that will turn on a pilot light after a time-delay period. Use a thumbwheel switch to vary the preset time-delay value of the timer.

4. Study the data compare program of Figure 10-49 and answer the following questions:
   a. Will the pilot light PL1 come on whenever switch S1 is closed? Why?
   b. Must switch S1 be closed to change the number stored in source *A* of the EQU instruction?
   c. What number or numbers need to be set on the thumbwheel in order to turn on the pilot light?

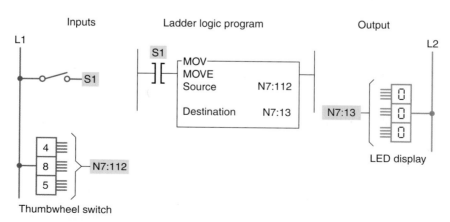

Figure 10-47   Program for Problem 1.

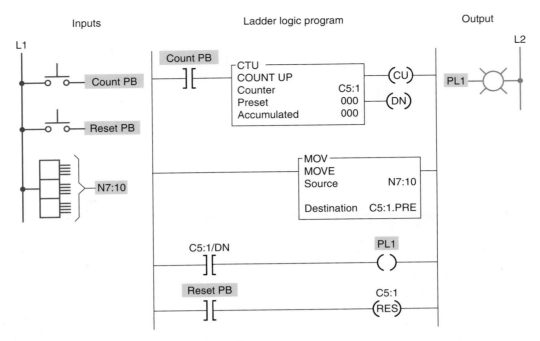

Figure 10-48 Program for Problem 2.

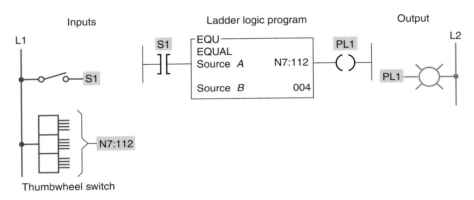

Figure 10-49 Program for Problem 4.

5. Study the data compare program in Figure 10-50 and answer the following questions:
   a. List the values for the thumbwheel switch that would allow the pilot light to turn on.
   b. If the value in the word N7:112 is 003 and switch S1 is open, will the pilot light turn on? Why?
   c. Assume that source *B* is addressed to the accumulated count of an up-counter. With S1 closed, what setting of the thumbwheel switch would be required to turn the pilot light off when the accumulated count reaches 150?

6. Write a program to perform the following:
   a. Turn on pilot light 1 (PL1) if the thumbwheel switch value is less than 4.
   b. Turn on pilot light 2 (PL2) if the thumbwheel switch value is equal to 4.
   c. Turn on pilot light 3 (PL3) if the thumbwheel switch value is greater than 4.
   d. Turn on pilot light 4 (PL4) if the thumbwheel switch value is less than or equal to 4.
   e. Turn on pilot light 5 (PL5) if the thumbwheel switch value is greater than or equal to 4.

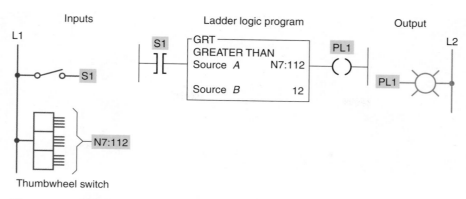

Inputs   Ladder logic program   Output

Figure 10-50 Program for Problem 5.

7. Write a program that will copy the value stored at address N7:56 into address N7:60.

8. Write a program that uses the mask move instruction to move only the upper 8 bits of the value stored at address I:2.0 to address O:2.1 and to ignore the lower 8 bits.

9. Write a program that uses the FAL instruction to copy 20 words of data from the integer data file, starting with N7:40, into the integer data file, starting with N7:80.

10. Write a program that uses the COP instruction to copy 128 bits of data from the memory area, starting at B3:0, to the memory area, starting at B3:8.

11. Write a program that will cause a light to come on only if a PLC counter has a value of 6 or 10.

12. Write a program that will cause a light to come on if a PLC counter value is less than 10 or more than 30.

13. Write a program for the following: The temperature reading from a thermocouple is to be read and stored in a memory location every 5 minutes for 4 hours. The temperature reading is brought in continuously and stored in address N7:150. File #7:200 is to contain the data from the last full 4-hour period.

11

Math Instructions

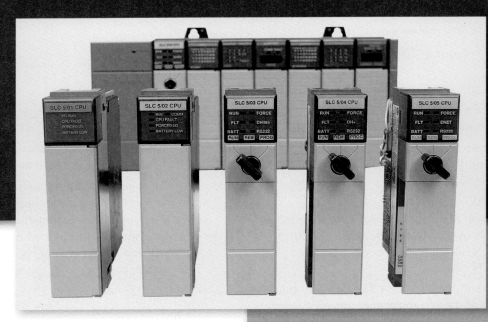

Image Used with Permission of Rockwell Automation, Inc.

Most PLCs have arithmetic function capabilities. Basic PLC math instructions include add, subtract, multiply and divide to calculate the sum, difference, product, and quotient of the content of word registers. The PLC is capable of doing many arithmetic operations per scan period for fast updating of data. This chapter covers the basic mathematical instructions performed by PLCs and their applications.

Chapter Objectives

After completing this chapter, you will be able to:

11.1 Analyze and interpret math instructions as they apply to a PLC program

11.2 Create PLC programs involving math instructions

11.3 Apply combinations of PLC arithmetic functions to processes

11.1 Math Instructions

Math instructions, like data manipulation instructions, enable the programmable controller to take on more of the qualities of a conventional computer. The PLC's math functions capability allows it to perform arithmetic functions on values stored in memory words or registers. For example, assume you are using a counter to keep track of the number of parts manufactured, and you would like to display how many more parts must be produced in order to reach a certain quota. This display would require the data in the accumulated value of the counter to be subtracted from the quota required. Other applications include combining parts counted, subtracting detected defects, and calculating run rates.

Depending on what type of processor is used, various math instructions can be programmed. The basic four mathematical functions performed by PLCs are:

Addition—The capability to add one piece of data to another.

Subtraction—The capability to subtract one piece of data from another.

Multiplication—The capability to multiply one piece of data by another.

Division—The capability to divide one piece of data by another.

Math instructions use the contents of two words or registers and perform the desired function. The PLC instructions for data manipulation (data transfer and data compare) are used with the math symbols to perform math functions. Math instructions are all output instructions. Figure 11-1 shows the Compute/Math menu tab for the SLC 500 PLC and its associated RSLogix software. The commands can be summarized as follows:

CPT (Compute)—Evaluates an expression and stores the result in the destination.

ADD (add)—Adds source *A* to source *B* and stores the result in the destination.

SUB (Subtract)—Subtracts source *B* from source *A* and stores the result in the destination.

MUL (Multiply)—Multiplies source *A* by source *B* and stores the result in the destination.

Figure 11-1 Compute/Math menu tab.

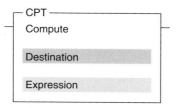

Figure 11-2 SLC 500 CPT (compute) instruction.

DIV (Divide)—Divides source *A* by source *B* and stores the result in the math register.

SQR (Square Root)—Calculates the square root of the source and places the integer result in the destination.

NEG (Negate)—Changes the sign of the source and places it in the destination.

TOD (To BCD)—Converts a 16-bit integer source value to BCD and stores it in the math register or the destination.

FRD (From BCD)—Converts a BCD value in the math register or the source to an integer and stores it in the destination.

Figure 11-2 shows the *CPT (compute)* instruction used with SLC 500 controllers. When CPT instruction is executed, then copy, arithmetic, logical, or conversion operation residing in the expression field of this instruction is performed and the result is sent to the destination. The execution time of a CPT instruction is longer than that of a single arithmetic operation and uses more instruction words.

11.2 Addition Instruction

Most math instructions take two input values, perform the specified arithmetic function, and output the result to an assigned memory location. For example, the *ADD* instruction performs the addition of two values stored in the referenced memory locations. How these values are accessed depends on the controller. Figure 11-3 shows the

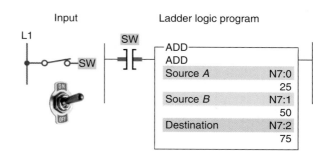

Figure 11-3 SLC 500 ADD instruction.

ADD instruction used with the SLC 500 controllers. The operation of the logic rung can be summarized as follows:

- When input switch SW is closed the rung will be true.
- The value stored at the source *A* address, N7:0 (25), is added to the value stored at the source *B* address, N7:1 (50).
- The answer (75) is stored at the destination address N7:2.
- Source *A* and source *B* can be either values or addresses that contain values, but *A* and *B* cannot both be constants.

The program of Figure 11-4 illustrates how the ADD instruction can be used to add the accumulated counts of two up-counters. This application requires a pilot light to come on when the sum of the counts from the two counters is equal to or greater than 350. The operation of the program can be summarized as follows:

- Source *A* of the ADD instruction is addressed to store the accumulated value of counter C5:0.
- Source *B* of the ADD instruction is addressed to store the accumulated value of counter C5:1.
- The value at source *A* is added to the value at source *B*, and the result (answer) is stored at destination address N7:1.
- Source *A* of the GEQ (greater than or equal) instruction is addressed to store the value of the destination address N7:1.
- Source *B* of the GEQ instruction contains the constant value of 350.

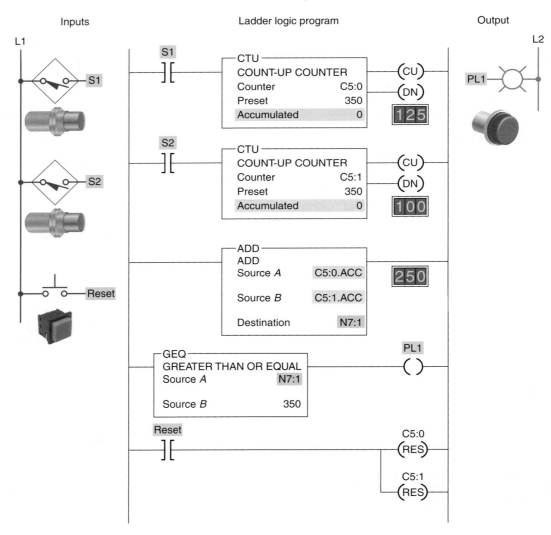

Figure 11-4 Counter program that uses the ADD instruction.

Status Table

	15	14	13	12	11	10	9	8	7	6	5	4	3	2	1	0
S2:0/	0	0	0	0	0	0	0	0	0	0	0	0	0	0	0	0
S2:1/	0	0	0	0	0	0	0	0	0	0	0	0	0	0	0	0
S2:2/	0	0	0	0	0	0	0	0	0	0	0	0	0	0	0	0
S2:3/	0	0	0	0	0	0	0	0	0	0	0	0	0	0	0	0
S2:4/	0	1	0	0	0	0	0	1	1	0	0	0	0	0	0	1
S2:5/	0	0	0	0	0	0	0	0	0	0	0	0	0	0	0	0

Address S2:0 Table: S2:Status

Figure 11-5 Processor status file S2.

- The GEQ instruction and PL1 output will be true whenever the accumulated values in the two counters are equal to or greater than the constant value 350.
- A reset button is provided to reset the accumulated count of both counters to zero.

When performing math functions, care must be taken to ensure that values remain in the range that the data table or file can store; otherwise, the overflow bit will be set. The arithmetic status bits for the SLC 500 controller are found in word 0, bits 0 to 3 of the processor status file S2 (Figure 11-5). After an instruction is executed, the arithmetic status bits in the status file are updated. The description of each bit can be summarized as follows:

Carry (C)—Address S2:0/0, is set to 1 when there is a carry in the ADD instruction or a borrow in the SUB instruction.

Overflow (O)—Address S2:0/1, is set to 1 when the result is too large to fit in the destination register.

Zero (Z)—Address S2:0/2, is set to 1 when the result of the subtract instruction is zero.

Sign (S)—Address S2:0/3, is set to 1 when the result is a negative number.

11.3 Subtraction Instruction

The *SUB (subtract)* instruction is an output instruction that subtracts one value from another and stores the result in the destination address. When rung conditions are true, the subtract instruction subtracts source *B* from source *A* and stores the result in the destination. Figure 11-6 shows the SUB instruction used with the SLC 500 controllers. The operation of the logic rung can be summarized as follows:

- When input switch SW is closed the rung will be true.
- The value stored at the source *B* address, N7:05 (322), is subtracted from the value stored at the source *A* address, N7:10 (520).

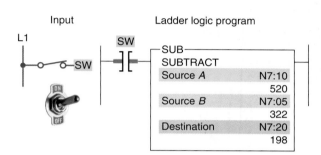

Figure 11-6 SLC 500 SUB (subtract) instruction.

- The answer (198) is stored at the destination address, N7:20.
- Source *A* and source *B* can be either values or addresses that contain values, but *A* and *B* cannot both be constants.

The program of Figure 11-7 shows how the SUB function can be used to indicate a vessel overfill condition. This application requires an alarm to sound when a supply system leaks 5 lb or more of raw material into the vessel after a preset weight of 500 lb has been reached. The operation of the program can be summarized as follows:

- When the start button is pressed, the fill solenoid (rung 1) and filling indicating light (rung 2) are turned on and raw material is allowed to flow into the vessel.
- The vessel has its weight monitored continuously by the PLC program (rung 3) as it fills.
- When the weight reaches 500 lb, the fill solenoid is de-energized and the flow is cut off.
- At the same time, the filling pilot light indicator is turned off and the full pilot light indicator (rung 3) is turned on.
- Should the fill solenoid leak 5 lb or more of raw material into the vessel, the alarm (rung 5) will energize and stay energized until the overflow level is reduced below the 5-lb overflow limit.

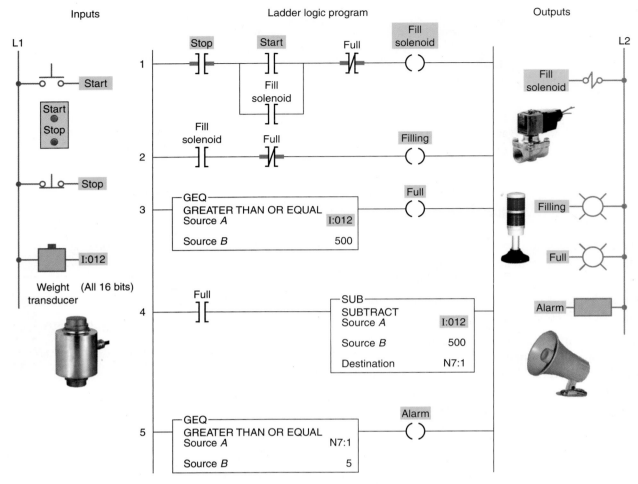

Figure 11-7 Vessel overfill alarm program.

11.4 Multiplication Instruction

The *multiply (MUL)* instruction is an output instruction that multiplies two values and stores the result in the destination address. Figure 11-8 shows the MUL instruction used with the SLC 500 controllers. The operation of the logic rung can be summarized as follows:

- When input switch SW is closed the rung will be true.

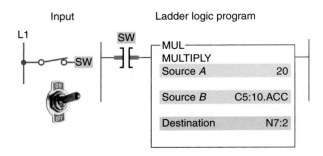

Figure 11-8 SLC 500 MUL (multiply) instruction.

- The data in source *A* (constant 20) will be multiplied by the data in source *B* (accumulated value of counter C5:10).
- The resultant answer is placed in the destination N7:2.
- Similar to previous math instructions, source *A* and *B* in multiplication instructions can be values (constants) or addresses that contain values, but *A* and *B* cannot both be constants.

The program of Figure 11-9 is an example of how MUL instruction calculates the product of two sources. The operation of the program can be summarized as follows:

- When input switch SW is closed the MUL instruction is executed.
- The value stored in source *A*, address N7:1 (123), is then multiplied by the value stored in source *B*, address N7:2(61).

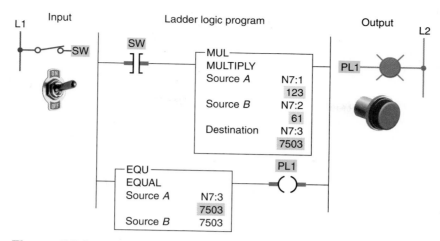

Figure 11-9 MUL instruction used to calculate the product of two sources.

- The product (7503) is placed into destination word N7:3.
- As a result, the equal instruction becomes true, turning output PL1 on.

The program of Figure 11-10 is an example of how the MUL instruction is used as part of an oven temperature control program. The operation of the program can be summarized as follows:

- The PLC calculates the upper and lower deadband, or off/on limits, about the set-point.
- Upper and lower temperature limits are set automatically at ±1 percent regardless of the set-point value.
- Set-point temperature is adjusted by means of the thumbwheel switch.
- The analog thermocouple interface module is used to monitor the current temperature of the oven.
- In this example, the set-point temperature is 400°F.
- Therefore, the electric heaters will be turned on when the temperature of the oven drops to less than 396°F and stay on until the temperature rises above 404°F.
- If the set-point is changed to 100°F, the deadband remains at ±1 percent, with the lower limit being 99°F and the upper limit being 101°F.
- The number stored in word N7:1 represents the upper temperature limit, and the number stored in word N7:2 represents the lower limit.

11.5 Division Instruction

The *divide (DIV)* instruction divides the value in source *A* by the value in source *B* and stores the result in the destination and math register. Figure 11-11 shows an example

of the DIV instruction. The operation of the logic rung can be summarized as follows:

- When input switch SW is closed the rung will be true.
- The data in source *A* (the accumulated value of counter C5:10) is then divided by the data in source *B* (the constant 2).
- The result is placed in the destination N7:3.
- If the remainder is 0.5 or greater, a roundup occurs in the integer destination.
- The value stored in the math register consists of the unrounded quotient (placed in the most significant word) and the remainder (placed in the least significant word).
- Some PLCs support the use of floating-decimal as well as integer (whole number) values. As an example, 10 divided by 3 may be expressed as 3.333333 (floating-decimal notation) or 3 with a remainder of 1.

The program of Figure 11-12 is an example of how the DIV instruction calculates the integer value that results from dividing source *A* by source *B*. The operation of the program can be summarized as follows:

- When input switch SW is closed the DIV instruction is executed.
- The value stored in source *A,* address N7:0 (120), is then divided by the value stored in source *B,* address N7:1 (4).
- The answer, 30, is placed in the destination address N7:5.
- As a result, the equal instruction becomes true, turning output PL1 on.

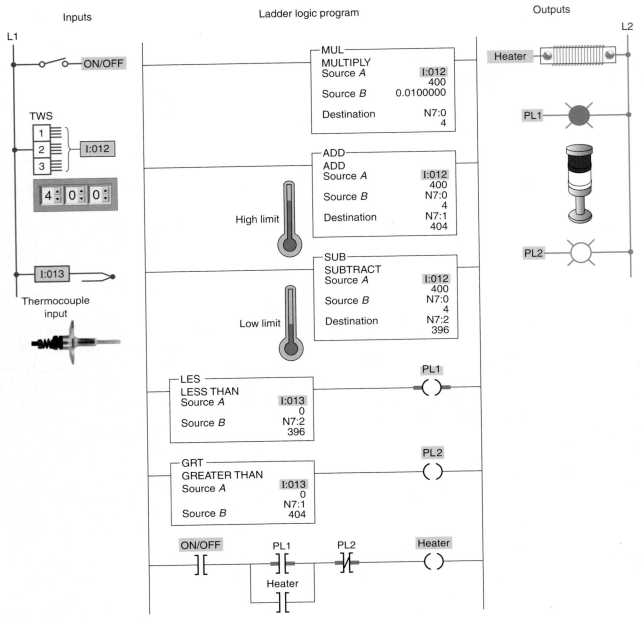

Figure 11-10 The MUL instruction used as part of a temperature control program.

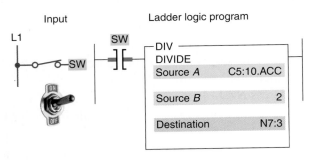

Figure 11-11 SLC 500 DIV (divide) instruction.

The program of Figure 11-13 is an example of how the DIV function is used as part of a program to convert Celsius temperature to Fahrenheit. The operation of the program can be summarized as follows:

- The thumbwheel switch connected to the input module indicates Celsius temperature.

- The program is designed to convert the recorded Celsius temperature in the data table to Fahrenheit values for display.

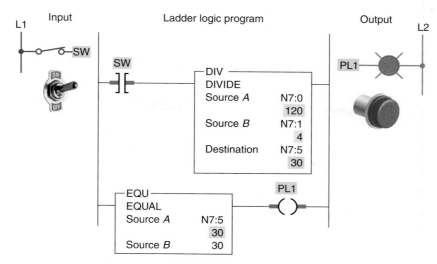

Figure 11-12 DIV instruction used to calculate the value that results from dividing source A by source B.

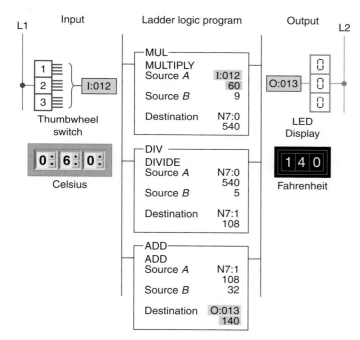

Figure 11-13 Program for converting Celsius temperature to Fahrenheit.

- The following conversion formula forms the basis for the program:

$$F = \left(\frac{9}{5} \times C\right) + 32$$

- In this example, a current temperature reading of 60°C is assumed.
- The MUL instruction multiplies the temperature (60°C) by 9 and stores the product (540) in address N7:0.

- Next, the DIV instruction divides 5 into the 540 and stores the answer (108) in address N7:1.
- Finally, the ADD instruction adds 32 to the value of 108 and stores the sum (140) in address O:13.
- Thus 60°C = 140°F.

11.6 Other Word-Level Math Instructions

The program of Figure 11-14 is an example of the *square root (SQR)* instruction. The operation of the logic rung can be summarized as follows:

- When input switch SW is closed the SQR instruction is executed.
- The number whose square root we want to determine (144) is placed in the source.
- The function calculates the square root and places it (12) in the destination.
- If the value of the source is negative, the instruction will store the square root of the absolute (positive) value of the source at the destination.

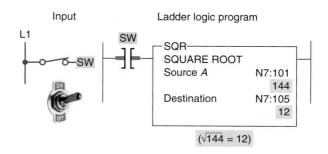

Figure 11-14 SLC 500 SQR (square root) instruction.

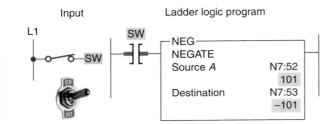

Figure 11-15 SLC 500 NEG (negate) instruction.

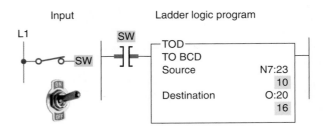

Figure 11-17 SLC 500 TOD (convert to BCD) instruction.

The program of Figure 11-15 is an example of the *negate (NEG)* instruction. This math function changes the sign of the source value from positive to negative. The operation of the logic rung can be summarized as follows:

- When input switch SW is closed the NEG instruction is executed.
- The positive value 101 stored at the source address N7:52 is negated to −101 and stored in destination address N7:53.
- Positive numbers will be stored in straight binary format, and negative numbers will be stored as 2's complement.

The program of Figure 11-16 is an example of the *clear (CLR)* instruction. The operation of the logic rung can be summarized as follows:

- When input switch SW is closed the CLR instruction is executed.
- Upon execution it sets all bits of a word to zero.
- In this example it changes the value of all bits stored in the destination address N7:22 to 0.

The *convert to BCD (TOD)* instruction is used to convert 16-bit integers into *binary-coded decimal (BCD)* values. This instruction could be used when transferring data from the processor (which stores data in binary format) to an external device, such as an LED display, that functions in BCD format. The program of

Figure 11-17 is an example of the TOD instruction. The operation of the logic rung can be summarized as follows:

- When input switch SW is closed the TOD instruction is executed.
- The binary bit pattern at the source address N7:23 is converted into a BCD bit pattern of the same decimal value at the destination address O:20.
- The source displays the value 10, which is the correct decimal value; however, the destination displays the value 16.
- The processor interprets all bit patterns as binary; therefore the value 16 is the binary interpretation of the BCD bit pattern.
- The bit pattern for 10 BCD is the same as the bit pattern for 16 binary.

The *convert from BCD (FRD)* instruction is used to convert binary-coded decimal (BCD) values to integer values. This instruction could be used to convert data from a BCD external source, such as a BCD thumbwheel switch, to the binary format in which the processor operates. The program of Figure 11-18 is an example of the FRD instruction. The operation of the logic rung can be summarized as follows:

- When input switch SW is closed the FDR instruction is executed.
- The BCD bit pattern stored at the source address I:30 is converted into a binary bit pattern of the same decimal value at the destination address, N7:24.

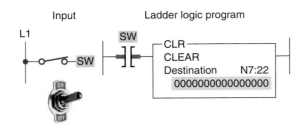

Figure 11-16 SLC 500 CLR (clear) instruction.

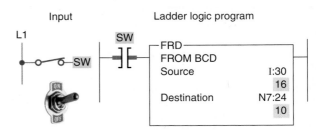

Figure 11-18 SLC 500 FRD (convert from BCD) instruction.

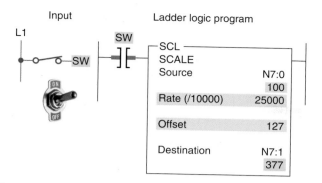

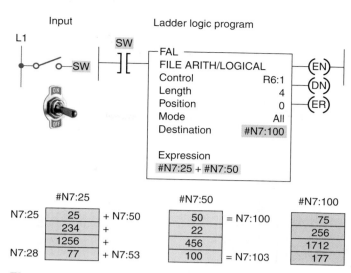

Figure 11-19 SLC 500 SCL (scale) instruction.

The *scale data (SCL)* instruction is used to allow very large or very small numbers to be enlarged or reduced by the rate value. When rung conditions are true, this instruction multiplies the source by a specified rate. The rounded result is then added to an offset value and placed in the destination. The program of Figure 11-19 is an example of the SCL instruction. The operation of the logic rung can be summarized as follows:

- When input switch SW is closed the SCL instruction is executed.
- The number 100 stored at the source address, N7:0, is multiplied by 25,000, divided by 10,000, and added to 127.
- The result, 377, is placed in the destination address, N7:1.

You can use SCL instruction to scale data from your analog module and bring it into the limits prescribed by the process variable or another analog module. For instance, you can use the SCL instruction to convert a 4–20 mA input signal to a PID process variable, or to scale an analog input to control an analog output.

11.7 File Arithmetic Operations

File arithmetic functions include file add, file subtract, file multiply, file divide, file square root, file convert from BCD, and file convert to BCD. The *file arithmetic and logic (FAL)* instruction can combine an arithmetic operation with file transfer. The arithmetic operations that can be implemented with the FAL are ADD, SUB, MULT, DIV, and SQR.

The *file add* function of the FAL instruction can be used to perform addition operations on multiple words. The program of Figure 11-20 is an example of the file add function of the FAL instruction. The operation of the logic rung can be summarized as follows:

- When input switch SW is closed the rung goes true and the expression tells the processor to add the

Figure 11-20 SLC 500 file add function of the FAL instruction.

data in file address N7:25 to the data stored in file address N7:50 and store the result in file address N7:100.

- The rate per scan is set at All, so the instruction goes to completion in one scan.

The program of Figure 11-21 is an example of the *file subtract* function of the FAL instruction. The operation of the logic rung can be summarized as follows:

- When input switch SW is closed the rung goes true and the processor subtracts a program constant (255) from each word of file address N10:0 and

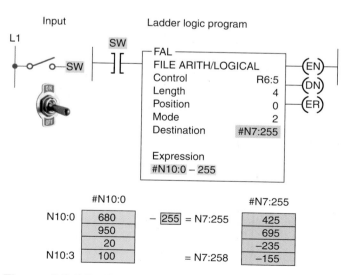

Figure 11-21 SLC 500 file subtract function of the FAL instruction.

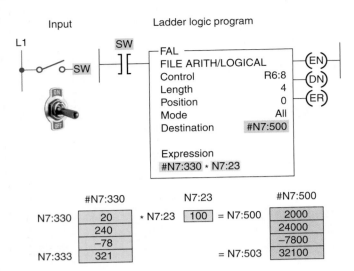

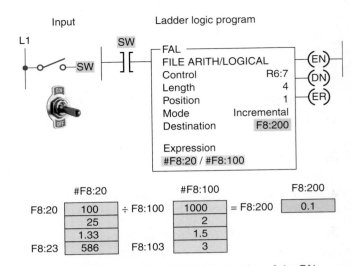

Figure 11-22 SLC 500 file multiply function of the FAL instruction.

Figure 11-23 SLC 500 file divide function of the FAL instruction.

stores the result at the destination file address, N7:255.

- The rate per scan is set at 2, so it will take 2 scans from the moment the instruction goes true to complete its operation.

The program of Figure 11-22 is an example of the *file multiply* function of the FAL instruction. The operation of the logic rung can be summarized as follows:

- When input switch SW is closed the rung goes true and the data in file address N7:330 is multiplied by the data in element address N7:23, with the result stored at the destination file address N7:500.

- The rate per scan is set at All, so the instruction goes to completion in one scan.

The program of Figure 11-23 is an example of the *file divide* function of the FAL instruction. The operation of the logic rung can be summarized as follows:

- When input switch SW is closed the rung goes true and the data in file address F8:20 is divided by the data in file address F8:100, with the result stored in element address F8:200.

- The mode is Incremental, so the instruction operates on one set of elements for each false-to-true transition of the instruction.

1. Explain the function of math instructions as applied to the PLC.

2. Name the four basic math functions performed by PLCs.

3. What standard format is used for PLC math instructions?

4. Would math instructions be classified as input or output instructions?

5. With reference to the instruction of Figure 11-24, what is the value of the number stored at source *B* if N7:3 contains a value of 60 and N7:20 contains a value of 80?

6. With reference to the instruction of Figure 11-25, what is the value of the number stored at the destination if N7:3 contains a value of 500?

7. With reference to the instruction of Figure 11-26, what is the value of the number stored at the destination if N7:3 contains a value of 40 and N7:4 contains a value of 3?

8. With reference to the instruction of Figure 11-27, what is the value of the number stored at the destination if N7:3 contains a value of 15 and N7:4 contains a value of 4?

9. With reference to the instruction of Figure 11-28, what is the value of the number stored at N7:20 if N7:3 contains a value of 2345?

10. With reference to the instruction of Figure 11-29, what will be the value of each of the bits in word B3:3 when the rung goes true?

11. With reference to the instruction of Figure 11-30, what is the value of the number stored at N7:101?

12. With reference to the instruction of Figure 11-31, list the values that will be stored in file #N7:10 when the rung goes true.

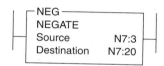

Figure 11-28   Instruction for Question 9.

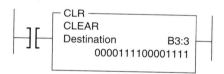

Figure 11-29   Instruction for Question 10.

Figure 11-30   Instruction for Question 11.

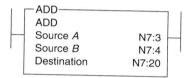

Figure 11-24   Instruction for Question 5.

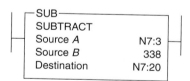

Figure 11-25   Instruction for Question 6.

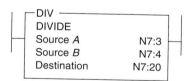

Figure 11-26   Instruction for Question 7.

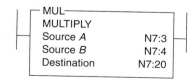

Figure 11-27   Instruction for Question 8.

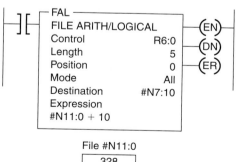

Figure 11-31   Instruction for Question 12.

1. Answer each of the following with reference to the counter program shown in Figure 11-32.
   a. Assume the accumulated count of counters C5:0 and C5:1 to be 148 and 36, respectively. State the value of the number stored in each of the following words at this point:
      (1) C5:0.ACC
      (2) C5:1.ACC
      (3) N7:1
      (4) Source *B* of the GEQ instruction
   b. Will output PL1 be energized at this point? Why?
   c. Assume the accumulated count of counters C5:0 and C5:1 to be 250 and 175, respectively. State the value of the number stored in each of the following words at this point:
      (1) C5:0.ACC
      (2) C5:1.ACC

   (3) N7:1
   (4) Source *B* of the GEQ instruction
   d. Will output PL1 be energized at this point? Why?

2. Answer each of the following with reference to the overfill alarm program shown in Figure 11-33.
   a. Assume that the vessel is filling and has reached the 300-lb point. State the status of each of the logic rungs (true or false) at this point.
   b. Assume that the vessel is filling and has reached the 480-lb point. State the value of the number stored in each of the following words at this point:
      (1) I:012
      (2) N7:1
   c. Assume that the vessel is filled to a weight of 502 lb. State the status of each of the logic rungs (true or false) for this condition.

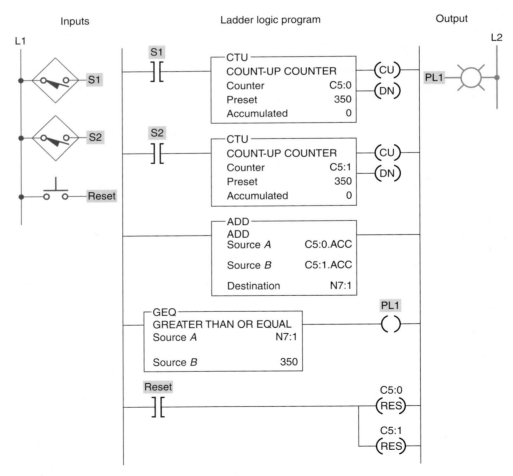

Figure 11-32 Program for Problem 1.

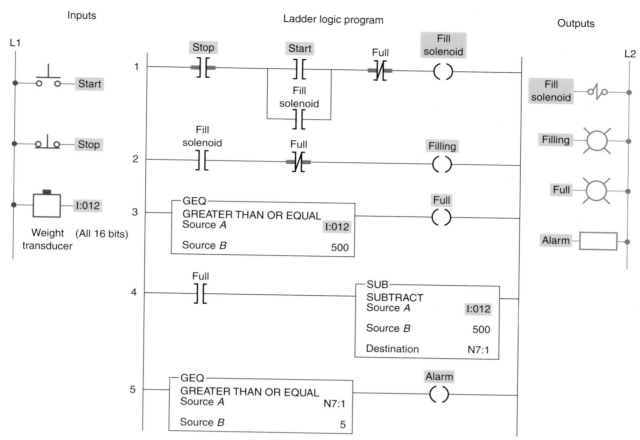

Figure 11-33 Program for Problem 2.

d. Assume that the vessel is filled to a weight of 510 lb. State the value of the number stored in each of the following words for this condition:
 (1) I:012
 (2) N7:1

e. With the vessel filled to a weight of 510 lb, state the status of each of the logic rungs (true or false).

3. Answer the following with reference to the temperature control program shown in Figure 11-34.
 a. Assume that the set-point temperature is 600°F. At what temperature will the electric heaters be turned on and off?
 b. Assume that the set-point temperature is 600°F and the thermocouple input module indicates a temperature of 590°F. What is the value of the number stored in each of the following words at this point?
 (1) I:012
 (2) I:013
 (3) N7:0
 (4) N7:1
 (5) N7:2

c. Assume that the set-point temperature is 600°F and the thermocouple input module indicates a temperature of 608°F. What is the status (energized or not energized) of each of the following outputs?
 (1) PL1
 (2) PL2
 (3) Heater

4. With reference to the Celsius to Fahrenheit conversion program shown in Figure 11-35, state the value of the number stored in each of the following words for a thumbwheel setting of 035:
 a. I:012
 b. N7:0
 c. N7:1
 d. O:013

5. Design a program that will add the values stored at N7:23 and N7:24 and store the result in N7:30 whenever input A is true, and then, when input B is true, will copy the data from N7:30 to N7:31.

6. Design a program that will take the accumulated value from TON timer T4:1 and display it on a 4-digit, BCD format set of LEDs. Use address

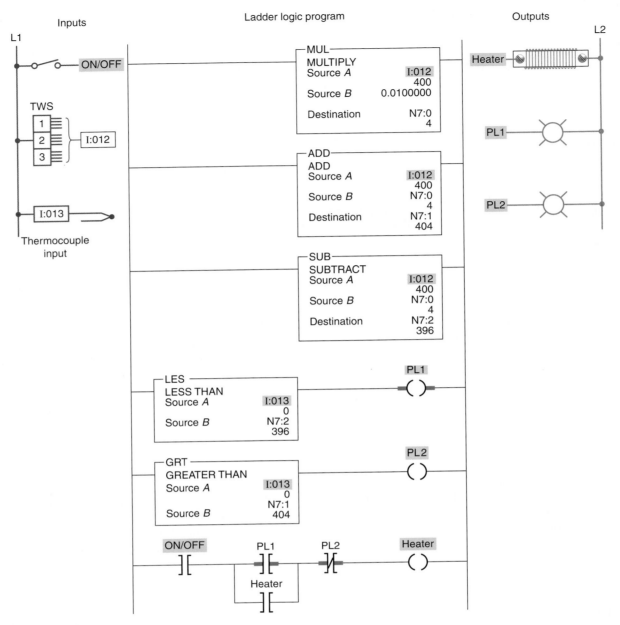

Figure 11-34 Program for Problem 3.

O:023 for the LEDs. Include the provision to change the preset value of the timer from a set of 4-digit BCD thumbwheels when input *A* is true. Use address I:012 for the thumbwheels.

7. Design a program that will implement the following arithmetic operation:
- Use a MOV instruction and place the value 45 in N7:0 and 286 in N7:1.
- Add the values together and store the result in N7:2.
- Subtract the value in N7:2 from 785 and store the result in N7:3.

- Multiply the value in N7:3 by 25 and store the result in N7:4.
- Divide the value in N7:4 by 35 and store the result in F8:0.

8. **a.** There are three part conveyor lines (1-2-3) feeding a main conveyor. Each of the three conveyor lines has its own counter. Construct a PLC program to obtain the total count of parts on the main conveyor.
   b. Add a timer to the program that will update the total count every 30 s.

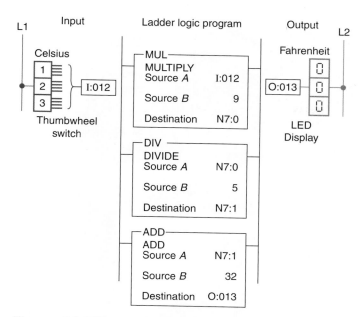

Figure 11-35 Program for Problem 4.

9. With reference to math instruction program shown in Figure 11-36, when the input goes true, what value will be stored at each of the following?
   a. N7:3
   b. N7:5
   c. F8:1

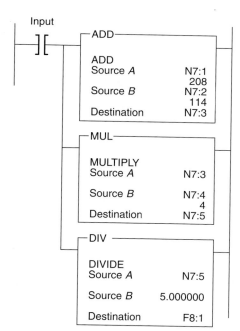

Figure 11-36 Program for Problem 9.

10. With reference to the math instruction program shown in Figure 11-37, when the input goes true, what value will be stored at each of the following?
    a. N7:3
    b. N7:4
    c. N7:5
    d. N7:6

11. Two part conveyor lines, A and B, feed a main conveyor line M. A third conveyor line, R, removes rejected parts a short distance away from the main conveyor. Conveyors A, B, and R have parts counters connected to them. Construct a PLC program to obtain the total parts output of main conveyor M.

12. A main conveyor has two conveyors, A and B, feeding it. Feeder conveyor A puts six-packs of canned soda on the main conveyor. Feeder conveyor B puts eight-packs of canned soda on the main conveyor. Both feeder conveyors have counters that count the number of *packs* leaving them. Construct a PLC program to give a *total can* count on the main conveyor.

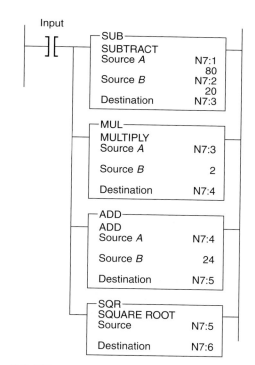

Figure 11-37 Program for Problem 10.

12

Sequencer and Shift Register Instructions

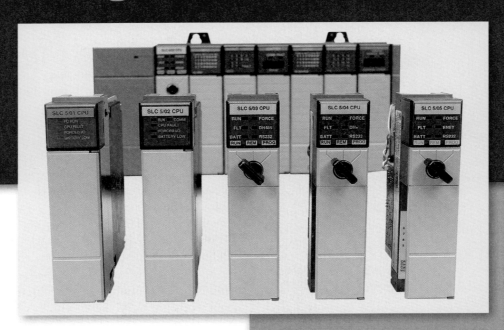

Image Used with Permission of Rockwell Automation, Inc.

This chapter explains how the PLC sequencer and shift register functions operate and how they can be applied to control problems. The sequencer instruction evolved from the mechanical drum switch, and it can handle complex sequencing control problems more easily than does the drum switch. Shift registers are often used to track parts on automated manufacturing lines by shifting either status or values through data files.

Chapter Objectives

After completing this chapter, you will be able to:

12.1 Identify and describe the various forms of mechanical sequencers and explain the basic operation of each

12.2 Interpret and explain information associated with PLC sequencer output, compare, and load instructions

12.3 Compare the operation of an event-driven and a time-driven sequencer

12.4 Describe the operation of bit and word shift registers

12.5 Interpret and develop programs that use shift registers

12.1 Mechanical Sequencers

Sequencer instructions are designed to operate much like the mechanical rotating cam limit switch shown in Figure 12-1. These mechanical type sequencers are often referred to as drum switches, rotary switches, stepper switches, or cam switches. They are often used to control machinery that has a repetitive cycle of operation.

Figure 12-2 illustrates the operation of a cam-operated sequencer switch. An electric motor is used to drive the cams. A series of leaf-spring mounted contacts interacts with the cam so that in different degrees of rotation of the cam, various contacts are closed and opened to energize and de-energize various electrical devices. As the cams rotate, load devices connected to the contacts can change from an on to an off state, from an off to an on state, or remain at the same state.

Figure 12-3 illustrates a typical mechanical drum-operated sequencer switch. The switch consists of a series of normally open contact blocks that are operated by pegs located on the motor-driven drum. The operation of this sequencer can be summarized as follows:

- Pegs are placed at specific locations around the circumference of the drum to operate the contact blocks.
- When the drum is rotated, contacts that align with the pegs will close, whereas the contacts where there are no pegs will remain open.
- The presence of a peg can be interpreted as logic 1, or on, and the absence of a peg as logic 0, or off.
- The equivalent sequencer data table illustrates the logic state for the first four steps of the drum cylinder.

Figure 12-1 Rotating cam limit switch.
Source: Image Used with Permission of Rockwell Automation, Inc.

Switch assembly   Enclosure   Symbol

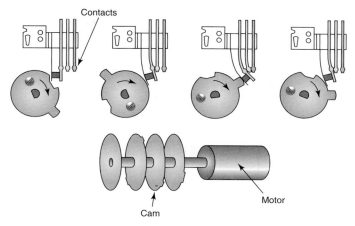

Contacts

Motor

Cam

Figure 12-2 Mechanical cam-operated sequencer.

- Each location where there was a peg is represented by a 1 (on), and the positions where there were no pegs are each represented by a 0 (off).

Sequencer switches are useful whenever a repeatable operating pattern is required. One example is the timed

Equivalent sequencer data table

0	1	0	1	0	1	1	0	0	0	1	0	1	0	1	0	4
1	0	0	0	0	0	0	0	0	1	0	0	0	1	0	0	3
0	1	1	1	0	0	1	0	1	0	1	0	1	0	1	0	2
1	1	1	1	0	0	1	1	0	0	1	0	1	0	1	1	1

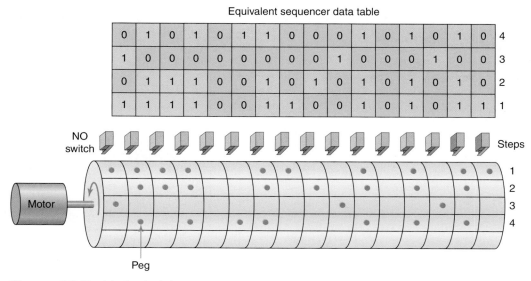

NO switch   Steps

Motor

Steps 1 2 3 4

Peg

Figure 12-3 Mechanical drum-operated sequencer switch.

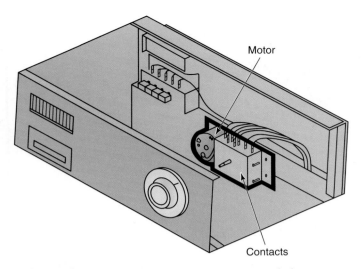

Figure 12-4 Dishwasher timed sequencer switch.

sequencer switch used in dishwashers to pilot the machinery through a wash cycle (Figure 12-4). The cycle is always the same with a fixed routine of actions at each step for a specific time to complete its specified task. The

domestic washing machine is another example of the use of a sequencer, as are dryers and similar time-clock controlled devices.

An example of the wiring and timing chart for a dishwasher that uses a cam-operated sequencer, commonly known as the timer, is shown in Figure 12-5. A synchronous motor drives a mechanical train that, in turn, drives a series of cam wheels. The operation of this sequencer can be summarized as follows:

- The timer motor operates continuously throughout the cycle of operation.
- The cam advances in time increments of 45 seconds in duration.
- The data timing chart shows the sequence of operation of the timer.
- A total of sixty 45 second steps are used to complete the 45 minute operating cycle.
- Numbers in the active devices column refer to control devices active during each step of the cycle.

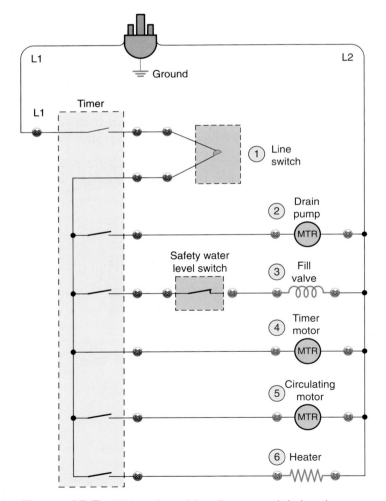

Machine function		Timer increment	Active devices
Off		0–1	
First prerinse	Drain	2	1 2 4
	Fill	3	1 3 4 5
	Rinse	4–5	1 4 5 6
	Drain	6	1 2 4 5
Prewash	Fill	7	1 3 4 5
	Wash	8–10	1 4 5 6
	Drain	11	1 2 4 5
Second prerinse	Fill	12	1 3 4 5
	Rinse	13–15	1 4 5 6
	Drain	16	1 2 4
Wash	Fill	17	1 3 4
	Wash	18–30	1 4 5 6
	Drain	31	1 2 4 5
First rinse	Fill	32	1 3 4 5
	Rinse	33–34	1 4 5 6
	Drain	35	1 2 4 5
Second rinse	Fill	36	1 3 4 5
	Rinse	37–41	1 4 5 6
	Drain	42	1 2 4 5
Dry	Dry	43–58	1 4 6
	Drain	59	1 2 4 6
	Dry	60	1 4 6

Figure 12-5 Dishwasher wiring diagram and timing chart.

12.2 Sequencer Instructions

PLC sequencer instructions replace the mechanical drum sequencer that is used to control machines that have a stepped sequence of repeatable operations. Programmed sequencers can perform the same specific on or off patterns of outputs that are continuously repeated with a drum switch, but with much more flexibility. Sequencer instructions simplify your ladder program by allowing you to use a single instruction or pair of instructions to perform complex operations. For example, the on/off operation of 16 discrete outputs can be controlled, using a sequencer instruction, with only one ladder rung. By contrast, the equivalent contact-coil ladder control arrangement would need 16 rungs in the program.

Depending on the PLC manufacturer, various sequencer instructions can be programmed. Figure 12-6 shows the **Sequencer** menu tab for the Allen-Bradley SLC 500 PLC and its associated RSLogix software. For the Allen-Bradley line of controllers, sequencer commands may include the following:

SQO (Sequencer Output)—Is an output instruction that uses a file to control various output devices.

SQI (Sequencer Input)—Is an input instruction that compares bits from an input file to corresponding bits from a source address. The instruction is true if all pairs of bits are the same.

SQC (Sequencer Compare)—Is an output instruction that compares bits from an input source file to corresponding bits from data words in a sequence file. If all pairs of bits are the same, then a bit in the control register is set to 1.

SQL (Sequencer Load)—Is an output instruction used to capture reference conditions by manually stepping the machine through its operating sequences. It transfers data from the input source module to the sequencer file. The instruction functions much like a file-to-word transfer instruction.

Figure 12-7 shows an example of an SQO (Sequencer Output) instruction. The SQO instruction reads data file elements (words) one at a time, applies a mask word to enable or disable bits from the current data file element, and transfers the masked data file element to a designated output.

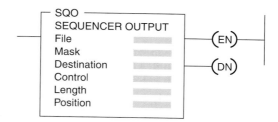

Figure 12-7 SQO (Sequencer Output) instruction.

Parameters that may be required to be entered in sequencer instructions can be summarized as follows:

File—Is the starting address for the registers in the sequencer file and you must use the indexed file indicator (#) for this address. The file contains the data that will be transferred to the destination address when the instruction undergoes a false-to-true transition. Each word in the file represents a position, starting with position 0 and continuing to the file length.

Mask—Is the bit pattern through which the sequencer instruction moves source data to the destination address. Recall that in the mask bit pattern, a 1 passes values while a 0 blocks the data flow. You use a mask register or file name when you want to change the mask pattern under program control. An **h** is placed behind the parameter to indicate that the mask is a hexadecimal number or a **B** to indicate binary notation. Decimal notation is entered without any indicator.

Source—Is the address of the input word or file from which the SQC and SQL instruction obtains data for comparison or input to its sequencer file.

Destination—Is the address of the output word or file to which the SQO moves the data from its sequencer file.

Control—Is the address that contains the parameters with control information for the instruction. The control register stores the status byte of the instruction, the length of the sequencer file, and the instantaneous position in the file as follows:

- The **enable bit (EN;** bit 15) is set by a false-to-true rung transition and indicates that the instruction is enabled. It follows the rung condition.
- The **done bit (DN;** bit 13) is set after the last word in the sequencer file is transferred. On the next false-to-true transition of the rung with the done bit set, the position pointer is reset to 1.
- The **error bit (ER;** bit 11) is set when the processor detects a negative position value, or a negative or zero length value.

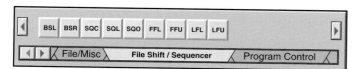

Figure 12-6 Sequencer menu tab.

Length—Is the number of steps of the sequencer file starting at position 1. Position 0 is the start-up position. The instruction resets (wraps) to position 1 at each cycle completion. The actual file length will be 1 plus the file length entered in the instruction.

Position—Indicates the step that is desired to start the sequencer instruction. The position is the word location or step in the sequencer file from which the instruction moves data. Any value up to the file length may be entered, but the instruction will always reset to 1 on the true-to-false transition after the instruction has operated on the last position. Before we start the sequence, we need a starting point at which the sequencer is in a neutral position. The start position is all zeros, representing this neutral position; thus, all outputs will be off in position 0.

To program a sequencer, binary information is first entered into the sequencer file or register made up of a series of consecutive memory words. The sequencer file is typically a bit file that contains one bit file word representing the output action required for each step of the sequence. Data are entered for each sequencer step according to the requirements of the control application. As the sequencer advances through the steps, binary information is transferred from the sequencer file to the output word.

To illustrate the purpose and function of the sequencer file we will examine the operation of the four-step sequence process shown in Figure 12-8. This sequencer is to be used to control traffic in two directions. The operation of the process can be summarized as follows:

- Six outputs are to be energized from one 16-point output module.
- Each light is controlled by one bit address of output word O:2.
- The first 6 bits are programmed to execute the following sequence of light outputs:
  - **Step 1:** Outputs O:2.0 (red) and O:2.5 (green) lights will be energized.
  - **Step 2:** Outputs O:2.0 (red) and O:2.4 (yellow) will be energized.
  - **Step 3:** Outputs O:2.2 (green) and O:2.3 (red) will be energized.
  - **Step 4:** Outputs O:2.1 (yellow) and O:2.3 (red) will be energized.

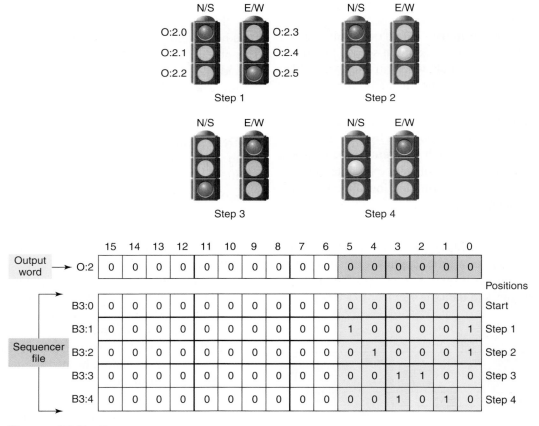

Figure 12-8 Four-step sequencer.

		15	14	13	12	11	10	9	8	7	6	5	4	3	2	1	0	
Output word	O:2	0	0	0	0	0	0	0	0	0	0	0	0	0	0	0	0	
																		Positions
	B3:0	0	0	0	0	0	0	0	0	0	0	0	0	0	0	0	0	Start
	B3:1	0	0	0	0	0	0	0	0	0	0	1	0	0	0	0	1	Step 1
Sequencer file	B3:2	0	0	0	0	0	0	0	0	0	0	0	1	0	0	0	1	Step 2
	B3:3	0	0	0	0	0	0	0	0	0	0	0	0	1	1	0	0	Step 3
	B3:4	0	0	0	0	0	0	0	0	0	0	0	0	1	0	1	0	Step 4

- Words B3:0, B3:1, B3:2, B3:3 and B3:4 make up the sequencer file.
- Binary information (1s and 0s) that reflects the desired on or off light status for each of the four steps is entered into each word of the sequencer file.
- Before starting the sequence, you need a starting point where the sequencer is in a neutral position. This is provided by the start position which is all zeros.

Due to the way in which the sequencer instruction operates, all output points must be on a single output module. When a sequencer operates on an entire output word, there may be outputs associated with the word that do *not* need to be controlled by the sequencer. In our example, bits 6 through 15 of output word O:2 are not used by the sequencer but could be used elsewhere in the program. To prevent the sequencer from controlling these bits of the output word, a mask word is used. The use of a mask word is illustrated in Figure 12-9. The operation of the mask can be summarized as follows:

- The mask word selectively screens out data from the sequencer word file to the output word.
- The hex number 003Fh is entered as the mask parameter.
- For each bit of output word O:2 that the sequencer is to control, the corresponding bit of the mask word must be set to 1.
- The arrows in the figure indicate the unmasked bits that are passed through the mask and into the designation address.
- The dashes in the bits of the designation address indicate that those bits remain unchanged in the designation location during the sequencing.
- These unchanged bits therefore can be used independently of the sequencer.

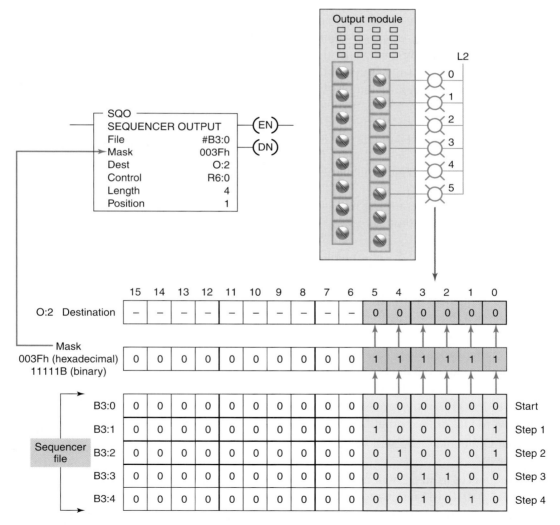

Figure 12-9 Sequencer moving data through a mask word.

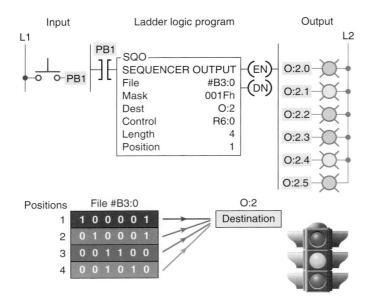

Figure 12-10 Sequencer moving data from a file to an output.

The sequencer output instruction requires preceding logic on the rung where it is located. When this logic goes from false to true, it triggers the sequencer to perform its functions. Only when the logic preceding the sequencer instruction makes the transition from false to true will it go through its functions of reading the data file, applying the mask, and transferring the masked data file to the output destination. After this cycle, it waits for another false-to-true occurrence of the preceding logic to increment to the next step.

Figure 12-10 illustrates how the sequencer moves data from a file to an output. The operation of the logic rung can be summarized as follows:

- Pushbutton PB is used to send false-to-true trigger signals to the sequencer output instruction.
- The position of the sequencer instruction is incremented by one for each false-to-true transition of the sequencer rung.
- Whenever PB is momentarily closed the sequencer is both enabled and advanced to the next position.
- When the sequencer is at step 1, the binary information in word B3:1 (100001) of the sequencer file is transferred into word O:2 of the output.
- As a result output O:2/0 and O:2/5 will be on and all the rest will be off.
- Advancing the sequencer to step 2 will transfer the data from word B3:2 (010001) into word O:2.
- As a result output O:2/0 and O:2/4 will be on and all the rest will be off.

- Advancing the sequencer to step 3 will transfer the data from word B3:3 (001100) into word O:2.
- As a result output O:2/2 and O:2/3 will be on and all the rest will be off.
- Advancing the sequencer to step 4 will transfer the data from word B3:4 (001010) into word O:2.
- As a result output O:2/1 and O:2/3 will be on and all the rest will be off.
- When the position parameter reaches 4 (the value in the length parameter) all words would have been moved so the DN (done bit) in the instruction is set to 1.
- On the next false-to-true transition of the rung, with done bit set, the position pointer is automatically reset to 1.

Sequencer instructions are usually retentive, and there can be an upper limit to the number of external outputs and steps that can be operated on by a single instruction. Many sequencer instructions reset the sequencer automatically to step 1 on completion of the last sequence step. Other instructions provide an individual reset control line or a combination of both.

12.3 Sequencer Programs

A sequencer program can be *event-driven* or *time-driven.* An event-driven sequencer operates similarly to a mechanical stepper switch that increments by one step for each pulse applied to it. A time-driven sequencer operates similarly to a mechanical drum switch that increments automatically after a preset time period.

A sequencer chart, such as the one shown in Figure 12-11, is a table that lists the sequence of operation of the outputs controlled by the sequencer instruction. These tables use a *matrix-style* chart format. A matrix is a two-dimensional, rectangular array of quantities. A time-driven sequencer chart usually indicates outputs on its horizontal axis and the time duration on its vertical axis. An event-driven sequencer indicates outputs on its horizontal axis and the input, or event, on its vertical axis.

An example of a time-driven sequencer with timed steps that are not all the same is shown in Figure 12-12. This sequencer program is used for automatic traffic light control at a four-way intersection. Output lights operate in a sequential fashion with variably timed steps. The system requires two SQO instructions: one for the light outputs and the other for the timed steps. Both SQOs have R6:0 for the control and 4 for the length. The first position is on for 25 seconds, the second for 5 seconds, the third for 25 seconds, and the fourth for 5 seconds.

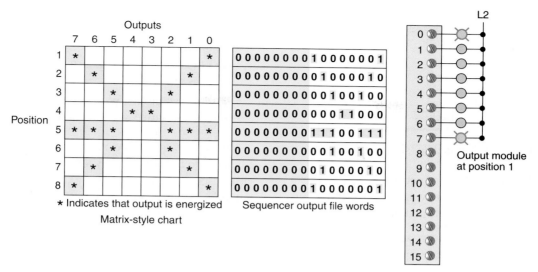

Figure 12-11 Sequencer chart.

The operation of the time-driven sequencer program can be summarized as follows:

- The bits controlling the traffic light outputs are stored in integer file #N7:0 of the first SQO instruction. The settings for the output bits for each position are entered and stored in binary table format as shown in Figure 12-13. Each word of the #N7:0 file is moved from the file by the program to the destination output word O:2 as previously described.

- The second SQO instruction sequencer file, #N7:10, contains the stored preset timer values 25, 5, 25, 5 seconds. These settings are stored in words N7:11, N7:12, N7:13, and N7:15 as illustrated in Figure 12-14. Each word of the #N7:10 file is moved by the program to the destination address T4:1.PRE, which is the preset value for the timer. The program moves information from this file to timer T4:1's preset. The mask allows the proper data to pass and blocks the unnecessary data.

- The timer cycles the two SQO instructions through their four states.

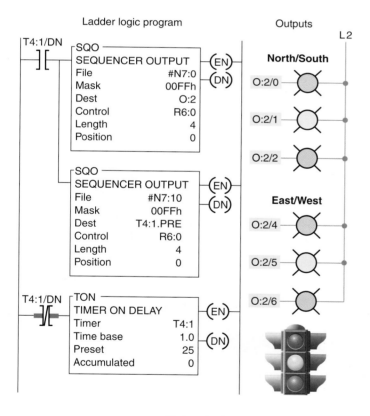

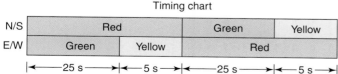

Figure 12-12 Time-driven sequencer output program.

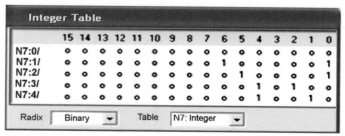

Figure 12-13 Sequencer file #N7:0 light cycle settings.

Integer Table	
	Value
N7:10	0
N7:11	25
N7:12	5
N7:13	25
N7:14	5
Radix	Decimal ▼

Figure 12-14   Sequencer file #N7:10 timer settings.

- Since both of the SQO instructions have R6:0 for control and 4 for length they are stepped in unison to provide a sequentially timed output.

An example of a time-driven sequencer program in which the time interval between sequencer steps is always a constant set value is shown in Figure 12-15. The operation of the program can be summarized as follows:

- The preset time of timer T4:0 is set for 3 seconds.
- The settings of the output bits for each sequencer position are entered and stored in bit file #B3:0.
- The timer is started by the closing switch SW and 3 seconds later the timer done bit is set to 1.
- As a result the timer done bit increments the SQO instruction to the next position and resets the timer.

- The destination is O:2 and all 16 bits of this word are used for outputs.
- The mask is FFFF hexadecimal or 1111111111111111 binary, which allows all 16 bits to pass through.
- As long as input SW is closed the program continues operating with 3 seconds between sequencer steps.

With an event-driven sequencer, the SQO instruction advances to the next step by an external pulsed input event rather than a preset time. An example of an event-driven sequencer is shown in Figure 12-16. The operation of the program can be summarized as follows:

- The sequencer SQO instruction uses two OR configured sensor switches (S1 and S2).
- Any one of the two parallel paths can make the SQO rung true.
- As each event occurs, that OR branch makes a false-to-true transition advancing the sequencer position.
- Data are copied from file #B3:0 at the bit locations through mask word, F0FF hex or 1111000011111111 binary, to the destination O:2. Mask bits are set to 1 to pass data and reset to 0 to mask data.

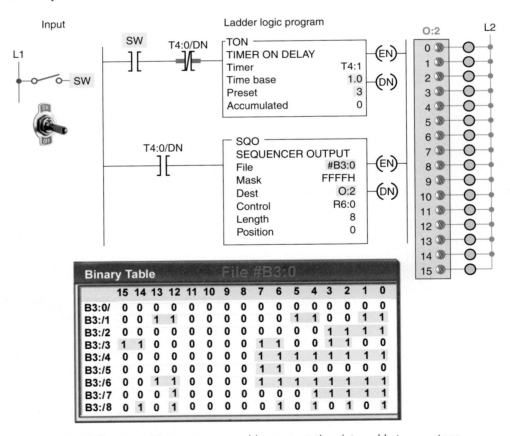

Figure 12-15   Time-driven sequencer with constant time interval between steps.

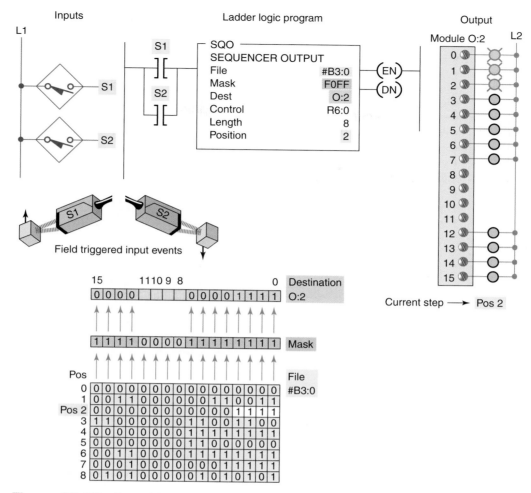

Figure 12-16 Event-driven sequencer output program.

- Once the position reaches the last position on the true-to-false transition of the instruction the position will reset to 1.
- Note that the data in O:2 match the data in position 2 in the file, except for the data in bits 8 through 11.
- Bits 8 through 11 may be controlled from elsewhere in the program; they are not affected by the sequencer instruction because of the 0 in these bit positions in the mask.

The sequencer *input* (SQI) instruction allows input data to be compared for equality against data stored in the sequencer file. For example, it can make comparisons between the states of input devices and their desired states: if the conditions match, the instruction is true.

The SQI instruction is an input instruction available in Allen-Bradley PLC-5 and ControlLogix controllers. An example of a PLC sequencer input instruction is shown in Figure 12-17. The entries in the instruction are similar to those in the sequencer output instruction, except that the destination is replaced by the source.

The operation of the program can be summarized as follows:

- The SQI instruction compares the input data in I:3 through the mask FFF0 with the data in the sequencer file N7:11 through N7:15 for equality.
- The specific data in the sequencer file used in the comparison is identified by the position parameter.
- When the unmasked source bits match those of the corresponding sequencer file word, the instruction goes true; otherwise, the instruction is false.
- In this example, the data at position 2 match the unmasked input data, so the SQI instruction is true, thus making the rung and output PL1 true.
- The input data can indicate the state of an input device, such as the combination of input switches shown in this example program.
- Anytime the combination of opened and closed switches is equal to the combination of 1s and 0s on a step in the sequencer reference file, the PL1 output of the sequencer becomes energized.

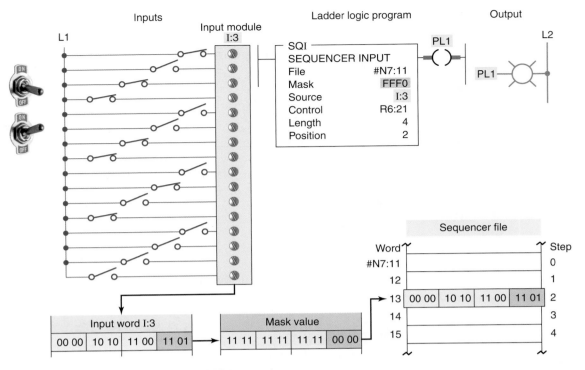

Figure 12-17 Sequencer input (SQI) instruction.

The SQI instruction uses a control register like the SQO instruction but does not have a done bit. In addition, the SQI instruction does not automatically increment its position each time its control logic makes a false-to-true transition at its input. If the SQI instruction is used alone, the position value must be changed by another instruction (such as the move instruction) to select a new input file value to compare against the value from the source address.

When the SQI is paired with an SQO instruction with identical control addresses the position is incremented by the SQO instruction for both. The program of Figure 12-18 illustrates the use of the sequencer input and sequencer output instructions in pairs to monitor and control, respectively, a sequential operation. The operation of the program can be summarized as follows:

- The same control address, length value, and position value is used for each instruction.
- The sequencer input instruction is indexed by the sequencer output instruction because both control elements have the same address, R6:5.
- This type of programming technique allows input and output sequences to function in unison, causing a specific output sequence to occur when a specific input sequence takes place.

The Allen-Bradley SLC 500's *sequencer compare (SQC)* instructions are similar but not identical to the SQI instruction. Differences between the two include:

- The SQC instruction is an output rather than an input instruction.
- The SQC instruction increments the position parameter
- The SQC instruction has an additional status bit—the *found bit (FD)*. When the source pattern matches the sequencer file word the FD is set to 1. It is zero under all other conditions.

Ladder logic program

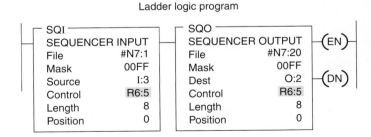

Figure 12-18 Sequencer input and sequencer output instructions used in pairs.

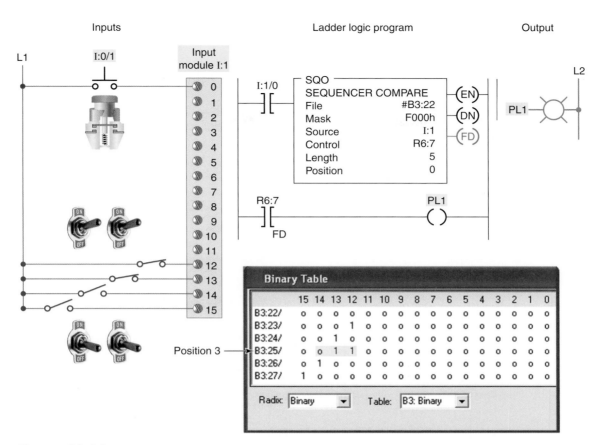

Figure 12-19 Sequencer compare (SQC) instruction program.

An example of an SLC 500 sequencer compare (SQC) instruction program is shown in Figure 12-19. The operation of the program can be summarized as follows:

- The data in the highest 4 bits of the source (I:1) are compared to the data in file #B3:22.
- In this example, the highest 4 bits in I:1 match the status of the highest 4 bits in B3:25 at step position 3.
- If the pushbutton input I:1/0 is true at this point, the found (FD) bit is set, which turns output PL1 on.
- Whenever the combination of opened and closed switches connected to I:1/12, I:1/13, I:1/14, and I:1/15 is equal to the combination of 1s and 0s on a step in the sequencer reference file and the input I:1/0 is true, the PL1 output will become energized.
- The mask (F000h) allows unused bits of the sequencer instruction to be used independently. In this example, unused bit I:1/0 is used for the conditional input of the sequencer compare rung.

The *sequencer load (SQL)* instruction is used to read the PLC input module and store the input data in the sequencer file. Loading input conditions for a large number of process steps is prone to errors. In such instances the sequencer load instruction can be used to load data into a sequencer file one step at a time. For example, a robot may be jogged manually through its sequence of operation, with its input devices read at each step. At each step, the status of the input devices is written to the data file in the sequencer compare instruction. As a result, the file is loaded with the desired input status at each step, and these data are then used for comparison with the input devices when the machine is run in automatic mode.

An example of an SLC 500 sequencer load (SQL) instruction program is shown in Figure 12-20. The operation of the program can be summarized as follows:

- The sequencer load instruction is used to load the file and does *not* function during the machine's normal operation.
- It replaces the manual loading of data into the file with the programming terminal.

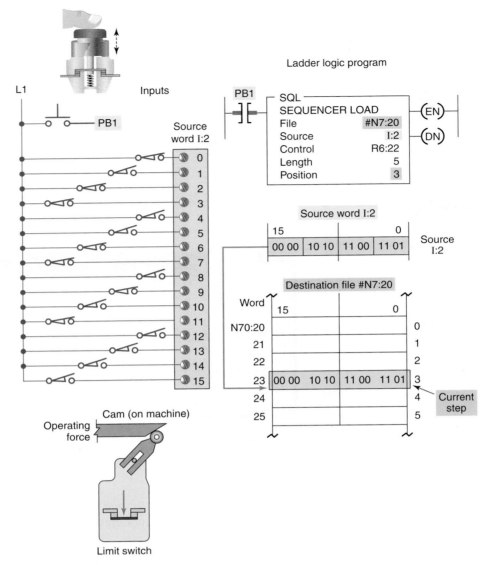

Figure 12-20 Sequencer load (SQL) instruction program.

- The sequencer load instruction *does not* use a mask. It copies data directly from the source address to the sequencer file.
- When the instruction goes from false to true, the instruction indexes to the next position and copies the data.
- When the instruction has operated on the last position and has a true-to-false transition, it resets to position 1.
- It transfers data in position 0 only if it is at position 0 and the instruction is true and the processor goes from the program to run mode.
- By manually jogging the machine through its cycle, the switches connected to input I:2 of the source can be read at each position and written into the file by momentarily pressing PB1. Otherwise, the data would have to be entered into the file manually.

12.4 Bit Shift Registers

The PLC not only uses a fixed pattern of register (word) bits, but also can easily manipulate and change individual bits. A bit *shift register* is a register that allows the shifting of bits through a single register or group of registers. The bit shift register shifts bits serially (from bit to bit) through an array in an orderly fashion.

A shift register can be used to simulate the movement, or *track* the flow, of parts and information. We use the shift register whenever we need to store the status of an event so that we can act on it at a later time. Shift registers can shift either status or values through data files. Common applications for shift registers include the following:

- Tracking of parts through an assembly line
- Controlling of machine or process operations

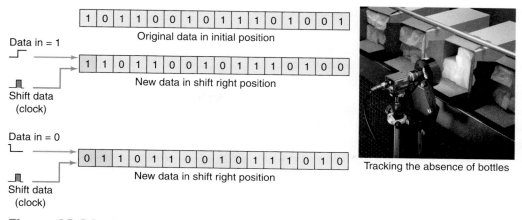

Original data in initial position

Data in = 1

New data in shift right position

Shift data
(clock)

Data in = 0

New data in shift right position

Shift data
(clock)

Tracking the absence of bottles

Figure 12-21 Basic concept of a shift register.

Source: Photo courtesy Omron Industrial Automation, **www.ia.omron.com**.

- Inventory control
- System diagnostics

Figure 12-21 illustrates the basic concept of a shift register. A shift pulse or clock causes each bit in the shift register to move 1 position to the right. At some point, the number of data bits fed into the shift register will exceed the register's storage capacity. When this happens, the first data bits fed into the shift register by the shift pulse are lost at the end of the shift register. Typically, data in the shift register could represent the following:

- Part types, quality, and size
- The presence or absence of parts
- The order in which events occur
- Identification numbers or locations
- A fault condition that caused a shutdown

You can program a shift register to shift status data either right or left, as illustrated in Figure 12-22, by shifting either status or values through data files. When you want to track parts on a status basis, use bit shift registers. Bit shift instructions will shift bit status from a source bit address, through a data file, and out to an unload bit, one bit at a time. There are two bit shift instructions: *bit shift left (BSL),* which shifts bit status from a lower address number to a higher address number through a data file, and *bit shift right (BSR),* which shifts data from a higher address number to a lower address number through a data file. Some PLCs provide a *circulating shift register* function, which allows you to repeat a pattern again and again.

When working with a bit shift register, you can identify each bit by its position in the register. Therefore, working with any bit in the register becomes a matter of identifying the position it occupies rather than the conventional word number/bit number addressing scheme.

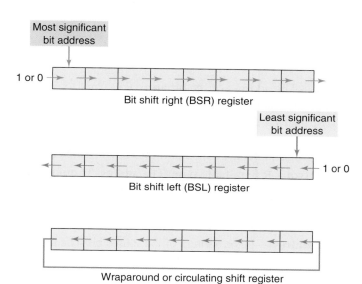

Most significant
bit address

1 or 0

Bit shift right (BSR) register

Least significant
bit address

1 or 0

Bit shift left (BSL) register

Wraparound or circulating shift register

Figure 12-22 Types of shift registers.

Figure 12-23 shows the **File Shift** menu tab and BSL and BSR instruction blocks that are part of the instruction set for the Allen-Bradley SLC 500 controllers. The commands can be summarized as follows:

BSL (Bit Shift Left)—Loads a bit of data into a bit array, shifts the pattern of data through the array

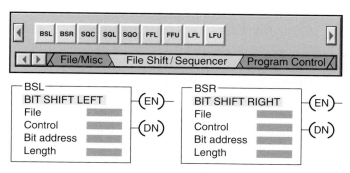

| BSL | BSR | SQC | SQL | SQO | FFL | FFU | LFL | LFU |

File/Misc   File Shift / Sequencer   Program Control

BSL
BIT SHIFT LEFT   (EN)
File
Control   (DN)
Bit address
Length

BSR
BIT SHIFT RIGHT   (EN)
File
Control   (DN)
Bit address
Length

Figure 12-23 Bit shift left and bit shift right instructions.

to the left, and unloads the last bit of data in the array.

BSR (Bit Shift Right)—Loads a bit of data into a bit array, shifts the pattern of data through the array to the right, and unloads the last bit of data in the array.

Shift registers are useful for tracking the status or identification of a part as it moves down an assembly line. The data file used for a shift register usually is the bit file because its data are displayed in binary format, making it easier to read. BSL and BSR are output instructions that load data into a bit array one bit at a time. The data are shifted through the array, then unloaded one bit at a time.

The BSL instruction has the same operands as the BSR instruction. The difference is the direction in which the bits are indexed. A bit shift instruction will execute when its input control logic goes from false to true. To program a bit shift instruction, you need to provide the processor with the following information:

File—The address of the bit array you want to manipulate. The address must start with the # sign and at bit 0 of the first word or element. Any remaining bits in the last word of the array cannot be used elsewhere in the program because the instruction invalidates them.

Control—R data-table type. The address is unique to the instruction and cannot be used to control any other instruction. It is a three-word element that consists of the status word, the length, and the position.

Bit address—Is the address of the source bit. The instruction inserts the status of this bit in either the first (lowest) bit position (for the BSL instruction) or the last (highest) bit position (for the BSR instruction) in the array.

Length—Indicates the number of bits to be shifted, or the file length, in bits. The status bits of the control word are the enable, done, error, and unload bits. Their functions can be summarized as follows:

- **Enable Bit (EN)**—The enable bit follows the instructions status and is set to 1 when the instruction is true.
- **Done Bit (DN)**—The done bit is set to 1 when the instruction has shifted all bits in the file one position. It resets to 0 when the instruction goes false.
- **Error Bit (ER)**—The error bit is set to 1 when the instruction has detected an error, which can happen when a negative number is entered in the length.
- **Unload Bit (UL)**—The unload bit's status is controlled by shifting of the last bit of the file into the unload bit when the instruction is executed. It is the bit location into which the status from the last bit in the file shifts when the instruction goes from false to true. When the next shift occurs, these data are lost, unless additional programming is done to retain the data.

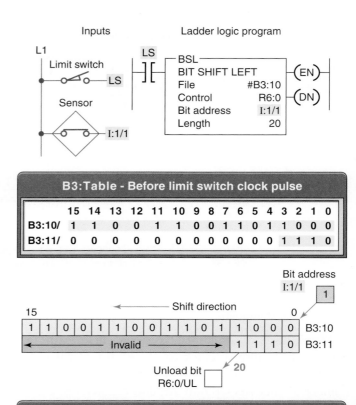

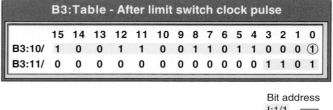

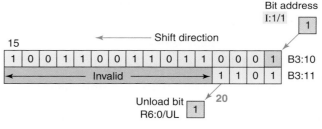

Figure 12-24   Bit shift left (BSL) instruction program.

An example of a bit shift left (BSL) instruction program is shown in Figure 12-24. The operation of the program can be summarized as follows:

- Momentary actuation of limit switch LS causes the BSL instruction to execute.
- When the rung goes from false to true, the enable bit is set and the data block is shifted to the left (to a higher bit number) one bit position.
- The specified bit, at sensor bit address I:1/1, is shifted into the first bit position, B3:10/0.
- The last bit is shifted out of the array and stored in the unload bit, R6:0/UL.

- The status that was previously in the unload bit is lost.
- All the bits in the unused portion of the last word of the file are invalid and should not be used elsewhere in the program.
- For wraparound operation, set the position of the bit address to the last bit of the array or to the UL bit, whichever applies.

An example of a bit shift right (BSR) instruction program is shown in Figure 12-25. The operation of the program can be summarized as follows:

- Before the rung goes from false to true, the status of bits in words B3:50 and B3:51 is as shown.
- The status of the bit address, I:3/5, is a 0, and the status of the unload bit, R6:1/UL, is a 1.
- When limit switch LS closes, the status of the bit address, I:3/5, is shifted into B3:51/7, which is the 24th bit in the file.
- The status of all the bits in the file is shifted one position to the right, through the length of 24 bits.
- The status of B3:50/0 is shifted to the unload bit, R6:1/UL. The status that was previously in the unload bit is lost.

An example of a bit BSL instruction program with wraparound operation is shown in Figure 12-26. The clock pulse input is a fixed regular 3 second pulse–generated on-delay timer T4:0. The operation of the program can be summarized as follows:

- Go to the data tables and set bit addresses B3:0/0, B3:0/1, B3:0/2 to logic 0 and bit address R6:0/UL to logic 1.
- When the PLC is then placed in run, bit B3:0/0 is set to logic 1 causing PL1 to turn on.
- Closing input switch SW starts timer T4:0 timing.
- After 3 seconds the timer done bit is set to reset the timer accumulated time to zero and shift the logic bit 1 to the left to B3:0/1.
- This causes PL1 to turn off and PL2 to turn on.
- After another 3 seconds, the timer done bit is set once again.
- The BSL instruction shifts the bits to the left once more and causes PL2 to turn off and PL3 to turn on.
- The process continues with each of the pilot lights turned on in sequence for 3 seconds.

A shift register is often used in material handling processes where some form of binary information must be synchronized with a moving part on a conveyor. The

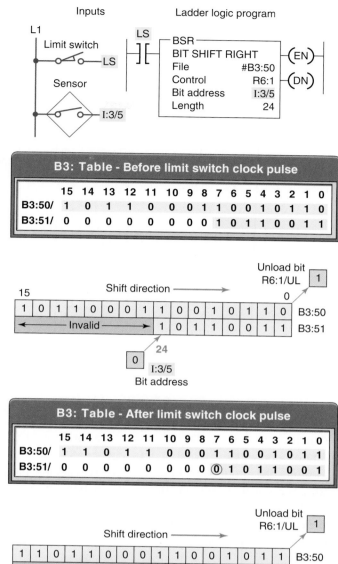

Figure 12-25 Bit shift right (BSR) instruction program.

binary information refers to any two conditions that can be assigned to the moving product—for example, the presence or absence of a part. As the part moves along the conveyer, some form of sensing device will determine which of these two categories the passing product falls into. Figure 12-27 illustrates cartons traveling on a conveyor being detected by a photoelectric sensor. The sensor that drives the data line on a shift register is fixed such that the beam detects the presence or absence of a carton. A logic 1 sensor condition state can indicate the presence of a carton, and a 0 the absence.

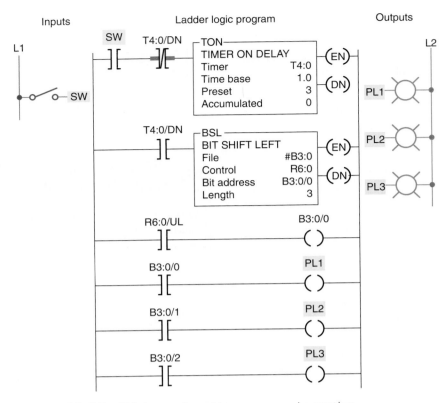

Figure 12-26 BSL instruction with a wraparound operation.

The process of Figure 12-28 illustrates a spray-painting operation controlled by a shift left register. As the parts pass along the production line, the shift register bit patterns represent the items on the conveyor hangers to be painted. Each file bit location represents a station on the line, and the status of the bit indicates whether or not a part is present at that station.

Figure 12-27 Cartons traveling on a conveyor being detected by a photoelectric sensor.
Source: Courtesy Banner Engineering Corp.

The program for the spray-painting operation is shown in Figure 12-29. Its operation can be summarized as follows:

- Limit switch LS1 is used to detect the hanger and limit switch LS2 the part.
- The pulse generated by the hanger-operated limit switch LS1 shifts the status of the data provided by part-detection limit switch LS2.
- The logic of this operation is such that when a part to be painted and a part hanger occur together at station 1 (indicated by simultaneous closing of LS2 and by LS1), logic 1 is input into the shift register at B3:0/0.
- This causes the SOL 1 rung to be true and the undercoat spray gun to energize.
- At station 5 a 1 appears in bit B3:0/5 of the shift register to make the SOL 2 rung true and topcoat spray gun energize.
- Logic 0 in the shift register indicates that the conveyor has no parts on it to be sprayed, and it therefore inhibits the operation of the spray guns.
- Counter C5:1 counts the parts as they enter the process and counter C5:2 as they exit.
- The count obtained by the two counters should be equal when no parts are being painted.

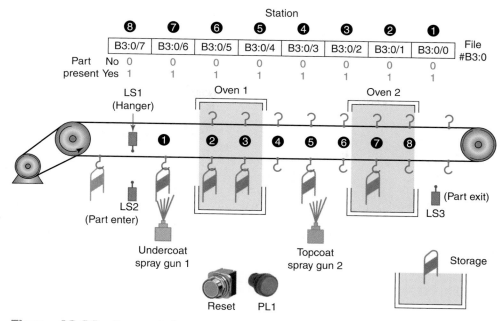

Figure 12-28 Spray-painting operation controlled by a shift left register.

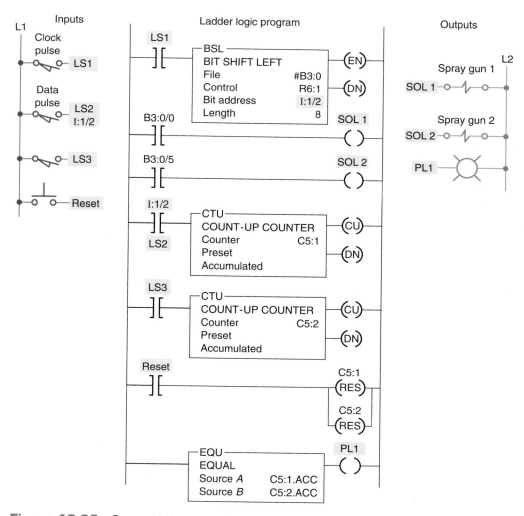

Figure 12-29 Spray-painting operation program.

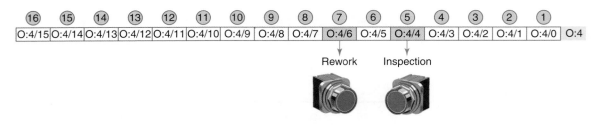

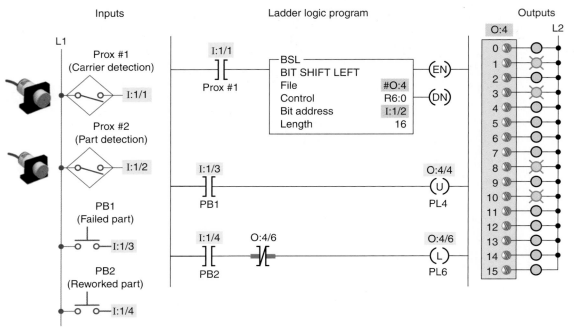

Figure 12-30 Program for tracking of carriers flowing through a 16-station machine.
Source: Photos courtesy Omron Industrial Automation, **www.ia.omron.com**.

- Whenever the two counts are equal in value the equal instruction executes to turn on pilot light PL1. This is an indication that the parts commencing the spray-painting run equal the parts that have completed it.

An example for a bit shift program used to keep track of carriers flowing through a 16-station machine is shown Figure 12-30. The operation of the program can be summarized as follows:

- Proximity switch 1 senses a carrier, and proximity switch 2 senses a part on the carrier.
- Clock pulse generated by carrier proximity switch I:1/1 shifts the status of the data provided by part detection proximity switch I:1/2.
- When a part and container are sensed together, indicated by simultaneous closing of I:1/2 and I:1/1; logic 1 is input into the shift register at output O4:0/0 to energize the pilot light connected to it.
- Remaining pilot lights connected to output module O:4 turn on in sequence as carriers with parts move through each station.

- They turn off or remain off as empty carriers move through.
- Station 5 is an inspection station where parts are examined.
- If the part fails, the inspectors push PB1 as they remove the part from the system, which turns output O:4/4 off.
- Rework parts can be added back into the system at station 7.
- When the operator puts a part on an empty carrier, he or she pushes PB2, turning output O:4/6 on to resume tracking.

12.5 Word Shift Operations

The *first in, first out (FIFO)* instructions are word shift operations that are similar to bit shift operations. Word shifting provides a simpler method of loading and unloading data into a file, usually called the *stack*. It is often used for tracking parts through an assembly line, where parts are represented by values that have a part number or

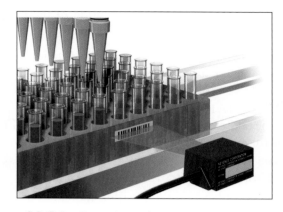

Figure 12-31 Barcode reader.
Source: Courtesy Keyence Canada Inc.

an assembly code. Figure 12-31 shows a barcode reader used for reading printed barcode data on boxes.

A bit shift register operates *synchronously* or in a serial fashion because information is shifted one bit at a time within a word or words. For every bit shifted in, one is shifted out. Data entered in a bit shift register must be shifted the length of the register (one position per shift pulse) before they are available to shift out.

A FIFO function operates *asynchronously*. Rather than shifting bits of information within a word it shifts the data from a complete word into a file or stack. Unlike the bit shift register, two separate shift pulses are required: one to shift data into the file (load) and one to shift data out of the file (unload). These two shift pulses operate independently (asynchronously) of each other. Data loaded in a FIFO can be immediately available for unload, regardless of length.

The FFL and FFU instruction are used in pairs. The FFL *loads* logic words into a user-created file called a FIFO stack. The FFU instruction is used to *unload* the words from the FIFO stack, in the same order as the words were entered. The first word entered is the first word out.

The SLC 500 FIFO load (FFL) instruction is shown in Figure 12-32. The parameters that are required to be entered in the instruction block are summarized as follows:

Source—Word address from which the data are entered into the FIFO file.

FIFO—Address of the file in which the data are entered. The address must start with a # sign.

Control—R data-table type and is the file address of the control structure. The status bits, stack length, and position are stored in this element.

Length—File length in words. Specifies the maximum number of words in the stack.

Position—Is the next available location where the instruction loads data into the stack. The first address in the stack is position 0. As each word is entered into the stack, the position counter, on both the FFL and FFU, will increment up by one. The stack is considered full when the position value equals the length. The status bits of the control word are the enable (EN), the done (DN), and the empty (EM) bits. Their functions can be summarized as follows:

- **Enable Bit (EN)**—The enable bit follows the instructions status and is set to 1 when the instruction is true.
- **Done Bit (DN)**—The done bit is set to 1 when the instruction's position equals the length. When the done bit is set, the FIFO is full and does not accept any more data. Also the data in the FIFO file are not overwritten when the instruction goes from false to true.
- **Empty Bit (EM)**—The empty bit is set to 1 when all the data have been unloaded from the FIFO file.

Figure 12-33 shows the SLC 500 FIFO unload (FFU) instruction. The following parameters need to be entered in the SLC 500 FFU instruction:

FIFO—Address of the file in which the data are entered. The address must start with a # sign. When *paired* with an FFL instruction, this address is the same as the address for the FFL.

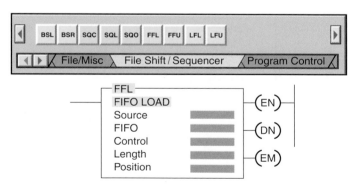

Figure 12-32 SLC 500 FIFO load (FFL) instruction.

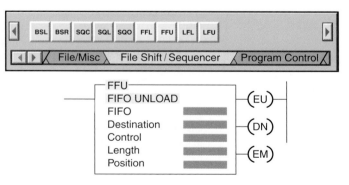

Figure 12-33 SLC 500 FIFO unload (FFU) instruction.

Destination—Address to which the FFU unloads data.

Control—R data-table type. It is a three-word element that consists of the status word, the length, and the position. When it is paired with the FFL, the control addresses are the same.

Length—File length in words. Specifies the maximum number of words in the stack.

Position—Next location from which data are unloaded when the instruction goes from false to true.

The status bits of the control word are the enable (EN), the done (DN), and the empty (EM) bits. The enable bit follows the instruction's status, the done bit is set when the instruction's position equals the length, and the empty bit is set when all the data have been unloaded from the FIFO file.

The program of Figure 12-34 is an example of how data are indexed in and out of a FIFO file using the FFL

and FFU instruction pair. The operation of the program can be summarized as follows:

- The FIFO load and FIFO unload instructions share the same control element, R6:0, which may not be used to control any other instructions.
- FIFO, #N7:12, is the address of the stack. The same address is programmed for the FFL and FFU instructions.
- Data enter the FIFO file from the source address, N7:10, on a false-to-true transition of input *A*.
- Data are placed at the position indicated in the instruction on a false-to-true transition of the FFL instruction, after which the position indicates the current number of data entries in the FIFO file.
- The FIFO file fills from the beginning address of the FIFO file and indexes to one higher address for each false-to-true transition of input *A*.

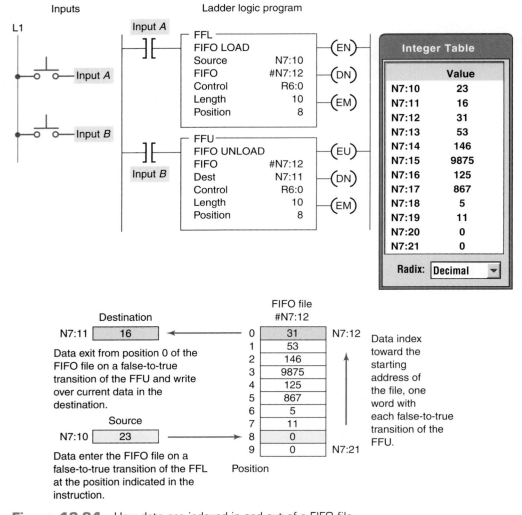

Figure 12-34 How data are indexed in and out of a FIFO file.

- A false-to-true transition of input *B* causes all data in the FIFO file to shift one position toward the starting address of the file, with the data from the starting address of the file shifting to the destination address, N7:11.

The FIFO instruction is often used for inventory control. One example is where different parts need to be removed from inventory to be used in production. Each part is assigned a unique code, which is loaded into a FIFO stack, and parts are removed in the order prescribed by the stack. This type of control ensures that the oldest part in the inventory is used first as the first part entered is the first part removed.

The opposite principle—where the last data to be stored are the first to be retrieved—is known as *LIFO (Last In, First Out)*. The LIFO instruction inverts the order of the data it receives by outputting the last data received first and the first data received last. A useful analogy is a pile of work on your desk. As new work arrives you drop it on the top of the stack. If your stack is LIFO, you pick your next job from the top of the pile. If your stack is FIFO, you pick your work from the bottom of the pile. Figure 12-35 shows how the FIFO and LIFO operations work for container stacking operations.

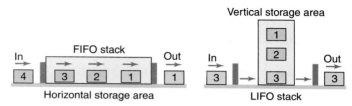

Figure 12-35 FIFO and LIFO container stacking operations.

The difference between FIFO and LIFO stack operation is that the LIFO instruction removes data in the reverse of the order they are loaded (last in, first out). An example of the LIFO instruction pair is shown in Figure 12-36 and the operation of this function can be summarized as follows:

- The load and unload of the LIFO stack operates similarly to that of the FIFO stack, except that the last word in the LIFO stack is the first word that is unloaded from the stack.

- Words can be added to the LIFO stack without disturbing the words already loaded on the stack.

- Otherwise, LIFO instructions operate the same as FIFO instructions.

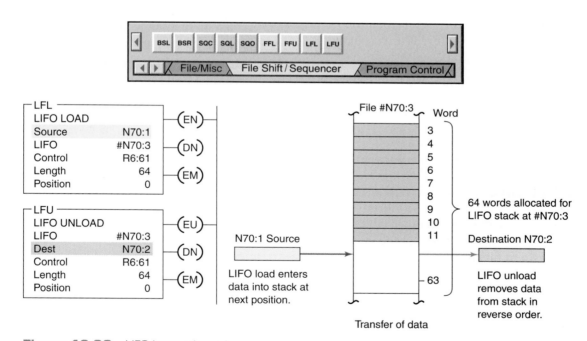

Figure 12-36 LIFO instruction pair.

1. Describe the operation of a drum switch.
2. What type of operations are sequencers most suitable for?
3. Why are PLC sequencers easier to program than PLC discrete outputs?
4. Answer the following with regard to an SLC 500 PLC sequencer output instruction:
   a. Where is the information for each sequencer step entered?
   b. What is the function of the output word?
   c. Explain the transfer of data that occurs as the sequencer is advanced through its various steps.
5. What is the function of the file of a sequencer?
6. What is the function of the mask in the sequencer instruction?
7. What is the relationship between the length and the position in a sequencer instruction?
8. What output and step programming limits may be placed on sequencer instructions?
9. Sequencer instructions are usually retentive. Explain what this means.
10. Compare the operation of an event-driven and a time-driven sequencer.
11. Explain the function of a sequencer input and compare instruction.
12. What is the difference between SQI and SQC instructions?

13. What is the purpose of using the SQI and SQO instruction in pairs?
14. What is the primary application in which an SQL instruction is used?
15. Explain the function of a sequencer load instruction.
16. How does a bit shift register manipulate individual bits?
17. List four common applications for bit shift registers.
18. When using a sensor as the input to the bit address of a BSL instruction, what is its function?
19. Compare the operation of the BSL and BSR bit shift instructions.
20. A bit shift register is said to operate in a synchronous manner. Explain what this means.
21. What is the function of the unload bit in a BSL instruction?
22. What is the function of the unload bit in a BSR instruction?
23. A first in, first out word shift register operates in an asynchronous manner. Explain what this means.
24. Why are both FFL and FFU instructions needed to perform a FIFO function?
25. Compare the operation a FIFO register and a LIFO register.

CHAPTER 12 PROBLEMS

1. Construct an equivalent sequencer data table for the four steps of the mechanical drum-operated sequencer drawn in Figure 12-37.

2. Answer the following with reference to the sequencer file #B3:0 shown in Figure 12-38:
   a. Assume that output bit addresses O:2/0 through O:2/15 are controlling associated output pilot

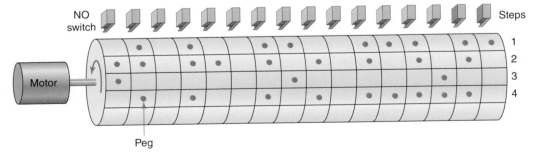

Figure 12-37 Drum-operated sequencer for Problem 1.

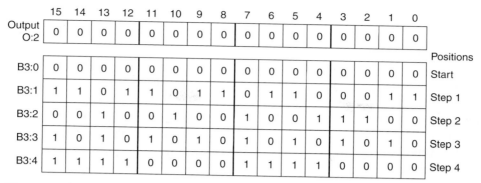

Figure 12-38 Sequencer file for Problem 2.

lights PL1 through PL16. State the status of each light for steps 1 through 4.

b. What output bit addresses could be masked? Why?

c. State the status of each bit of output word O:2 for step 3 of the sequencer cycle.

3. Answer each of the following with reference to the timer-driven sequencer program shown in Figure 12-39:

a. How many bit outputs are controlled by this sequencer?

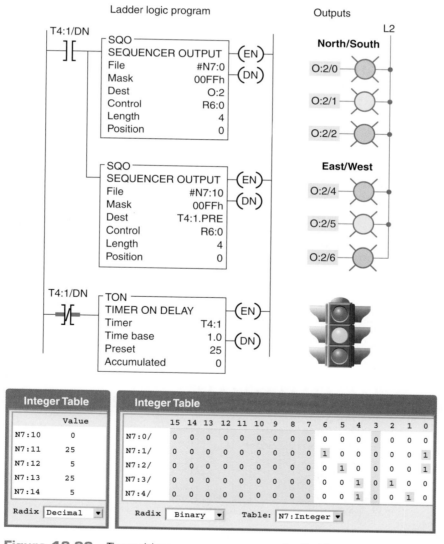

Figure 12-39 Timer-driven sequencer program for Problem 3.

b. What is the address of the word that controls the outputs?

c. What is the address of the sequencer file that sets the states for the outputs?

d. What is the address of the sequencer file that contains the preset timer values?

e. For what length of time is the red light programmed to be on?

f. For what length of time is the green light programmed to be on?

g. For what length of time is the yellow light programmed to be on?

h. What is the time required for one complete cycle of the sequencer?

i. Assume that the decimal value stored in N7:13 is changed to 35. Outline the changes that this new value will have on the timing of the traffic lights.

4. Answer each of the following with reference to the event-driven sequencer program shown in Figure 12-40:

a. When does the sequencer advance to the next step?

b. Assume that the sequencer is at position 2, as shown; what bit outputs will be on?

c. Assume that the sequencer is stepped to position 8; what bit outputs will be on?

d. Assume that the sequencer is at position 8 and a true-to-false transition of one of the inputs occurs. What happens as a result?

5. Using whatever PLC sequencer output instruction you are most familiar with, develop a program that will operate the cylinders in the desired sequence. The time between each step is to be 3 seconds. The desired sequence of operation will be as follows:

- All cylinders to retract.
- Cylinder 1 advance.
- Cylinder 1 retract and cylinder 3 advance.
- Cylinder 2 advance and cylinder 5 advance.
- Cylinder 4 advance and cylinder 2 retract.
- Cylinder 3 retract and cylinder 5 retract.
- Cylinder 6 advance and cylinder 4 retract.
- Cylinder 6 retract.
- Sequence to repeat.

6. Using whatever PLC sequencer output instruction you are most familiar with, develop a program to implement an automatic car-wash process. The process is to be event-driven by the vehicle, which activates

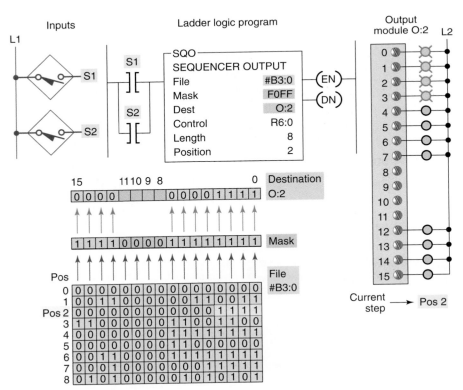

Figure 12-40 Event-driven sequencer program for Problem 4.

various limit switches (LS1 through LS6) as it is pulled by a conveyor chain through the car-wash bay. Design the program to operate the car wash in the following manner:

- The vehicle is connected to the conveyor chain and pulled inside the car-wash bay.
- LS1 turns on the water input valve.
- LS2 turns on the soap release valve, which mixes with the water input valve to provide a wash spray.
- LS3 shuts off the soap valve, and the water input valve remains on to rinse the vehicle.
- LS4 shuts off the water input valve and activates the hot wax valve, if selected.
- LS5 shuts off the hot wax valve and starts the air-blower motor.
- LS6 shuts off the air blower. The vehicle exits the car wash.

7. A product moves continuously down an assembly line that has four stations, as shown in Figure 12-41.
    - The product enters the inspection zone, where its presence is sensed by the proximity switch.
    - The inspector examines it and activates a reject button if the product fails inspection.

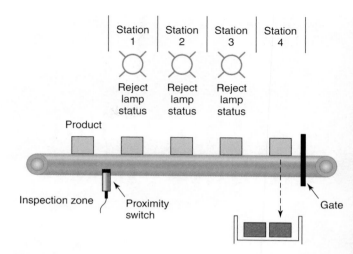

Figure 12-41 Assembly line program for Problem 7.

- If the product is defective, reject status lights come on at stations 1, 2, and 3 to tell the assembler to ignore the part.
- When a defective part reaches station 4, a diverter gate is activated to direct that part to a reject bin.
- Using whatever PLC bit shift register you are most familiar with, develop a program to implement this process.

PLC system. The master control relay provides a means of de-energizing the entire circuit that *is not* dependent on software. The internally programmed MCR of a PLC is not sufficient to meet safety requirements. The hardwired MCR is connected to interrupt power to the I/O rack in the event of an emergency, but still allow power to be maintained at the processor. Figure 13-3 shows the typical wiring for an AC power distribution with a master control

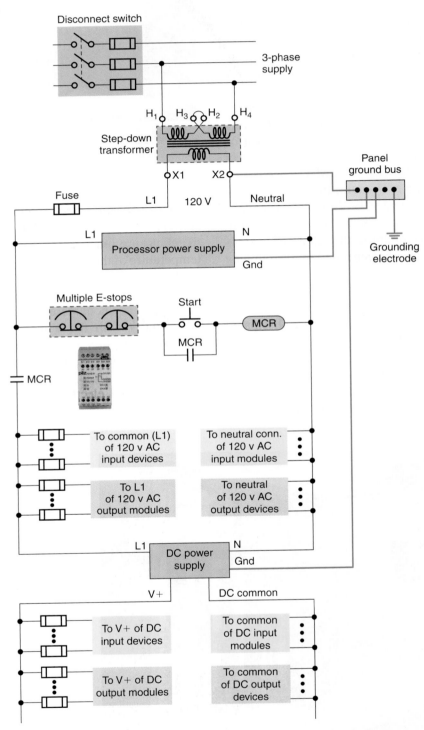

Figure 13-3 Typical wiring for an AC power distribution with a master control relay.
Source: Courtesy Pilz GmbH & Co. KG.

relay. The operation of the circuit can be summarized as follows:

- A power disconnect switch is provided so that, when required, the PLC can be serviced with the power off.
- The step-down transformer provides isolation from the main power distribution system and decreases the voltage to the 120 volts required for the controller power supplies and DC power supplies.
- The momentary start button is pressed to energize the master control relay.
- Pressing any one of the emergency-stop switches de-energizes the master control relay and thus de-energizes the I/O devices.
- Power to the processor of the PLC remains on so status LEDs can continue to provide up-to-date information.
- Emergency stop buttons use normally closed contacts wired in series for fail-safe operation. In the event a wire is broken or comes off a terminal, the MCR relay is de-energized and power is removed.

13.2 Electrical Noise

Electrical noise, also called electromagnetic interference, or EMI, is unwanted electrical signals that produce undesirable effects and otherwise disrupt the control system circuits. EMI may be either radiated or conducted. *Radiated* noise originates from a source and travels through the air while *conducted* noise travels on an actual conductor, such as a power line.

When the PLC is operated in a noise-polluted industrial environment, special consideration should be given to possible electrical interference. To increase the operating noise margin, the controller should be located away from noise-generating devices such as large AC motors and high-frequency welders. Malfunctions resulting from noise are temporary occurrences of operating errors that can result in hazardous machine operation in certain applications. Noise usually enters through input, output, and power supply lines. Noise may be coupled into these lines by an electrostatic field or through electromagnetic induction. The following reduce the effect of electrical interference:

- Manufacturer design features
- Proper mounting of the controller within an enclosure
- Proper equipment grounding

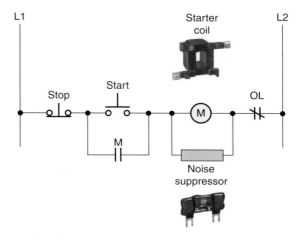

Figure 13-4 Motor starter noise suppression.
Source: Image Used with Permission of Rockwell Automation, Inc.

- Proper routing of wiring
- Proper suppression added to noise-generating devices

Noise suppression is normally needed for inductive loads such as relays, solenoids, and motor starters when operated by hard contact devices such as pushbuttons or selector switches. When inductive loads are switched off, high transient voltages are generated that if not suppressed can reach several thousand volts. Figure 13-4 illustrates a typical noise suppression circuit that is used to suppress the high voltage spikes generated when a motor starter coil is de-energized.

Lack of surge suppression on inductive loads may contribute to processor faults and sporadic operation. RAM can be corrupted (lost), and I/O modules can appear faulty or can reset themselves. When inductive devices are energized or de-energized, they can cause an electrical pulse to be back-fed into the PLC system. The back-fed pulse, when entering the PLC system, can be mistaken by the PLC for a computer pulse. It takes only one false pulse to create a malfunction of the orderly flow of PLC operational sequences.

Proper routing of field power and signal wiring to the PLC enclosure as well as inside the enclosure helps to cut down on electrical noise. The following are some general guidelines for PLC wire routing:

- Use the shortest possible wire runs for I/O signals.
- When possible, conductors that are run from the PLC enclosure to another location should be in a metal conduit as the metal can serve as a shield against EMI.
- *Never* run signal wiring and power wiring in the same conduit.
- Segregate I/O wiring by signal type. Route AC and DC I/O signal wires in separate wireways.

Figure 13-5 Heat-shrinkable wire identification sleeves.
Source: Photo courtesy Tyco Electronics, **www.tycoelectronics.com**.

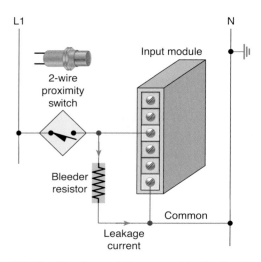

Figure 13-6 Bleeder resistor connection for input sensors.

- Low-level signal conductors such as thermocouples and serial communications should be run as shielded twisted pair and routed separately.
- A fiber optic system, which is totally immune to all kinds of electrical interference, can also be used for signal wiring.

An important part of a PLC installation is clearly identifying each wire to be connected and the terminal to which it is connected. A reliable labeling method, such as the heat-shrinkable wire identification sleeves shown in Figure 13-5, should be used to label each wire. Wiring connectors for input/output modules usually includes spaces for labels used for identifying each I/O address and device connected. Proper wire and terminal identification will simplify the installation and aid in troubleshooting and maintenance.

13.3 Leaky Inputs and Outputs

Many electronic devices with transistor or triac outputs exhibit a small leakage current even when in the off state that may need to be considered when they are connected to PLC input modules. This so-called leakage is typically exhibited by two-wire proximity, photoelectric, and other such sensors. Often, the leaky input will only cause the module's input indicator to flicker. However, a large enough leakage current can activate the input circuit, creating a false input signal.

A common solution to the problem of leaky input current is to connect a bleeder resistor across or in parallel with the input, as shown in Figure 13-6. The bleeder resistor acts as an additional lower resistance load, which allows the leakage current to flow through the lower resistance path. Typically a 10 kΩ to 20 kΩ resistor is used to solve the problem.

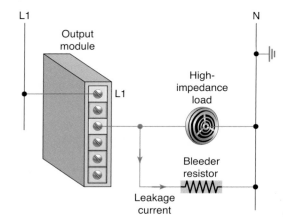

Figure 13-7 Bleeder resistor connection for a high-impedance output.

Leakage current may also occur with the solid-state switch used in many output modules. Problems similar to that encountered with input modules can be created when a high-impedance load device is used with these modules. For example, a PLC output might supply a sound alert device as illustrated in Figure 13-7. In this case the leakage current could cause continuous false or intermittent operation. A resistor can be connected as shown to bleed off this current. An isolation relay could also be used to solve this type of problem.

13.4 Grounding

Proper grounding is an important safety measure in all electrical installations. The authoritative source on grounding requirements for a PLC installation is the National Electrical Code. The NEC specifies the types

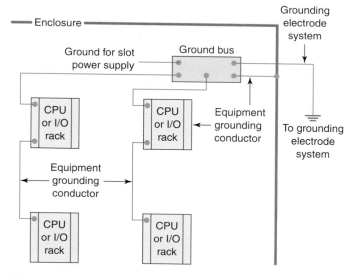

Figure 13-8 PLC grounding system.

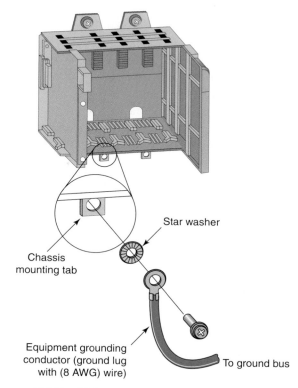

Figure 13-9 Make ground connections using a star washer.

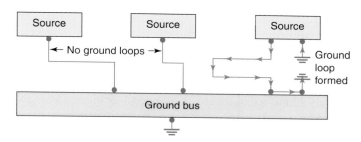

Figure 13-10 Formation of ground loops.

of conductors, color codes, and connections necessary for safe grounding of electrical components. In addition, most manufacturers provide detailed information on the proper grounding methods to use in an enclosure.

Figure 13-8 illustrates a PLC grounding system. A properly installed grounding system will provide a low-impedance path to earth ground. The complete PLC installation, including enclosures, CPU and I/O chassis, and power supplies are all connected to a single low-impedance ground. These connections should exhibit low DC resistance and low high-frequency impedance. A central ground bus bar is provided as a single point of reference inside the enclosure to which all chassis and power supply equipment grounding conductors are connected. The ground bus is then connected to the building's earth ground.

In the event of a high value of ground current, the temperature of the conductor could cause the solder to melt, resulting in interruption of the ground connection. Therefore the grounding path must be permanent (no solder), continuous, and able to conduct safely the ground-fault current in the system with minimal impedance. Paint or other nonconductive material should be scraped away from the area where a chassis makes contact with the enclosure. The minimum ground wire size should be No. 12 AWG stranded copper for PLC equipment grounds and No. 8 AWG stranded copper for enclosure backplane grounds. Ground connections should be made with a star washer between the grounding wire and lug and metal enclosure surface, as illustrated in Figure 13-9.

Ground loops can cause problems by adding or subtracting current or voltage from input signal devices. A ground loop circuit can develop when each device's ground is tied to a different earth potential thereby allowing current to flow between the grounds, as illustrated in Figure 13-10. If a varying magnetic field passes through one of these ground loops, a voltage is produced and current flows in the loop. The receiving device is unable to differentiate between the wanted and unwanted signals and, thus, can't accurately reflect actual process conditions. Certain connections require shielded cables to help reduce the effects of electrical noise coupling. Each shield should be grounded at one end only, as a shield grounded at both ends forms a ground loop.

13.5 Voltage Variations and Surges

The power supply section of the PLC system is built to sustain line fluctuations and still allow the system to function within its operating range. If voltage fluctuations exceed this range, then a system shutdown will occur. In areas where excessive line voltage variation or extended brownouts are anticipated, installing a *constant voltage (CV) transformer* may be required to minimize nuisance shutdowns of the PLC.

Isolation transformers are used in some PLC systems to isolate the PLC from electrical disturbances generated by other equipment connected to the distribution system. Although the PLC is designed to operate in harsh environments, other equipment may generate considerable amounts of interference that may result in intermittent disturbances in normal operation. A normal practice is to place the PLC power supply and I/O devices on a separate transformer that may also serve as a step-down transformer to reduce the incoming voltage to the desired level.

When current in an inductive load is interrupted or turned off, a very high voltage spike is generated. This high voltage can be reduced or eliminated through suppression techniques which absorb the inductive induced voltage. Generally, output modules designed to drive inductive loads include suppression networks built in as part of the module circuit.

An additional external suppression device is recommended if an output module is used to control devices such as relays, solenoids, motor starters, or motors. The suppression device is wired in parallel (directly across) and as close as possible to the load device. The suppression components must be rated appropriately to suppress the switching transient characteristic of the particular inductive device. Figure 13-11 illustrates how a diode is connected to suppress DC inductive

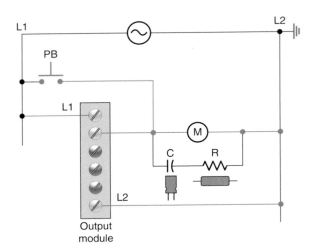

Figure 13-12 RC snubber circuit connected to suppress AC loads.

loads. The operation of the circuit can be summarized as follows:

- The diode is connected in reverse-bias across the solenoid load.
- In normal operation, the electric current can't flow through the diode, so it flows through the solenoid coil.
- When voltage to the solenoid is switched off a voltage opposite in polarity to the original applied voltage is generated by the collapsing magnetic field.
- The induced voltage creates a current flow through the diode bleeding off the high voltage spike.

Figure 13-12 illustrates how an RC (resistor/capacitor) snubber circuit is connected for suppressing AC load devices. The operation of the circuit can be summarized as follows:

- The voltage peak, which occurs at the instant the current path to the coil is opened, is safely short-circuited by the RC network.
- The resistor and capacitor connected in series slows the rate of rise of the transient voltage.
- The voltage across the capacitor cannot change instantaneously, so a decreasing transient current will flow through it for a small fraction of a second, allowing the voltage to increase more slowly when the circuit is opened.

The *metal oxide varistor (MOV)* surge suppressor, shown in Figure 13-13, is the most popular surge protection device. It functions in a manner similar to two zener

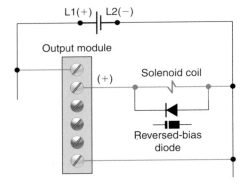

Figure 13-11 Diode connected to suppress DC inductive loads.

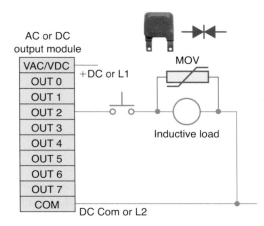

Figure 13-13 Metal oxide varistor (MOV) surge suppressor.

diodes connected back-to-back. The operation of a MOV can be summarized as follows:

- The device acts as an open circuit until the voltage across it in either direction exceeds its rated value.
- Any greater voltage peak instantly makes the device act like a short circuit that bypasses this voltage away from the rest of the circuit.

13.6 Program Editing and Commissioning

After you have entered the rungs for your program, you may need to modify them. *Editing* is simply the ability to make changes to an existing program through a variety of editing functions. Using the editing function, instructions and rungs can be added or deleted; addresses, data, and bits can be changed. Again, the editing format varies with different manufacturers and PLC models.

Today, most PLC programming software is Microsoft Windows based, so if you are familiar with Windows and know how to point and click with a mouse, you should have no problem editing a program. In general, both instructions and rungs are selected simply by clicking on them with the left mouse button. Double clicking with the left mouse button allows you to edit an instruction's address, whereas right clicking displays a pop-up menu of related editing commands. If you want to include additional explanation of a symbol or address, you can place an address description on your ladder rung directly above the symbol. To add a page or rung comment, right click on the rung number to which you wish to add the page or rung comment.

Preparing a control process for start-up, also called *commissioning,* involves a series of tests to ensure that the

PLC, the ladder logic program, the I/O devices, and all associated wiring operate according to specifications. Before commissioning any control system, you should have a good understanding of how the control system operates and how the various components interact. The following are general steps to be followed when commissioning a PLC system:

- Before applying power to the PLC or the input devices, disconnect or otherwise isolate any output device that could potentially cause damage or injury. Typically this precaution would pertain to outputs that cause movement such as starting a motor or operating a valve.
- Apply power to the PLC and input devices. Measure the voltage to verify that rated voltage is being applied.
- Examine the PLC's status indicator lights. If power is properly applied, the power indicator should be on, and there should be no fault indication. If the PLC does not power up properly, it may be faulty. PLCs rarely fail, but if they do fail, it usually happens immediately upon powering up.
- Verify that you have communication with the PLC via the programming device that is running the PLC programming software.
- Place the PLC in a mode that prevents it from energizing its output circuits. Depending on the make of the PLC, this mode may be called *disable, continuous test,* or *single-scan* mode. This mode will allow you to monitor input devices, execute the program, and update the output image file while keeping the output circuits de-energized.
- Manually activate each input device, one at a time, to verify that the PLC's input status lights turn on and off as expected. Monitor the associated condition instruction to verify that the input device corresponds to the correct program address and that the instruction turns true or false as expected.
- Manually test each output. One way you can do this is by applying power to the terminal where the output device is wired. This test will check the output field device and its associated wiring.
- After verifying all inputs, outputs, and program addresses, verify all preset values for counters, timers, and so on.
- Reconnect any output devices that may have been disconnected and place the PLC in the run mode. Test the operation of all emergency stop buttons and the total system operation.

13.7 Programming and Monitoring

When you program a PLC, several instruction entry modes are available, depending on the manufacturer and the model of the unit. A personal computer, with appropriate software, is generally used to program and monitor the program in the PLC. Additionally, it makes possible *offline programming,* which involves writing and storing the program in the personal computer without its being connected to the PLC and later downloading it to the PLC. Figure 13-14 illustrates how programs are downloaded and uploaded from and to the computer.

With *online programming* the program can be modified, the modifications can be tested, and finally they can be accepted or rejected while the PLC is running. However, offline programming is the safest manner in which to edit a program because additions, changes, and deletions do not affect the operation of the system until downloaded to the PLC.

Many manufacturers provide a *continuous test mode* that causes the processor to operate from the user program without energizing any outputs. This mode allows the control program to be executed and debugged while the outputs are disabled. A check of each rung can be done by monitoring the corresponding output rung on the programming device. A *single-scan* test mode may also be available for debugging the control logic. This mode causes the processor to complete a single scan of the user program each time the single-scan key is pressed with no outputs being energized.

An online programming mode permits the user to change the program during machine operation. As the PLC controls its equipment or process, the user can add, change, or delete control instructions and data values as desired. Any modification made is executed immediately on entry of the instruction. Therefore, the user should assess in advance all possible sequences of machine operation that will result from the change. Online programming should be done only by experienced personnel who understand fully the operation of the PLC they are dealing with and the machinery being controlled. If at all possible, changes should be made offline to provide a safe transition from existing programming to new programming.

Two useful monitoring tools provided with PLC programming packages are data monitor and cross reference. *Data monitoring* functions allow you to monitor and/or modify specified program variables. The *cross reference* function allows you to search each instance of a particular address.

The data monitor feature allows you to display data from any place in the data table. Depending on the PLC, the data monitor function can be used to do the following:

- View data within an instruction
- Store data or values for an instruction prior to use
- Set or reset values and/or bits during a debug operation for control purposes
- Change the radix or data format

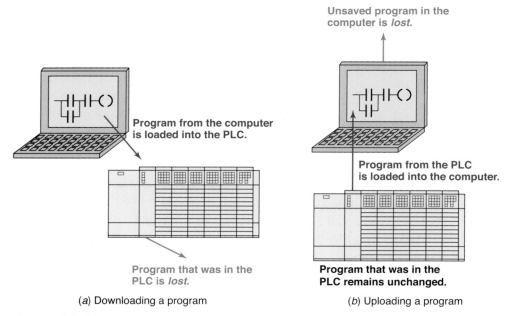

(a) Downloading a program

Program from the computer is loaded into the PLC.

Program that was in the PLC is *lost.*

Unsaved program in the computer is *lost.*

Program from the PLC is loaded into the computer.

Program that was in the PLC remains unchanged.

(b) Uploading a program

Figure 13-14 Downloading and uploading PLC program.

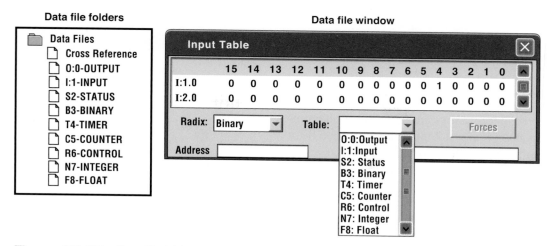

Figure 13-15 Data file folder and window.

Figure 13-15 shows the data file folder and window for the Allen-Bradley SLC 500 PLC and its associated RSLogix software. The data file folder allows the user to determine the status of I/O files as well as the status file (S2), binary file (B3), timer file (T4), counter file (C5), control file (R6), integer file (N7), and the floating-point file (F8). Always be careful when manipulating data using the data monitor function. Changing data could affect the program and turn output devices on or off.

When troubleshooting a PLC, it may be necessary to locate each instance of a particular address in the ladder program. The cross reference function searches all program files to locate each instance of the selected address. A user can then trace the operation by finding all the places where a particular output coil or contact with the same address is used in the program. Figure 13-16 shows a sample cross reference report for the Allen-Bradley SLC 500 PLC and its associated RSLogix software. Its contents can be summarized as follows:

- The report contains all the addresses used in the program.
- Addresses are displayed in the same order as the data table files.
- The address that the search was performed for (O:2/1) is highlighted.
- The description for each address is displayed.
- Listing includes the instruction type, program file, and rung number for each address.
- Each occurrence of the address is displayed, starting with program file 2 and rung 0.

The *contact histogram* function allows you to view the transition history (the on and off states) of a data table value.

The status of the bit(s) (on or off) and the length of time the bit(s) remained on or off (in hours, minutes, seconds, and hundredths of a second) are displayed. In a contact histogram file, the accumulated time indicates the total time that the histogram function was running. The delta time of the contact histogram indicates the elapsed time between the changes in states. Contact histograms are extremely useful

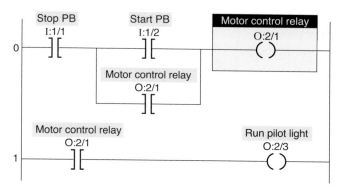

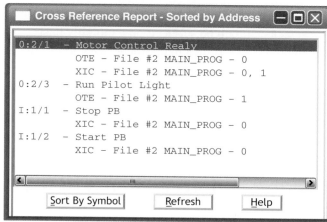

Figure 13-16 Sample cross reference report.

for detecting intermittent problems, either hardware- or logic-related. By tracking the status and time between status changes, you can detect different types of problems.

13.8 Preventive Maintenance

The biggest deterrent to PLC system faults is a proper preventive maintenance program. Although PLCs have been designed to minimize maintenance and provide trouble-free operation, there are several preventive measures that should be looked at regularly.

Many control systems operate processes that must be shut down for short periods for product changes. The following preventive maintenance tasks should be carried out during these short shutdown periods:

- Any filters that have been installed in enclosures should be cleaned or replaced to ensure that clear air circulation is present inside the enclosure.

- Dust or dirt accumulated on PLC circuit boards should be cleaned. If dust is allowed to build up on heat sinks and electronic circuitry, an obstruction of heat dissipation could occur and cause circuit malfunction. Furthermore, if conductive dust reaches the electronic boards, a short circuit could result and cause permanent damage to the circuit board. Ensuring that the enclosure door is kept closed will prevent the rapid buildup of these contaminants.

- Connections to the I/O modules should be checked for tightness to ensure that all plugs, sockets, terminal strips, and module connections are making connections and that the module is installed securely. Loose connections may result not only in improper function of the controller but also in damage to the components of the system.

- All field I/O devices should be inspected to ensure that they are adjusted properly. Circuit boards dealing with process control analogs should be calibrated every 6 months. Other devices, such as sensors, should be serviced on a monthly basis. Field devices in the environment, which have to translate mechanical signals into electrical, may gum up, get dirty, crack, or break—and then they will no longer trip at the correct setting.

- Care should be taken to ensure that heavy noise- or heat-generating equipment is not moved too close to the PLC.

- Check the condition of the battery that backs up the RAM memory in the CPU (Figure 13-17). Most CPUs have a status indicator that shows whether the battery's voltage is sufficient to back up the memory stored in the PLC. If a battery module is to be

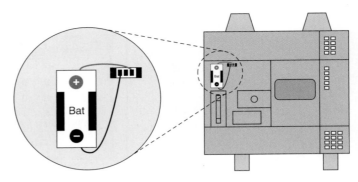

Figure 13-17 CPU backup memory battery.

replaced, it must be replaced with exactly the same type of battery module.

- Stock commonly needed spare parts. Input and output modules are the PLC components that fail most often.

- Keep a master copy of operating programs used.

To avoid injury to personnel and to prevent equipment damage, connections should always be checked with power removed from the system. In addition to disconnecting electrical power, all other sources of power (pneumatic and hydraulic) should be de-energized before someone works on a machine or process controlled by a PLC. Most companies use lockout and tagout procedures, shown in Figure 13-18, to make sure that equipment does not operate while maintenance and repairs are conducted. A personnel protection tag is placed on the power source for the equipment and the PLC, and it can be removed only by the person who originally placed the tag. In addition to the tag, a lock is also attached so that equipment cannot be energized.

Figure 13-18 Lockout/tagout devices.
Source: Photo courtesy Panduit Corporation, **www.panduit.com**.

13.9 Troubleshooting

In the event of a PLC fault, you should employ a careful and systematic approach to troubleshoot the system to resolve the problem. PLCs are relatively easy to troubleshoot because the control program can be displayed on a monitor and watched in real time as it executes. If a control system has been operating, you can be fairly confident of the accuracy of the program logic. For a system that has never worked or is just being commissioned, programming errors should be considered.

When a problem occurs, the first step in the troubleshooting procedure is to identify the problem and its source. The source of a problem can generally be narrowed down to the processor module, I/O hardware, wiring, machine inputs or outputs, or ladder logic program. Once a problem is recognized, it is usually quite simple to deal with. The following sections will deal with troubleshooting these potential problem areas.

Processor Module

The processor is responsible for the *self-detection* of potential problems. It performs error checks during its operation and sends status information to indicators that are normally located on the front of the processor module. You can diagnose processor faults or obtain more detailed information about the processor by accessing the processor status through programming software. Figure 13-19 shows sample diagnostics LEDs found on a processor module. What they indicate can be summarized as follows:

RUN (Green)

- On steady indicates that the process is in the RUN mode.
- Flashing during operation indicates that the process is transferring a program from RAM to the memory module.
- Off indicates that processor is in a mode other than RUN.

FLT (Red)

- Flashing at power-up indicates that the processor has not been configured.

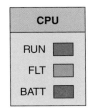

Figure 13-19 Processor diagnostics LEDs.

- Flashing during operation indicates a major error either in the processor, chassis, or memory.
- On steady indicates that a fatal error is present (no communications).
- Off indicates there are no errors.

BATT (Red)

- On steady indicates the battery voltage has fallen below a threshold level, or the battery is missing or not connected.
- Off indicates that the battery is functional.

The processor then monitors itself continually for any problems that might cause the controller to execute the user program improperly. Depending on the controller, a set of fault relay contacts may be available. The fault relay is controlled by the processor and is activated when one or more specific fault conditions occur. The fault relay contacts are used to disable the outputs and signal a failure.

Most PLCs incorporate a *watchdog timer* to monitor the scan process of the system. The watchdog timer is usually a separate timing circuit that must be set and reset by the processor within a predetermined period. The watchdog timer circuit monitors how long it takes the CPU to complete a scan. If the CPU scan takes too long, a watchdog major error will be declared. PLC user manuals will show how to apply this function.

The PLC processor hardware is not likely to fail because today's microprocessors and microcomputer hardware are very reliable when operated within the stated limits of temperature, moisture, and so on. The PLC processor chassis is typically designed to withstand harsh environments.

Input Malfunctions

If the controller is operating in the RUN mode but output devices do not operate as programmed, the faults could be associated with any of the following:

- Input and output wiring between field devices and modules
- Field device or module power supplies
- Input sensing devices
- Output actuators
- PLC I/O modules
- PLC processor

Narrowing down the problem source can usually be accomplished by comparing the actual status of the suspect I/O with controller status indicators. Usually each input or output device has at least two status indicators. One of these indicators is on the I/O module; the other indicator is provided by the programming device monitor.

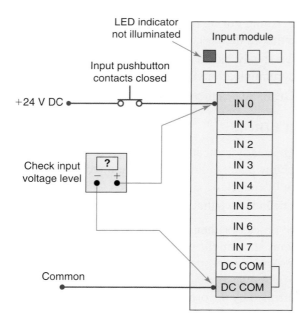

Figure 13-20 Checking for input malfunctions.

The circuit of Figure 13-20 illustrates how to check for discrete input malfunctions. The steps taken can be summarized as follows:

- When input hardware is suspected to be the source of a problem, the first check is to see if the status indicator on the input module illuminates when it is receiving power from its corresponding input device (e.g., pushbutton, limit switch).
- If the status indicator on the input module does *not* illuminate when the input device is on, take a voltage measurement across the input terminal to check for the proper voltage level.
- If the voltage level is correct, then the input module should be replaced.
- If the voltage level is not correct, power supply, wiring, or input device may be faulty.

If the programming device monitor does not show the correct status indication for a condition instruction, the input module may not be converting the input signal properly to the logic level voltage required by the processor module. In this case, the input module should be replaced. If a replacement module does not eliminate the problem and wiring is assumed to be correct, then the I/O rack, communication cable, or processor should be suspected. Figure 13-21 shows a typical input device troubleshooting guide. This guide reviews condition instructions and how their true/false status relates to external input devices.

Input device troubleshooting guide				
Input device condition	Input module status indicator	Monitor display status indicator ⊣⊢	⊣N⊦	Possible fault(s)
Closed — ON 24 V DC input	ON	True	False	None - correct indications
Open — OFF 0 V DC input	OFF	False	True	None - correct indications
Closed — ON 24 V DC input	ON	False	True	Sensor condition, input voltage, status indicator are correct. Ladder instructions have incorrect indications. Input module or processor fault.
Closed — ON 0 V DC input	OFF	False	True	Status indicator and instructions agree but not with the sensor condition. Open field device or wiring.
Open — OFF 0 V DC input	OFF	True	False	Sensor condition, input voltage, status indicator are correct. Ladder instructions have incorrect indications. Input module or processor fault.
Open — OFF 24 V DC input	ON	True	False	Input voltage, status indicator, and ladder instructions agree but not with sensor condition. Short circuit in the field device or wiring.

Figure 13-21 Input troubleshooting guide.

Output Malfunctions

In addition to the logic indicator, some output modules incorporate either a blown fuse indicator or a power indicator or both. A blown fuse indicator indicates the status of the protective fuse in the output circuit, while a power indicator shows that power is being applied to the load.

Electronic protection, as shown in Figure 13-22, is also used to provide protection for the modules from short-circuit and overload current conditions. The protection is based on a thermal cut-out principle. In the event of a short-circuit or overload current condition on an output channel, that channel will limit current within milliseconds after its thermal cut-out temperature has been reached. All other channels continue to operate as directed by the processor.

When an output does not energize as expected, first check the output module blown fuse indicator. Many output modules have each output fused. This indicator will normally illuminate only when the output circuit corresponding to the blown fuse is energized. If this indicator is illuminated, correct the cause of the malfunction and replace the blown fuse in the module.

Figure 13-23 shows a typical discrete output module troubleshooting guide. In general, the following items should be noted when troubleshooting discrete output modules:

- If the blown fuse indicator is not illuminated (fuse OK), then check to see if the output device is responding to the LED status indicator.
- An output module's logic status indicator functions similarly to an input module's status indicator. When it is on, the status LED indicates that the module's logic circuitry has recognized a command from the processor to turn on.
- If an output rung is energized, the module status indicator is on, and the output device is not responding, then the wiring to the output device or the output device itself should be suspected.

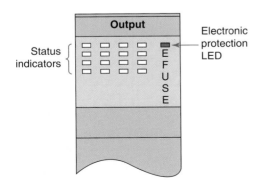

Figure 13-22 Electronic output module protection.

- If, according to the programming device monitor, an output device is commanded to turn on but the status indicator is off, then the output module or processors may be at fault.
- Check voltage at output; if incorrect, power supply, wiring, or output device may be faulty.

Ladder Logic Program

Many PLC software programs offer various software checks used to verify program logic. Figure 13-24 shows a sample of verifying program errors using RSLogix 500 software. Selecting **edit** then **verify project** will check the program for errors. The sample shows what the error message might look like.

The ladder logic program itself is not likely to fail, assuming that the program was at one time working correctly. A hardware fault in the memory IC that holds the ladder logic program could alter the program, but this is a PLC hardware failure. If all other possible sources of trouble have been eliminated, the ladder logic program should be reloaded into the PLC from the master copy of the program. Make sure the master copy of the program is up to date before you download it to the PLC.

Start program troubleshooting by identifying which outputs operate properly and which outputs do not. Then trace back from the output on the nonfunctioning rung and examine the logic to determine what may be preventing the output from energizing. Common logic errors include:

- Programming an examine if closed instruction instead of an examine if open (or vice versa)
- Using an incorrect address in the program

Although the ladder logic program is not likely to fail, the process may be in a state that was unaccounted for in the original program and thus is not controlled properly. In this case, the program needs to be modified to include this new state. A careful examination of the description of the control system and the ladder logic program can help identify this type of fault.

The force on and force off instructions allow you to turn specific bits on or off for testing purposes. Figure 13-25 illustrates how forces are identified as being enabled or disabled in RSLogix 500 software. Forcing lets you simulate operation or control an output device. For example, forcing a solenoid valve on will tell you immediately whether the solenoid is functional when the program is bypassed. If it is, the problem must be related to the software and not the hardware. If the output fails to respond when forced, either the actual output module is causing the problem or the solenoid itself is malfunctioning. *Take all necessary precautions to protect personnel and equipment during forcing.*

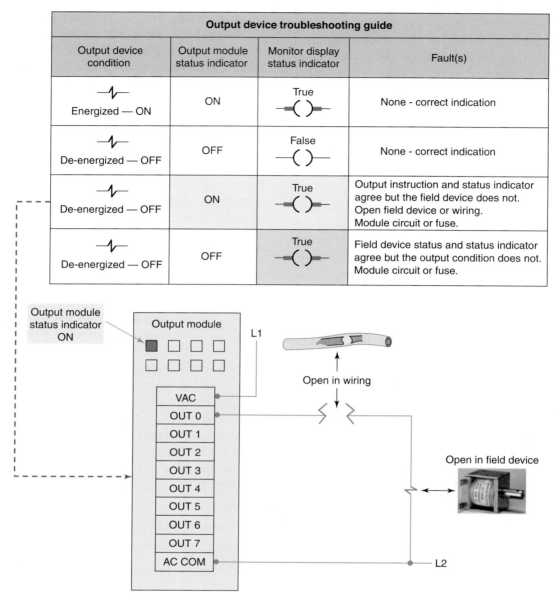

Output device troubleshooting guide			
Output device condition	Output module status indicator	Monitor display status indicator	Fault(s)
Energized — ON	ON	True —( )—	None - correct indication
De-energized — OFF	OFF	False —( )—	None - correct indication
De-energized — OFF	ON	True —( )—	Output instruction and status indicator agree but the field device does not. Open field device or wiring. Module circuit or fuse.
De-energized — OFF	OFF	True —( )—	Field device status and status indicator agree but the output condition does not. Module circuit or fuse.

Figure 13-23 Output troubleshooting guide.

Source: Photo courtesy Guardian Electric, **www.guardian-electric.com.**

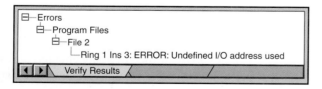

Figure 13-24 Sample of verifying program errors.

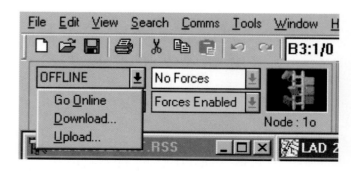

Figure 13-25 Indication of enabled forces.

Certain diagnostic instructions may be included as part of a PLC's instruction set for troubleshooting purposes. The *temporary end (TND)* instruction, shown in Figure 13-26, is used when you want to change the amount of logic scanned to progressively debug your program. The

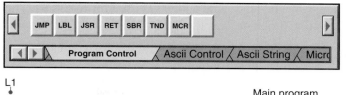

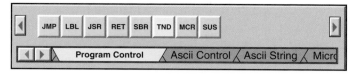

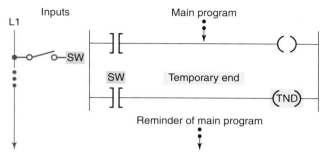

Figure 13-26 TND (temporary end) diagnostic instruction.

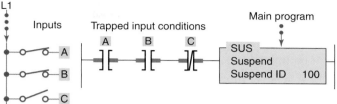

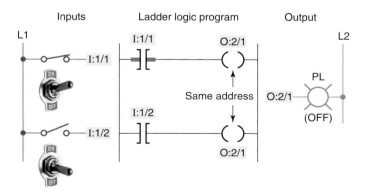

Figure 13-27 SUS (suspend) diagnostic instruction.

operation of this output instruction can be summarized as follows:

- The instruction operates only when its rung conditions are true and stops the processor from scanning any logic beyond the TND instruction.
- When the processor encounters a true TND rung, it resets the watchdog timer (to 0), performs an I/O update, and begins running the ladder program at the first instruction in the main program.
- If the TND rung is false, the processor continues the scan until the next TND instruction or the END statement.
- By inserting the TND instruction at different locations in the program you can test parts of the program sequentially until the entire program has been tested.
- Once the troubleshooting process has been completed, any remaining TND instructions are removed from the program.

The *suspend (SUS)* instruction, shown in Figure 13-27, is used to trap and identify specific conditions for program debugging and system troubleshooting. The operation of this output instruction can be summarized as follows:

- When the rung is true, this instruction places the controller in the *suspend* or *idle* mode.
- The suspend ID, in this case 100, must be selected by the programmer and entered in the instruction.
- When the SUS instruction executes, the ID number 100 is written in word 7 (S:7) of the status file.
- If multiple suspend instructions are present, then this will indicate which SUS instruction was active.
- The suspend file (program or subroutine number identifying where the executed SUS instruction resides) is placed in word 8 (S:8) of the status file.

- All ladder logic outputs are de-energized, but other status files have the data present when the suspend instruction was executed.

Most PLC system faults occur in the field wiring and devices. The wiring between the field devices and the terminals of the I/O modules is a likely place for problems to occur. Faulty wiring and mechanical connection problems can interrupt or short the signals sent to and from the I/O modules.

The sensors and actuators connected to the I/O of the process can also fail. Mechanical switches can wear out or be damaged during normal operation. Motors, heaters, lights, and sensors can also fail. Input and output field devices must be compatible with the I/O module to ensure proper operation.

When an instruction does not seem to be working correctly, the problem may be an addressing conflict caused by the *same address* being used for two or more coil instructions in the same program. As a result, multiple rung conditions can control the same output coil, making troubleshooting more difficult. In the case of duplicate outputs, the monitored rung may be true; but if a rung farther down in the ladder diagram is false, the PLC will keep the output off. The program of Figure 13-28 illustrates what

Figure 13-28 Program with the same address used for two coils.

happens when the same address is used for two coils. The resulting problem scenario can be summarized as follows:

- The problem is turning input switch I:1/1 on *will not* turn on PL output O:2/1 as it appears to be programmed.
- The root of the problem lies in the fact that the PLC scans the program from left to right and top to bottom.
- Whenever input switch I:1/1 is true (closed) and input switch I:1/2 is false (open) output O:2/1 will be off.
- This is because when the PLC updates the outputs it does so based on the status of input I:1/2.
- Regardless of whether input I:1/1 is open or closed the output reacts only to the status of input switch I:1/2.

When a problem occurs, the best way to proceed is to try to logically identify the devices or connections that could be causing the problem rather than arbitrarily checking every connection, switch, motor, sensor, I/O module, and so on. First, observe the system in operation and try to describe the problem. Using these observations and the description of the control system, you should identify the possible sources of trouble. Compare the logic status of the hardwired inputs and outputs to their actual state, as illustrated in Figure 13-29. Any disagreements indicate malfunctions as well as their approximate location.

Some of your troubleshooting can be accomplished by interpreting the status indicators on the I/O modules. The key is to know whether the status indicators are telling you that there is a fault or that the system is normal. Often PLC manufacturers supply a troubleshooting guide, map, or tree that presents a list of observed problems and their possible sources. Figure 13-30 shows a sample troubleshooting tree for a discrete output module. Figures 13-31 and 13-32 are samples of input and output troubleshooting guides.

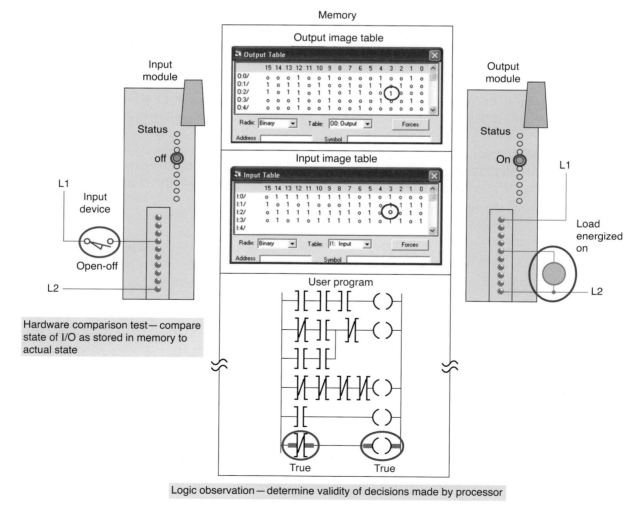

Figure 13-29 General methods of troubleshooting.

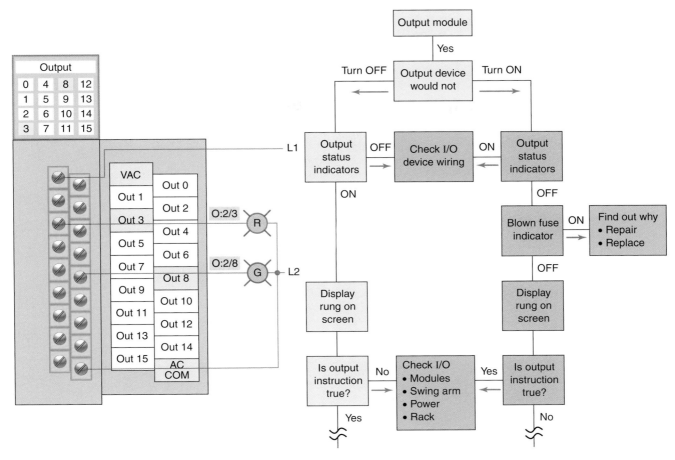

Figure 13-30 Troubleshooting tree for a discrete output module.

If Your Input Circuit LED Is . . .	And Your Input Device Is . . .	And	Probable Cause
ON Input: 0 4 8 12 / 1 5 9 13 / [2] 6 10 14 / 3 7 11 15	On/Closed/Activated	Your input device will not turn off.	Device is shorted or damaged.
		Your program operates as though it is off.	Input circuit wiring or module.
			Input is forced off in program.
	Off/Open/Deactivated	Your program operates as though it is on and/or the input circuit will not turn off.	Input device off-state leakage current exceeds input circuit specification.
			Input device is shorted or damaged.
			Input circuit wiring or module.
OFF Input: 0 4 8 12 / 1 5 9 13 / [2] 6 10 14 / 3 7 11 15	On/Closed/Activated	Your program operates as though it is off and/or the input circuit will not turn on.	Input circuit is incompatible.
			Low voltage across the input.
			Input circuit wiring or module.
			Input signal turn-on time too fast for input circuit.
	Off/Open/Deactivated	Your input device will not turn on.	Input device is shorted or damaged.
		Your program operates as though it is on.	Input is forced on in program.
			Input circuit wiring or module.

Figure 13-31 Input troubleshooting guide.

If Your Output Circuit LED Is . . .	And Your Output Device Is . . .	And	Probable Cause
ON Output: 0 4 **8** 12 / 1 5 9 13 / 2 6 10 14 / 3 7 11 15	On/Energized	Your program indicates that the output circuit is off or the output circuit will not turn off. —( )—	Programming problem: - Check for duplicate outputs and addresses. - If using subroutines, outputs are left in their last state when not executing subroutines. - Use the force function to force output off. If this does not force the output off, output circuit is damaged. If the output does force off, then check again for logic/programming problem.
			Output is forced on in program.
			Output circuit wiring or module.
	Off/De-energized	Your output device will not turn on and the program indicates that it is on. —( )—	Low or no voltage across the load.
			Output device is incompatible: check specifications and sink/source compatibility (if dc output).
			Output circuit wiring or module.
OFF Output: 0 4 **8** 12 / 1 5 9 13 / 2 6 10 14 / 3 7 11 15	On/Energized	Your output device will not turn off and the program indicates that it is off. —( )—	Output device is incompatible.
			Output circuit off-state leakage current may exceed output device specification.
			Output circuit wiring or module.
			Output device is shorted or damaged.
	Off/De-energized	Your program indicates that the output circuit is on or the output circuit will not turn on. —( )—	Programming problem: - Check for duplicate outputs and addresses. - If using subroutines, outputs are left in their last state when not executing subroutines. - Use the force function to force output on. If this does not force the output on, output circuit is damaged. If the output does force on, then check again for logic/programming problem.
			Output is forced off in program.
			Output circuit wiring or module.

Figure 13-32 Output troubleshooting guide.

13.10 PLC Programming Software

You must establish a way for your personal computer (PC) software to communicate with the programmable logic controller (PLC) processor. Making this connection is known as *configuring* the communications. The method used to configure the communications varies with each brand of controller. In Allen-Bradley controllers, *RSLogix* software is required to develop and edit ladder programs. A second software package, *RSLinx,* is needed to monitor PLC activity, download a program from your PC to your PLC, and upload a program from your PLC into your PC. You cannot download multiple projects to the PLC and then run them when required. The PLC will accept only one program at a time, but the program can consist of multiple subroutine files which can be conditionally called from the main program.

RSLinx software is available in multiple packages to meet the demand for a variety of cost and functionality requirements. This software package is used as the driver between your PC and PLC processor. A *driver* is a computer program that controls a device. For example, you must have the correct printer driver installed in your PC

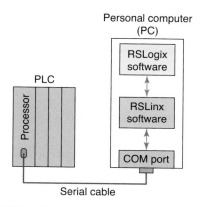

Figure 13-33 Direct PC-to-PLC software connection.

in order to be able to print a word-processing document created on your PC. RSLinx works much like the printer driver for RSLogix software. The RSLinx program must be opened and drivers configured before communications can be established between a PC and a PLC that is using RSLogix software.

RSLinx allows RSLogix to communicate through an interface cable to the PLC processor. The simplest connection between a PC and a PLC is a point-to-point direct connection through the computer serial port, as illustrated in Figure 13-33. A serial cable is used to connect to your PC's COM 1 or COM 2 port and to the PLC processor's serial communications port. With RSLinx software you can auto-configure the serial connection and thus automatically find the proper serial port configuration.

Two important aspects of the communication link must be considered, namely, the RS-232 standard and the communications protocol. The *RS-232 standard* specifies a function for each of the wires inside the standard

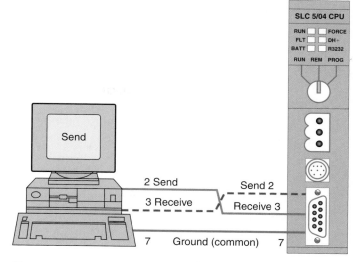

Figure 13-34 Serial wiring connection.

communications cable and their associated pins. *Communications protocol* is a standardized method for transmitting data and/or establishing communications between different devices.

Minimum configuration for two-way communications requires the use of only three connected wires, as shown in Figure 13-34. For ease of connection, the RS-232 standard specifies that computer devices have male connectors and that peripheral equipment have female connectors. Direct communication between two computers, such as a PC and a PLC, does not involve intermediate peripheral equipment. Therefore, a serial null-modem type cable must be used for the connection because both the PC and the PLC processor use pin 2 for data output and pin 3 for data input.

1. Why are PLCs installed within an enclosure?
2. What methods are used to keep enclosure temperatures within allowable limits?
3. State two ways in which electrical noise may be coupled into a PLC control system.
4. List three potential noise-generating inductive devices.
5. Describe four ways in which careful wire routing can help cut down on electrical noise.
6. **a.** What type of input field devices and output modules are most likely to have a small leakage current flow when they are in the off state? Why?
   b. Explain how an input bleeder resistor reduces leakage current.
7. Summarize the basic grounding requirements for a PLC system.
8. Under what condition can a ground loop circuit be developed?
9. When line voltage variations to the PLC power supply are excessive, what can be done to solve the problem?
10. What operating state will cause an inductive load to generate a very high voltage spike?
11. Explain how a diode is connected to suppress a DC inductive load.
12. Explain how a MOV suppresses an AC inductive load.
13. What is the purpose of PLC editing functions?
14. What is involved with commissioning a PLC system?
15. **a.** Compare offline and online programming.
    b. Which method is safer? Why?
16. List four uses for the data monitor function.

17. What information is provided by the cross reference function?
18. What information is provided by the contact histogram function?
19. List five preventive maintenance tasks that should be carried out on the PLC installation regularly.
20. Outline the general procedure followed to lock out and tag a PLC installation.
21. Typically, what does each of the following processor diagnostic light states indicate?
    a. RUN light is off.
    b. Fault light is off.
    c. BATTERY light is on.
22. When a processor comes equipped with a fault relay, what are the relay contacts used for?
23. Explain the function of a watchdog timer circuit.
24. A PLC is operating in the RUN mode but output devices do not operate as programmed. List five faults that could be responsible for this condition.
25. What is the verify results function used for?
26. A fast-acting solenoid-operated gate is suspected of not functioning properly when energized and de-energized by the PLC program. Explain how you would use the force function to check its operation.
27. What happens when the processor encounters a temporary end instruction?
28. Explain the function of the suspend instruction.
29. In what negative ways can faulty wiring and connections affect signals sent to and from the I/O modules?
30. The same address is used for two coil instructions within the same PLC program. What will happen as a consequence of this?
31. Compare the uses for RSLogix and RSLinx programming software.

1. The enclosure door of a PLC installation is not kept closed. What potential problem could this create?
2. A fuse is blown in an output module. Suggest two possible reasons why the fuse blew.
3. Whenever a crane located over a PLC installation is started from a standstill, temporary malfunction of the PLC system occurs. What is one likely cause of the problem?

4. During the static checkout of a PLC system, a specific output is forced on by the programming device. If an indicator other than the expected one turns on, what is the probable problem?

5. The input device to a module is activated, but the LED status indicator does not come on. A check of the voltage to the input module indicates that no voltage is present. Suggest two possible causes of the problem.

6. An output is forced on. The module logic light comes on, but the field device does not work. A check of the voltage on the output module indicates the proper voltage level. Suggest two possible causes of the problem.

7. A specific output is forced on, but the LED module indicator does not come on. A check of the voltage at the output module indicates a voltage far below the normal on level. What is the first thing to check?

8. An electronic-based input sensor is wired to a high-impedance PLC input and is falsely activating the input. How can this problem be corrected?

9. An LED logic indicator is illuminated, and according to the programming device monitor, the processor is not recognizing the input. If a replacement module does not eliminate the problem, what two other items should be suspected?

10. a. A normally open field limit switch examined for an on state normally cycles from on to off five times during one machine cycle. How could you tell by observing the LED status light that the limit switch is functioning properly?
    b. How could you tell by observing the programming device monitor that the limit switch is functioning properly?
    c. How could you tell by observing the LED status light whether the limit switch was stuck open?
    d. How could you tell by observing the programming device monitor whether the limit switch was stuck open?
    e. How could you tell by observing the LED status light if the limit switch was stuck closed?

f. How could you tell by observing the programming device monitor if the limit switch was stuck closed?

11. Assume that prior to putting a PLC system into operation, you want to verify that each *input device* is connected to the correct input terminal and that the input module or point is functioning properly. Outline a method of carrying out this test.

12. Assume that prior to putting a PLC system into operation, you want to verify that each *output device* is connected to the correct output terminal and that the output module or point is functioning properly. Outline a method of carrying out this test.

13. With reference to the ladder logic program of Figure 13-35, add instructions to modify the program to ensure that the second pump_2 does not run while pump_1 is running. If this condition occurs, the program should suspend operation and enter code identification number 100 into S2:7.

14. The program of Figure 13-36 is supposed to execute to sequentially turn PL1 off for 5 seconds and on for 10 seconds whenever input A is closed.
    a. Examine the ladder logic and describe how the circuit would operate as programmed.
    b. Troubleshoot the program and identify what needs to be changed to have it operate properly.

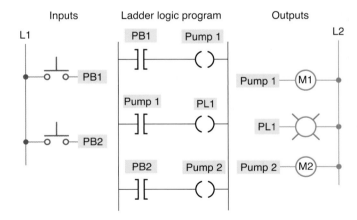

Figure 13-35 Program for Problem 13.

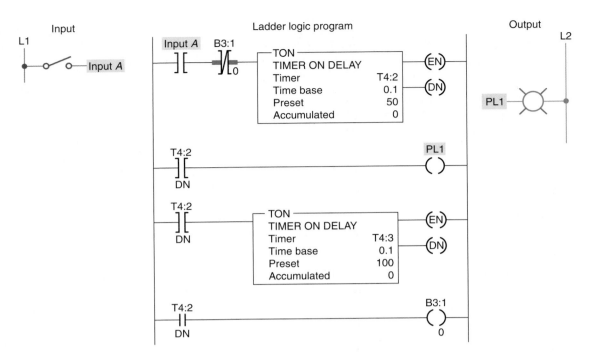

Figure 13-36 Program for Problem 14.

14

Process Control, Network Systems, and SCADA

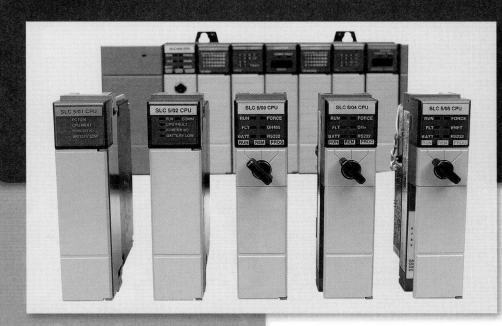

Image Used with Permission of Rockwell Automation, Inc.

Chapter Objectives

After completing this chapter, you will be able to:

14.1 Discuss the operation of continuous process, batch production, and discrete manufacturing processes

14.2 Compare individual, centralized, and distributive control systems

14.3 Explain the functions of the major components of a process control system

14.4 Describe the various functions of electronic HMI screens

14.5 Recognize and explain the functions of the control elements of a closed-loop control system

14.6 Explain how on/off control works

14.7 Explain how PID control works

This chapter introduces the kinds of industrial processes that can be PLC controlled. SCADA is such a process. Different types of control systems are used for complex processes. These control systems may be PLCs, but other controllers include robots, data terminals, and computers. For these controllers to work together, they must communicate. This chapter will discuss the different kinds of industrial processes and the means by which they communicate.

14.1 Types of Processes

Process control is the automated control of a process. Such systems typically deal with analog signals from sensors. The ability of a PLC to perform math functions and utilize analog signals makes it ideally suited for this type of control. Manufacturing is based on a series of processes being applied to raw materials. Typical applications of process control systems include automobile assembly, petrochemical production, oil refining, power generation, and food processing.

A *continuous process* is one in which raw materials enter one end of the system and the finished product comes out the other end of the system; the process itself runs continuously. Figure 14-1 shows a continuous process used in an automotive engine assembly line. Parts are mounted sequentially, in an assembly-line fashion, through a series of stations. Assembly and adjustments are carried out by both automated machine and manual operations.

In *batch processing,* there is no flow of product material from one section of the process to another. Instead, a set amount of each of the inputs to the process is received in a batch, and then some operation is performed on the batch to produce a product. Products produced using the batch process include food, beverages, pharmaceutical products, paint, and fertilizer. Figure 14-2 shows an example of a batch process. Three ingredients are mixed together, heated, and then stored. Recipes are the key to producing batches as each batch may have different characteristics by design.

Discrete manufacturing is characterized by individual or separate unit production. With this manufacturing process, a series of operations produces a useful output product. Discrete manufacturing systems typically deal with digital inputs to PLCs that cause motors and robotic devices to be activated. The work piece is normally a discrete part that must be handled on an individual basis. Making

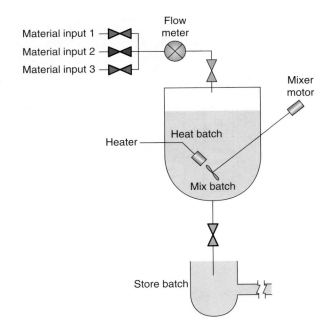

Figure 14-2   Batch process.

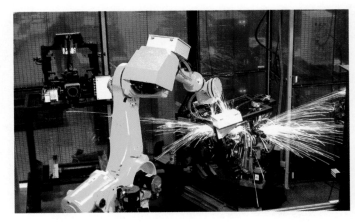

Figure 14-3   Discrete manufacturing.
Source: Courtesy Automation IG.

car interiors, as illustrated in Figure 14-3, is one example of discrete manufacturing.

Possible control configurations include individual, centralized, and distributed. *Individual control* is used to control a single machine. This type of control does not normally require communication with other controllers. Figure 14-4 shows an individual control application for a cut to length operation. The operator enters the feed length and batch count via the interface control panel and then presses the start button to initiate the process. Stock lengths vary so the operator needs to select the length and the number of pieces to be cut.

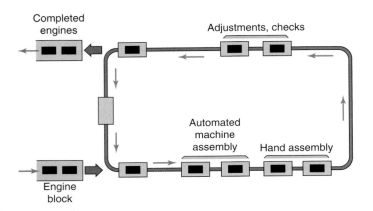

Figure 14-1   Continuous process.

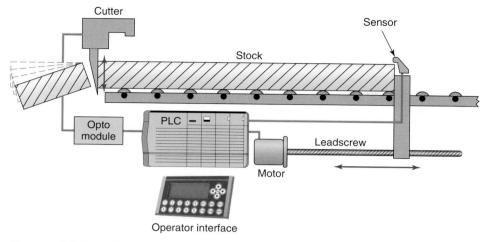

Figure 14-4 Individual control.

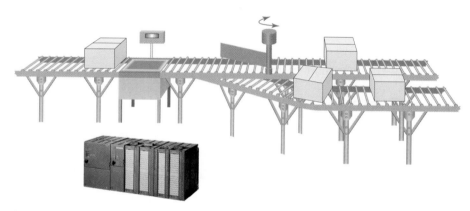

Figure 14-5 Centralized control.
Source: Courtesy Siemens.

Centralized control is used when several machines or processes are controlled by one central controller. The control layout uses a single, large control system to control many diverse manufacturing processes and operations, as illustrated in Figure 14-5. The main features of centralized control can be summarized as follows:

- Each individual step in the manufacturing process is handled by a central control system controller.
- No exchange of controller status or data is sent to other controllers.
- If the main controller fails, the whole process stops.

A *distributive control system (DCS)* is a network-based system. Distributive control involves two or more PLCs communicating with each other to accomplish the complete control task, as illustrated in Figure 14-6. Each PLC controls different processes locally and the PLCs are constantly exchanging information through the communications link and reporting on the status of the process. The

main features of a distributive control system can be summarized as follows:

- Distributive control permits the distribution of the processing tasks among several controllers.
- Each PLC controls its associated machine or process.
- High-speed communication among the computers is done through CAT-5 or CAT-6 twisted pair wires, single coaxial cables, fiber optics, or the Ethernet.
- Distributive control drastically reduces field wiring and heightens performance because it places the controller and I/O close to the machine process being controlled.
- Depending on the process, one PLC failure would not necessarily halt the complete process.
- DCS is supervised by a host computer that may perform monitoring/supervising functions such as report generation and storage of data.

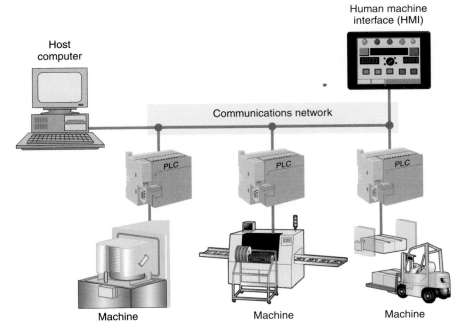

Figure 14-6   Distributive control system (DCS).

14.2   Structure of Control Systems

Process control normally applies to the manufacturing or processing of products in industry. In the case of a programmable controller, the process or machine is operated and supervised under the control of the user program. The major components of a process control system include the following:

Sensors

- Provide inputs from the process and from the external environment
- Convert physical information such as pressure, temperature, flow rate, and position into electrical signals

Human Machine Interface (HMI)

- Allows human inputs through various types of programmed switches, controls, and keypads to set up the starting conditions or alter the control of a process

Signal Conditioning

- Involves converting input and output signals to a usable form
- May include signal-conditioning techniques such as amplification, attenuation, filtering, scaling, A/D and D/A converters

Actuators

- Convert system output electrical signals into physical action

- Process actuators that include flow control valves, pumps, positioning drives, variable speed drives, clutches, brakes, solenoids, stepping motors, and power relays

Controller

- Makes the system's decisions based on the input signals
- Generates output signals that operate actuators to carry out the decisions

Human machine interface (HMI) equipment provides a control and visualization interface between a human and a process (Figure 14-7). HMIs allow operators to control,

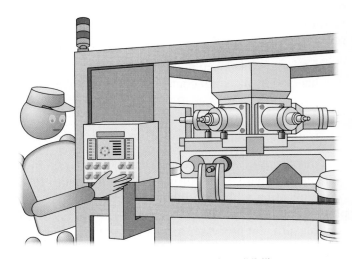

Figure 14-7   Human machine interface (HMI).

monitor, diagnose, and manage the application. Depending on the requirements and complexity of the process, the operator may be required to:

- Stop and start the process.
- Operate the controls and make the adjustments required for the process and monitor its progress.
- Detect abnormal situations and undertake corrective action.

Graphic HMI terminals offer electronic interfacing in a wide variety of sizes and configurations. They replace traditional wired panels with a touch screen with graphical representations of switches and indicators. Types of graphical display screens include the following:

Operational Summary—used to monitor the process.

Configuration/Setup—textual in nature used to detail process parameters.

Alarm Summary—provides a list of time-stamped active alarms.

Event History—presents a time-stamped list of all significant events that have occurred in the process.

Trend Values—displays information on process variables, such as flow, temperature, and production rate, over a period of time.

Manual Control—generally available only to maintenance personnel and meant to bypass parts of the automatic control system.

Diagnostics—used by maintenance personnel to diagnose equipment failures.

Graphic terminals come fully packaged with hardware, software, and communications. Figure 14-8 shows the Allen-Bradley family of PanelView graphic terminals. The setup varies with the vendor. In general, the tasks required to develop an HMI application include:

- Establish a communication link with the PLCs
- Create the tag addresses database

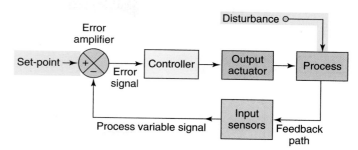

Figure 14-9   Closed-loop control system.

- Edit and create graphical objects on the screens
- Animate the objects

Most control systems are closed loop in that they utilize feedback in which the output of a process affects the input control signal. A closed-loop system measures the actual output of the process and compares it to the desired output. Adjustments are made continuously by the control system until the difference between the desired and actual output falls within a predetermined tolerance.

Figure 14-9 illustrates an example of a closed-loop control system. The actual output is sensed and fed back to be subtracted from the set-point input that indicates what output is desired. If a difference occurs, a signal to the controller causes it to take action to change the actual output until the difference is 0. The operation of the component parts are as follows:

Set-point—The input that determines the desired operating point for the process.

Process Variable—Refers to the feedback signal that contains information about the current process status.

Error Amplifier—Determines whether the process operation matches the set-point. The magnitude and polarity of the error signal will determine how the process will be brought back under control.

Controller—Produces the appropriate corrective output signal based on the error signal input.

Figure 14-8   PanelView graphic terminals.
Source: Image Used with Permission of Rockwell Automation, Inc.

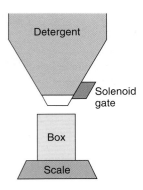

Figure 14-10 Container-filling process.

Output Actuator—The component that directly affects a process change. Examples are motors, heaters, fans, and solenoids.

The process shown in Figure 14-10 is an example of a closed-loop continuous control process used to automatically fill box containers to a specified weight of detergent. An empty box is moved into position and filling begins. The weight of the box and contents is monitored. When the actual weight equals the desired weight, filling is halted.

Operation and block diagrams for the container-filling process are shown in Figure 14-11. The operation of the process can be summarized as follows:

- A sensor attached to the scale weighing the container generates the voltage signal or digital code

that represents the weight of the container and contents.

- The sensor signal is subtracted from the voltage signal or digital code that has been input to represent the desired weight.

- As long as the difference between the input signal and feedback signal is greater than 0, the controller keeps the solenoid gate open.

- When the difference becomes 0, the controller outputs a signal that closes the gate.

Virtually all feedback controllers determine their output by observing the error between the set-point and a measurement of the process variable. Errors can occur when an operator changes the set-point or when a disturbance or a load on the process changes the process variable. The controller's role is to eliminate the error automatically.

14.3 On/Off Control

With *on/off controllers* the final control element is either on or off—one for the occasion when the value of the measured variable is above the set-point and the other for the occasion when the value is below the set-point. The controller will never keep the final control element in an intermediate position. Controlling activity is achieved by the period of on-off cycling action.

Figure 14-12 shows a system using on/off control in which a liquid is heated by steam. The operation of the process can be summarized as follows:

- If the liquid temperature goes below the set-point, the steam valve opens and the steam is turned on.

- When the liquid temperature goes above the set-point, the steam valve closes and the steam is shut off.

- The on/off cycle will continue as long as the system is operating.

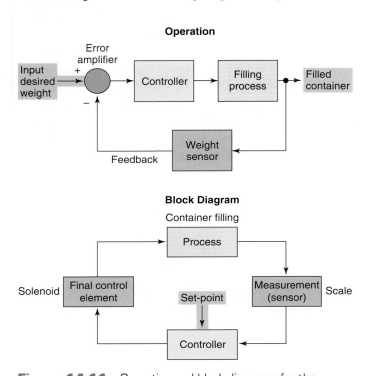

Figure 14-11 Operation and block diagrams for the container-filling process.

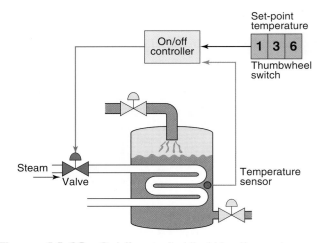

Figure 14-12 On/off controlled liquid heating system.

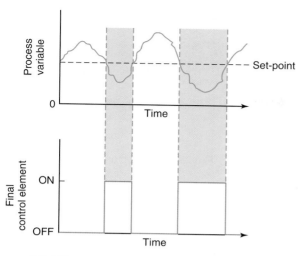

Figure 14-13 On/off control response.

Figure 14-13 illustrates the control response for an on/off temperature controller. The action of the control response can be summarized as follows:

- The output turns on when the temperature falls below the set-point and turns off when the temperature reaches the set-point.
- Control is simple, but overshoot and cycling about the set-point can be disadvantageous in some processes.
- The measured variable will oscillate around the set-point at an amplitude and frequency that depend on the capacity and time response of the process.
- Oscillations may be reduced in amplitude by increasing the sensitivity of the controller. This increase will cause the controller to turn on and off more often, a possibly undesirable result.
- On/off control is used when a more precise control is unnecessary.

A *deadband* is usually established around the set-point. The deadband of the controller is usually a selectable value that determines the error range above and below the set-point that will not produce an output as long as the process variable is within the set limits. The inclusion of deadband eliminates any hunting by the control device around the set-point. Hunting occurs when minor adjustments of the controlled position are continually made due to minor fluctuations.

14.4 PID Control

Proportional controllers are designed to eliminate the hunting or cycling associated with on/off control. They allow the final control element to take intermediate positions between on and off. Proportioning action permits *analog control* of the final control element to vary the amount of energy to the process, depending on how much the value of the measured variable has shifted from the desired value.

A proportional controller allows tighter control of the process variable because its output can take on any value between fully on and fully off, depending on the magnitude of the error signal. Figure 14-14 shows an example of a motor-driven analog proportional control valve used as a final control element. The action of the control valve actuator can be summarized as follows:

- The actuator receives an input current between 4 mA and 20 mA from the controller.
- In response, it provides linear control of the valve.
- A value of 4 mA corresponds to a minimum value opening (often 0) and 20 mA corresponds to a maximum value opening (full scale).
- The 4 mA lower limit allows the system to detect opens. If the circuit is open, 0 mA would result, and the system can issue an alarm.
- Because the signal is a current, it is unaffected by reasonable variations in connecting wire resistance and is less susceptible to noise pickup from other signals than is a voltage signal.

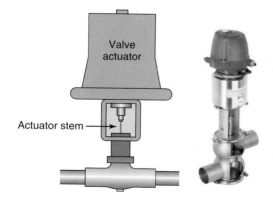

Actuator current (mA)	Valve response (% open)
4	0
6	12.5
8	25
10	37.5
12	50
14	62.5
16	75
18	87.5
20	100

Figure 14-14 Motor-driven analog proportional control valve.
Source: Courtesy GEA Tuchenhagen.

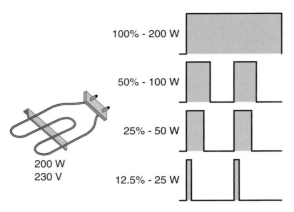

Figure 14-15 Time proportioning of a heater element.

Proportioning action can also be accomplished by turning the final control element on and off for short intervals. This *time proportioning* (also known as *pulse width modulation*) varies the ratio of on time to off. Figure 14-15 shows an example of time proportioning used to produce varying wattage from a 200 watt heater element as follows:

- To produce 100 watts the heater must be on 50% of the time.
- To produce 50 watts the heater must be on 25% of the time.
- To produce 25 watts the heater must be on 12.5% of the time.

Proportioning action occurs within a proportional band around the set-point. The table of Figure 14-16 is an example of the proportional band for a heating application with a set-point of 500°F and a proportional band of 80°F (±40°F). Proportioning action can be summarized as follows:

- Outside proportional band, the controller functions as an on/off unit, with the output either fully on (below the band) or fully off (above the band).

Time proportional				4–20 mA proportional	
Percent on	On time (seconds)	Off time (seconds)	Temp. (°F)	Output level	Percent output
0.0	0.0	20.0	over 540	4 mA	0.0
0.0	0.0	20.0	540.0	4 mA	0.0
12.5	2.5	17.5	530.0	6 mA	12.5
25.0	5.0	15.0	520.0	8 mA	25.0
37.5	7.5	12.5	510.0	10 mA	37.5
50.0	10.0	10.0	500.0	12 mA	50.0
62.5	12.5	7.5	490.0	14 mA	62.5
75.0	15.0	5.0	480.0	16 mA	75.0
87.5	17.5	2.5	470.0	18 mA	87.5
100.0	20.0	0.0	460.0	20 mA	100.0
100.0	20.0	0.0	under 460	20 mA	100.0

Figure 14-16 Proportional band for a heating application.

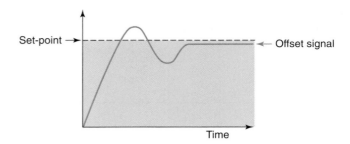

Figure 14-17 Proportional control steady-state error.

- Within the proportional band the output is turned on and off in the ratio of the measurement difference from the set-point.
- At the set-point (the midpoint of the proportional band), the output on:off ratio is 1:1; that is, the on time and off time are equal.
- If the temperature is further from the set-point, the on and off times vary in proportion to the temperature difference.
- If the temperature is below the set-point, the output will be on longer; if the temperature is too high, the output will be off longer.

In theory, a proportional controller should be all that is needed for process control. Any change in system output is corrected by an appropriate change in controller output. Unfortunately, the operation of a proportional controller leads to a steady-state error known as *offset,* or *droop.* This steady-state error is the difference between the attained value of the controller and the required value that results in an offset signal that is slightly lower than the set-point value, as illustrated in Figure 14-17. Depending on the PLC application, this offset may or may not be acceptable.

The process of Figure 14-18 illustrates what effect a proportional control steady-state error might have on a tank-filling operation. It may require an operator to make a small adjustment (manual reset) to bring the controlled variable to the set-point on initial start-up, or whenever

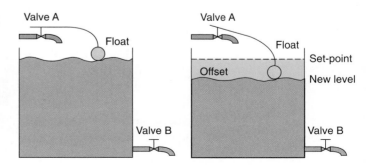

Figure 14-18 Proportional control tank-filling operation.

the process conditions change significantly. The operation can be summarized as follows:

- When valve B opens liquid flows out and the level in the tank drops.
- This causes the float to lower, opening valve A and allowing more liquid in.
- This process continues until the level drops to a point at which the float is low enough to open valve A, thus allowing the same input flow as output flow.
- Due to the steady-state error, the level will stabilize at a new lower level, not at the desired set-point.

Proportional control is often used in conjunction with integral control and/or derivative control.

- The *integral action,* sometimes termed reset action, responds to the size and time duration of the error signal. An error signal exists when there is a difference between the process variable and the set-point, so the integral action will cause the output to change and continue to change until the error no longer exists. Integral action eliminates steady-state error. The amount of integral action is measured as minutes per repeat or repeats per minute, which is the relationship between changes and time.
- The *derivative action* responds to the speed at which the error signal is changing—that is, the greater the error change, the greater the correcting output. The derivative action is measured in terms of time.

Proportional plus integral (PI) control combines the characteristics of both types of control. A step change in the set-point causes the controller to respond proportionally, followed by the integral response, which is added to the proportional response. Because the integral mode determines the output change as a function of time, the more integral action found in the control, the faster the output changes. This action can be summarized as follows:

- To eliminate the offset error, the controller needs to change its output until the process variable error is zero.
- Reset integral control action changes the controller output by the amount needed to drive the process variable back to the set-point value.
- The new equilibrium point after reset action is at point "C."
- Since the proportional controller must always operate on its proportional band, the proportional band must be shifted to include the new point "C."
- A controller with reset integral control does this automatically.

Rate action (derivative control) acts on the error signal just like reset does, but rate action is a function of the rate of change rather than the magnitude of error. Rate action is applied as a change in output for a selectable time interval, usually stated in minutes. Rate-induced change in controller output is calculated from the derivative of the error. Input change, rather than proportional control error change, is used to improve response. Rate action quickly positions the output, whereas proportional action alone would eventually position the output. In effect, rate action puts the brakes on any offset or error by quickly shifting the proportional band. *Proportional plus derivative (PD) control* is used in process control systems with errors that change very rapidly. By adding derivative control to proportional control, we obtain a controller output that responds to the error's rate of change as well as to its magnitude.

PID control is a feedback control method that combines proportional, integral, and derivative actions. The proportional action provides smooth control without hunting. The integral action automatically corrects offset. The derivative action responds quickly to large external disturbances. The PID controller is the most widely used type of process controller. When combined into a single control loop the proportional, integral and derivative modes complement each other to reduce the system error to zero faster than any other controller. Figure 14-19 shows the block diagram of a PID control loop, the operation of which can be summarized as follows:

- During setup, the set-point, proportional band, reset (integral), rate (derivative), and output limits are specified.
- All these can be changed during operation to tune the process.
- The integral term improves accuracy, and the derivative reduces overshoot for transient upsets.
- The output can be used to control valve positions, temperature, flow metering equipment, and so on.

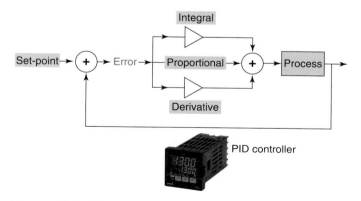

Figure 14-19 PID control loop.

- PID control allows the output power level to be varied.
- As an example, assume that a furnace is set at 50°C.
- The heater power will increase as the temperature falls below the 50°C set-point.
- The lower the temperature the higher the power.
- PID has the effect of gently turning the power down as the signal gets close to the set-point.

The long-term operation of any system, large or small, requires a mass-energy balance between input and output. If a process were operated at equilibrium at all times, control would be simple. Because change does occur, the critical parameter in process control is time, that is, how long it takes for a change in any input to appear in the output. System time constants can vary from fractions of a second to many hours. The PID controller has the ability to tune its control action to specific process time constants and therefore to deal with process changes over time. PID control changes the amount of output signal in a *mathematically* specified way that accounts for the amount of error and the rate of signal change.

Either programmable controllers can be fitted with input/output modules that produce PID control, or they will already have sufficient mathematical functions to allow PID control to be carried out. PID is essentially an equation that the controller uses to evaluate the controlled variable. Figure 14-20 illustrates how a programmable logic controller can be used in the control of a PID loop. The operation of the PID loop can be summarized as follows:

- The process variable (pressure) is measured and feedback is generated.

- The PLC program compares the feedback to the set-point and generates an error signal.
- The error is examined by the PID loop calculation in three ways: with proportional, integral, and derivative methodology.
- The controller then issues an output to correct for any measured error by adjustment of the position of the variable flow outlet valve.

The *response* of a PID loop is the rate at which it compensates for error by adjusting the output. The PID loop is adjusted or tuned by changing the proportional gain, the integral gain, and/or the derivative gain. A PID loop is normally tested by making an abrupt change to the set-point and observing the controller's response rate. Adjustments can then be made as follows:

- As the proportional gain is increased, the controller responds faster.
- If the proportional gain is too high, the controller may become unstable and oscillate.
- The integral gain acts as a stabilizer.
- Integral gain also provides power, even if the error is zero (e.g., even when an oven reaches its set-point, it still needs power to stay hot).
- Without this base power, the controller will droop and hunt for the set-point.
- The derivative gain acts as an anticipator.
- Derivative gain is used to slow the controller down when change is too fast.

Basically, PID controller tuning consists of determining the appropriate values for the gain (proportional band), rate (derivative), and reset time (integral) tuning

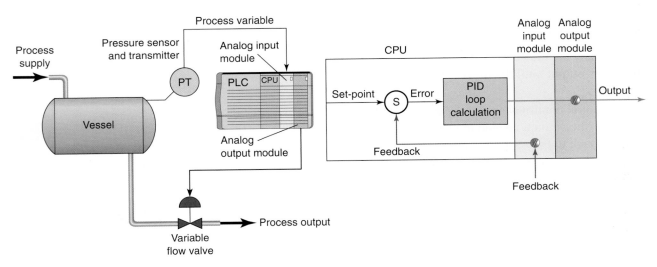

Figure 14-20 PLC control of a PID loop.

parameters (control constants) that will give the control required. Depending on the characteristics of the deviation of the process variable from the set-point, the tuning parameters interact to alter the controller's output and produce changes in the value of the process variable. In general, three methods of controller tuning are used:

Manual

- The operator estimates the tuning parameters required to give the desired controller response.
- The proportional, integral, and derivative terms must be adjusted, or tuned, individually to a particular system using a trial-and-error method.

Semiautomatic or Autotune

- The controller takes care of calculating and setting PID parameters.
  - Measures sensor output
  - Calculates error, sum of error, rate of change of error
  - Calculates desired power with PID equations
  - Updates control output

Fully Automatic or Intelligent

- This method is also known in the industry as fuzzy logic control.
- The controller uses artificial intelligence to readjust PID tuning parameters continually as necessary.
- Rather than calculating an output with a formula, the fuzzy logic controller evaluates rules. The first step is to "fuzzify" the error and change-in-error from continuous variables into linguistic variables, like "negative large" or "positive small." Simple if-then rules are evaluated to develop an output. The resulting output must be de-fuzzified into a continuous variable such as valve position.

The PID programmable controller output instruction uses closed-loop control to automatically control physical properties such as temperature, pressure, liquid level, or flow rate of process loops. Figure 14-21 shows the PID output instruction and setup screen associated with the

Allen-Bradley SLC 500 instruction set. The PID instruction is straightforward: it takes one input and controls one output. Normally, the PID instruction is placed on a rung without conditional logic. The output remains at its last value when the rung goes false. A summary of the basic information that is entered into the instruction is as follows:

Control Block—File that stores the data required to operate the instruction.

Process Variable—The element address that stores the process input value.

Control Variable—The element address that stores the output of the PID instruction.

Setup Screen—Instruction on which you can double-click to bring up a display that prompts you for other parameters you must enter to fully program the PID instruction.

14.5 Motion Control

A motion control system provides precise positioning, velocity, and torque control for a wide range of motion applications. PLCs are ideally suited for both linear and rotary motion control applications. *Pick and Place* machines are used in the consumer products industry for a wide variety of product transfer applications. The machine takes a product from one point to another. One example is the transfer of a product to a moving conveyor belt as illustrated in Figure 14-22.

A basic PLC motion control system consists of a controller, a motion module, a servo drive, one or more motors with encoders, and the machinery being controlled. Each motor controlled in the system is referred to as an

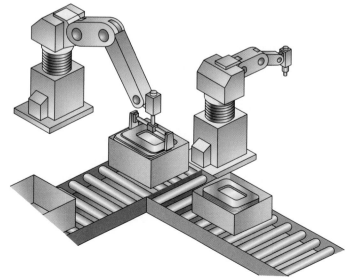

Figure 14-22 Pick and Place machine.

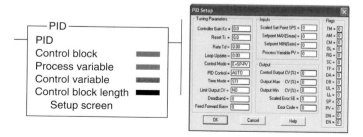

Figure 14-21 PID output instruction and setup screen.

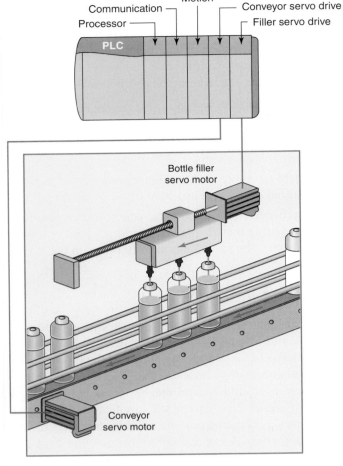

Figure 14-23 Bottle-filling motion control process.

axis of motion. Figure 14-23 illustrates a bottle-filling motion control process. This application requires two axes of motion: the motor operating the bottle filler mechanism and the motor controlling the conveyor speed. The role of each control component can be summarized as follows:

Programmable Logic Controller

- The controller stores and executes the user program that controls the process.
- This program includes motion instructions that control axis movements.
- When the controller encounters a motion instruction it calculates the motion commands for the axis.
- A motion command represents the desired position, velocity, or torque of the servo motor at the particular time the calculations take place.

Motion Module

- The motion module receives motion commands from the controller and transforms them into a compatible form the servo drive can understand.

- In addition it updates the controller with motor and drive information used to monitor drive and motor performance.

Servo Drive

- The servo drive receives the signal provided by the motion module and translates this signal into motor drive commands.
- These commands can include motor position, velocity, and/or torque.
- The servo drive provides power to the servo motors in response to the motion commands.
- Motor power is supplied and controlled by the servo drive.
- The servo drive monitors the motor's position and velocity by use of an encoder mounted on the motor shaft. This feedback information is used within the servo drive to ensure accurate motor motion.

Servo Motor

- The servo motors represent the axis being controlled.
- The servo motors receive electrical power from their servo drive which determines the motor shaft velocity and position.
- The filler motor must accelerate the filler mechanism in the direction the bottles are moving, match their speed, and track the bottles.
- After the bottles have been filled, the filler motor has to stop and reverse direction to return the filler mechanism to the starting position to begin the process again.

A robot is simply a series of mechanical links driven by servo motors. The basic industrial robot widely used today is an *arm* or *manipulator* that moves to perform industrial operations. Figure 14-24 illustrates the motion of

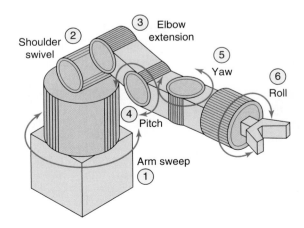

Figure 14-24 Six-axis robot arm.

a six-axis robot arm. Each axis of the robot arm is fundamentally a closed-loop servo control system. The wrist is the name usually given to the last three joints on the robot's arm. Going out along the arm, these wrist joints are known as the pitch joint, yaw joint, and roll joint. There are two types of controller setups that can be used to control an industrial robot—PLC- and PC-based systems. Depending on the difficulty of the task the robotic system will be performing, you may need a PLC or just a robot controller.

14.6 Data Communications

Data communications refers to the different ways that PLC microprocessor-based systems talk to each other and to other devices. The two general types of communications links that can be established between the PLC and other devices are point-to-point links and network links. Figure 14-25 illustrates a *point-to-point* serial communications link. Serial communications is used with devices such as printers, operator workstations, motor drives, bar code readers, computers, or another PLC. Serial communications interfaces are either built into the processor module or come as separate modules. A serial module installed in each controller is normally all that is required for two PLCs of the same manufacturer to establish a point-to-point link.

As control systems become more complex, they require more effective communications schemes between the system components. A *local area network* or *LAN* is a system that interconnects data communications components within a limited geographical area, typically

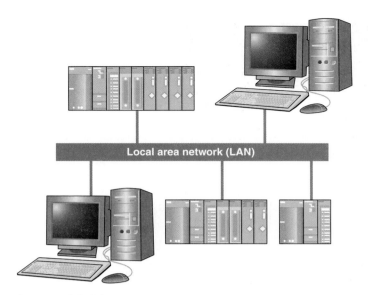

Figure 14-26 Local area network (LAN) communication link.

no more than one or two miles. Figure 14-26 illustrates a LAN communication link. Network communications supports communication among multiple PLCs and other devices. PLC networks allow:

- Sharing of information such as the current state of status bits among PLCs that may determine the action of one another.
- Monitoring of information from a central location.
- Programs to be uploaded or downloaded from a central location.
- Several PLCs to operate in unison to accomplish a common goal.

Transmission media are the cable through which data and control signals flow on a network. The transmission media used in data communications systems include coaxial cable, twisted pair, or fiber optics (Figure 14-27). Each cable has different electrical capabilities and may be more or less suitable to a specific environment or network type. Not all networks transmit information through cable. Wireless Wi-Fi Ethernet networks, such as the DF1

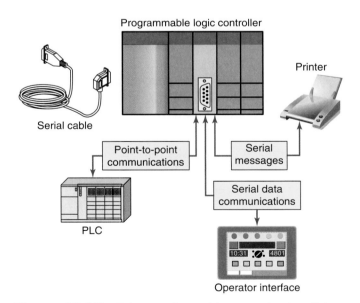

Figure 14-25 Point-to-point serial communications link.

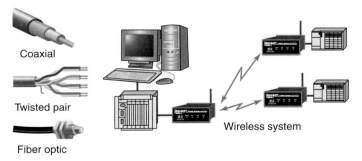

Figure 14-27 Transmission media.

Radio Modem, communicate through radio waves, which are transmitted through the air.

In industrial applications, LANs have most often been used as the communication system for distributed control systems (DCS). Recall that a DCS system uses individual controllers to control the subsystems of a machine or process. This approach contrasts with centralized control in which a single controller governs the entire operation. A second major use of local area networks is that of supervisory control and data acquisition (SCADA). A LAN allows data collection and processing for a group of controllers to be accomplished using one host computer as the central point for collecting data.

There are three general levels of functionality of industrial networks. Figure 14-28 shows an illustration of the three levels, which can be summarized as follows:

Device Level—The device level involves various sensor and actuator devices of machines and processes. These may include devices such as sensors, switches, drives, motors, and valves.

Control Level—The control level would be the networks industrial controllers are on. This level may include controllers such as PLCs and robot controllers. Communications on the control level includes sharing I/O and program data between controllers.

Information Level—The information level is a plantwide network typically composed of the company's business networks and computers. This level may include scheduling, sales, management, and corporatewide information.

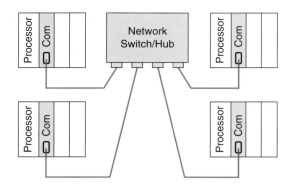

Figure 14-29 Star topology network.

Each device connected on a network is known as a *node* or *station*. As signals travel along a network cable, they degrade and become distorted in a process that is called attenuation. If a cable is long enough, the attenuation will finally make a signal unrecognizable. A *repeater* is a device that amplifies a signal to its original strength in order to enable its signals to travel further. Different network types will have different specifications for cable length and type without a repeater.

Network topology is the physical layout of devices on a network formed by the network cables when nodes are attached. The *star topology* illustrated in Figure 14-29 and its operation can be summarized as follows:

- A network controller switch or hub is connected to several PLC network nodes.

- Currently, most Ethernet networks use switches rather than hubs. A switch performs the same basic function as a hub but effectively increases the speed, size, and data handling capacity of the network.

- The configuration allows for bidirectional communication between switch/hub and each PLC.

- All transmission must be between the switch/hub and the PLCs because the network controller hub controls all communication.

- All transmissions must be sent to the switch/hub, which then sends them to the correct PLC.

- One problem with the star topology is that if the switch/hub goes down, the entire LAN is down.

- This type of system works best when information is transmitted primarily between the main controller and remote PLCs. However, if most communication is to occur between PLCs, the operation speed is affected.

- Also, the star system can use substantial amounts of communication conductors to connect all remote PLCs to one central location.

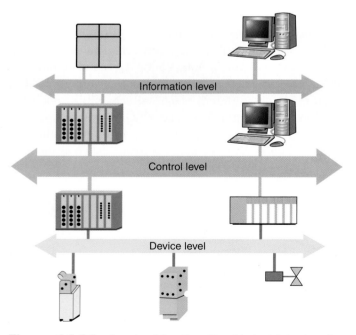

Figure 14-28 Levels of functionality of industrial networks.

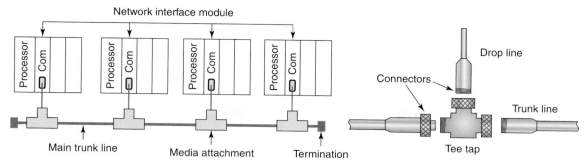

Figure 14-30 Bus topology network.

Bus topology, illustrated in Figure 14-30, is a network configuration in which all stations are connected in parallel with the communication medium and all stations receive information from every other station on the network. The operation of a bus topology network can be summarized as follows:

- Uses a single bus trunk cable to which individual PLC nodes are attached by a cable drop that taps off the main cable.
- Each PLC is interfaced to the bus using a network interface module that is attached using a drop cable or connector.
- Due to the nature of the bus technology, and the way the data are transmitted on the network, each end of the bus must be terminated with a terminating resistor.
- As the data move along the total bus, each PLC node is listening for its own node identification address and accepts only information sent to that address.
- Because of the simple linear layout, bus networks require less cable than all other topologies.
- No single station controls the network and stations can communicate freely to one another.
- Bus networks are very useful in distributive control systems, because each station or node has equal independent control capability and can exchange information at any given time.
- Another advantage of the bus network is that you can add or remove stations from the network with a minimum amount of system reconfiguration.
- This network's main disadvantage is that all the nodes rely on a common bus trunk line, and a break in that common line can affect many nodes.

I/O bus networks can be divided into two categories: device bus networks and process bus networks. *Device bus networks* interface with low-level information devices such as pushbuttons and limit switches that primarily

transmit data relating to the on/off state of the device and its operational status. Device bus networks can be further classified as bit-wide or byte-wide buses. Device bus networks that include discrete devices as well as small analog devices are called *byte-wide bus networks.* These networks can transfer 50 or more bytes of data at a time. Device bus networks that interface only with discrete devices are called *bit-wide bus networks.* Bit-wide networks transfer less than 8 bits of information to and from simple discrete devices.

Process bus networks are capable of communicating several hundred bytes of data per transmission. The majority of devices used in process bus networks are analog, whereas most devices used in device bus networks are discrete. Process bus networks connect with high-level information devices such as smart process valves and flowmeters, which are typically used in process control applications. Process buses are slower because of their large data packet size. Most analog control devices are used in controlling such process variables as flow and temperature, which are typically slow to respond.

A *protocol* is a set of rules that two or more devices must follow if they are to communicate with each other. Protocols are to computers what language is to humans. This book is in English, and to understand it, you must be able to read English. Similarly, for two devices on a network to successfully communicate, they must both understand the same protocols.

A network protocol defines how data is arranged and coded for transmission on a network. In the past, communications networks were often proprietary systems designed to a specific vendor's standards; users were forced to buy all their control components from a single supplier. This is because of the different communications protocols, command sequences, error-checking schemes, and communications media used by each manufacturer. Today, the trend is toward open network systems based on international standards developed through industry associations.

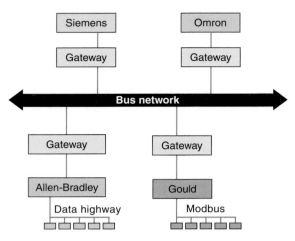

Figure 14-31 Translating from one network-access scheme to another.

Gateways (Figure 14-31) make communication possible between different architectures and protocols. They repackage and convert data going from one network to another network so that the one can understand the other's application data. Gateways can change the format of a message so that it will conform to the application program at the receiving end of the transfer. If network-access translation is their only function, the interfaces are known as *bridges*. If the interface also adjusts data formats or performs data transmission control, then it is called a *gateway*.

A bus topology network requires some method of controlling a particular device's access to the bus. An *access method* is the manner in which a PLC accesses the network to transmit information. Network access control ensures that data are transmitted in an organized manner preventing the occurrence of more than one message on the network at a time. Although many access methods exist, the most common are token passing, collision detection, and polling.

In a *token passing* network, a node can transmit data on the network only when it has possession of a token. A token is simply a small packet that is passed from node to node as illustrated in Figure 14-32. When a node finishes transmitting messages, it sends a special message to the

next node in the sequence, granting it the token. The token passes sequentially from node to node, allowing each an opportunity to transmit without interference. Tokens usually have a time limit to prevent a single node from tying up the token for a long period of time.

Ethernet networks use a *collision detection* access control scheme. With this access method, nodes listen for activity on the network and transmit only if there are no other messages on the network. On Ethernet networks there is the possibility that nodes will transmit data at the same time. When this happens a collision is detected. Each node that had sent out a message will wait a random amount of time and will resend its data if it does not detect any network activity.

The access method most often used in master/slave protocols is *polling*. The master/slave network is one in which a master controller controls all communications originating from other controllers. This configuration is illustrated in Figure 14-33 and consists of several slave controllers and one master controller. Its operation can be summarized as follows:

- The master controller sends data to the slave controllers.
- When the master needs data from a slave, it will *poll* (address) the slave and wait for a response.
- No communication takes place without the master initiating it.
- Direct communication among slave devices is not possible.
- Information to be transferred between slaves must be sent first to the network master unit, which will,

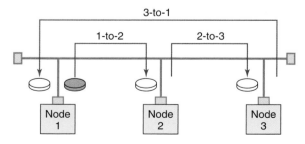

Figure 14-32 Example of token passing.

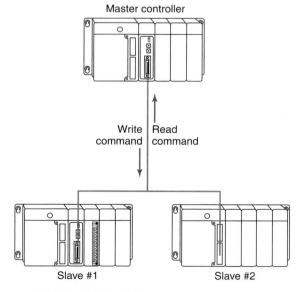

Figure 14-33 Master/slave network.

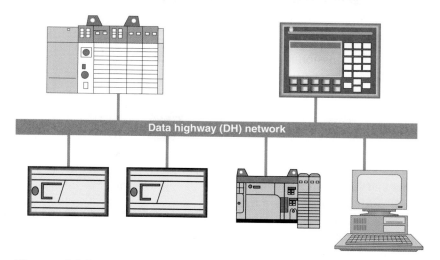

Figure 14-34 Peer-to-peer network.

in turn, retransmit the message to the designated slave device.

- Master/slave networks use two pairs of conductors. One pair of wires is used for the master to transmit data and the slave to receive them. On the other pair, the slaves transmit and the master receives.

A peer-to-peer network has a distributive means of control, as opposed to a master/slave network in which one node controls all communications originating from other nodes. The Allen-Bradley Data Highway, shown in Figure 14-34, is an example of a peer-to-peer network of programmable controllers and computers linked together to form a data communication system. The operation of the network can be summarized as follows:

- Peer-to-peer networks use the token passing media access method.
- Each device has the ability to request use of, and then take control of, the network for the purpose of transmitting information to or requesting information from other network devices.
- Each device is identified by an address.
- When the network is operating, the token passes from one device to the next sequentially.
- The device that is transmitting the token also knows the address of the next station that will receive the token.
- Each device receives the packet information and uses it, if needed.
- Any additional information that the node has will be sent in a new packet.

There are two methods of transmitting PLC digital data: parallel and serial transmission. In *parallel* data transmission, all bits of the binary data are transmitted

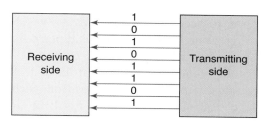

Figure 14-35 Parallel data transmission.

simultaneously, as illustrated in Figure 14-35. Parallel transmission of data can be summarized as follows:

- Eight transmission lines are required to transmit the 8-bit binary number.
- Each bit requires its own separate data path and all bits of a word are transmitted at the same time.
- Parallel data transmission is less common but faster than serial transmission.
- A common example of parallel data transmission is the connection between a computer and a printer.

In *serial* transmission one bit of the binary data is transferred at a time, as illustrated in Figure 14-36. Serial transmission of data can be summarized as follows:

- In serial transmission, bits are sent sequentially on the same channel (wire) which reduces costs for wire but also slows the speed of transmission.

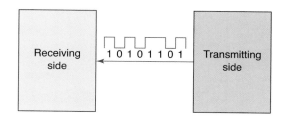

Figure 14-36 Serial data transmission.

- Serial data can be transmitted effectively over much greater distances than can parallel data.
- Each data word in the serial transmission must be denoted with a known start bit sequence followed by the data bits that contain the intelligence of the data transmission and a stop bit.
- An extra bit, termed a *parity bit,* may be used to provide some error-detecting ability.

A *duplex* communication system is a system composed of two connected devices that can communicate with one another in both directions at the same time. A *half-duplex* system provides for communication in both directions, but only one direction at a time (not simultaneously). Half-duplex transmission is use for master/slave communications. *Full-duplex* transmission allows the transmission of data in both directions simultaneously and can be used for peer-to-peer communications.

The different networking schemes replace traditional point-to-point hardwiring. Network control of systems minimizes the amount of wiring that needs to be done. With traditional wiring multiple wires from each device, fed through control cabinets, often result in large wire bundles running through the system. Due to the sheer volume of wires, installation time is considerable and troubleshooting is complex. If a network is used all devices can be directly connected to a single transmission media cable.

High-speed industrial networking technologies offer a variety of methods for connecting devices. PLC network configurations may be either open or proprietary (vendor-unique). Following is an overview of some of the industrial communication technologies that play a critical role in today's control systems.

Data Highway

The Allen-Bradley Data Highway networks, Data Highway Plus (DH+) and DH-485, are proprietary communications networks. They use peer-to-peer communication implementing token passing. The medium is shielded twisted pair cable. Figure 14-37 shows the DH+ network

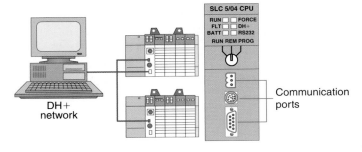

Figure 14-37 Data Highway network connection.

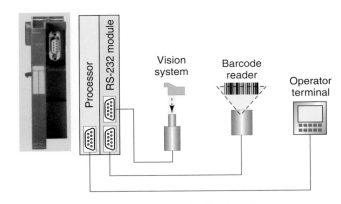

Figure 14-38 Serial communication interface.
Source: Courtesy Siemens.

connection for a SLC 5/04 controller. The three-pin Phoenix connector is used to form the network transmission media.

Serial Communication

Serial data communication is implemented using standards such as RS-232, RS-422, and RS-485. The RS in the standard's name means *recommended standard* that specifies the electrical, mechanical, and functional characteristics for serial communications. Serial communication interfaces are either built into the processor module or come as a separate communications interface module, as illustrated in Figure 14-38. The simplest type of connection is the RS-232 serial port. The RS interfaces are used to connect to devices such as vision systems, barcode readers, and operator terminals that must transfer quantities of data at a reasonably high rate between the remote device and the PLC. The RS-232 type of serial transmission is designed to communicate between one computer and one controller and is usually limited to lengths up to 50 feet. RS-422 and RS-485 serial transmission types are designed to communicate between one computer and multiple controllers, have a high level of noise immunity, and are usually limited to lengths of 650 feet (for RS-485) or 1650 feet (for RS-422).

DeviceNet

DeviceNet is an open device-level network. It is relatively low speed but efficient at handling the short messages to and from I/O modules. As PLCs have become more powerful, they are being required to control an increasing number of I/O field devices. Therefore, at times it may not be practical to separately wire each sensor and actuator directly into I/O modules. Figure 14-39 shows a comparison between conventional and DeviceNet I/O systems. Conventional systems have racks of inputs and outputs with each I/O device wired back to the controller. The DeviceNet protocol dramatically reduces costs

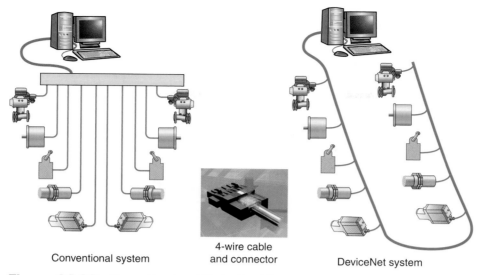

Conventional system

4-wire cable
and connector

DeviceNet system

Figure 14-39 Conventional and DeviceNet I/O systems.
Source: Photo courtesy Omron Industrial Automation, **www.ia.omron.com**.

by integrating all I/O devices on a 4-wire trunk network with data and power conductors in the same cable. This direct connectivity reduces costly and time-consuming wiring.

The basic function of a DeviceNet I/O bus network is to communicate information with, as well as supply power to, the field devices that are connected to the bus. The PLC drives the field devices directly with the use of a *network scanner* instead of I/O modules, as illustrated in Figure 14-40. The scanner module communicates with DeviceNet devices over the network to:

- Read inputs from a device.
- Write outputs to a device.
- Download configuration data.
- Monitor a device's operational status.

The scanner module communicates with the controller to exchange information which includes:

- Device I/O data
- Status information
- Configuration data

DeviceNet also has the unique feature of having power on the network. This allows devices with limited power requirements to be powered directly from the network, further reducing connection points and physical size.

DeviceNet uses the Common Industrial Protocol, called *CIP*, which is strictly object oriented. Each object has attributes (data), services (commands), and behavior (reaction to events). Two different types of objects are defined in the CIP specification: communication objects and application-specific objects. A DeviceNet network can support up to 64 nodes and the network end-to-end distance is variable, based on network speed. Figure 14-41 shows an example of a typical layout of the trunk wiring for a DeviceNet network. Communications data is carried over two wires with a second pair of wires carrying power.

The field devices that are connected to the network contain intelligence in the form of microprocessors or

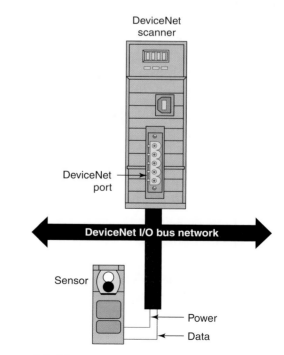

Figure 14-40 DeviceNet network scanner.

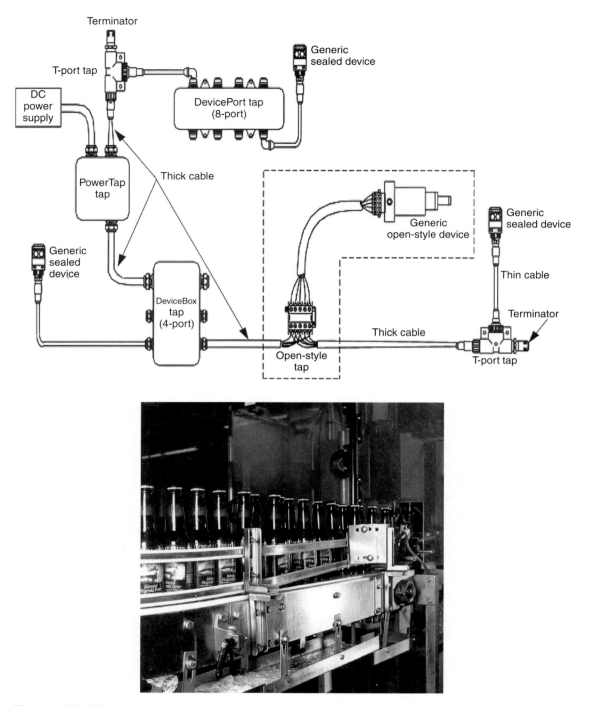

Figure 14-41 Layout of a DeviceNet network.
Source: Image Used with Permission of Rockwell Automation, Inc.

other circuits. These devices can communicate not only the on/off status of field devices but also diagnostic information about their operating state. For example, you can detect via the network that a photoelectric sensor is losing margin because of a dirty lens, and you can correct the situation before the sensor fails to detect an object. A limit switch can report the number of motions it has performed, which may be an indication that it has reached the end of its operating life and thus requires replacement.

ControlNet

ControlNet is positioned one level above DeviceNet. It uses the Common Industrial Protocol (CIP) to combine the functionality of an I/O network and a peer-to-peer

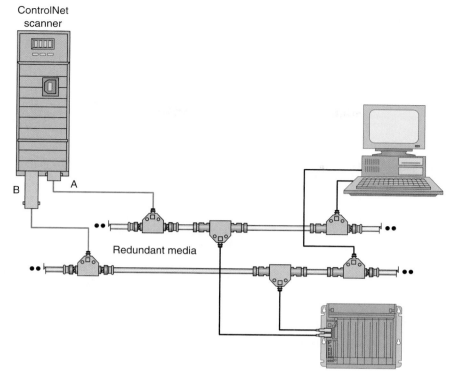

Figure 14-42   ControlNet network with redundant media installed.

network providing high-speed performance for both functions. This open high-speed network is highly deterministic and repeatable. *Determinism* is the ability to reliably predict when data will be delivered, and *repeatability* ensures that transmit times are constant and unaffected by devices connecting to, or leaving, the network. Electronic device data sheets (EDS-Files) are required for each ControlNet device. During the setup phase the ControlNet scanner must configure each device according to the EDS-Files. The ControlNet layout shown in Figure 14-42 has a *redundant media* option in which two separate cables are installed to guard against failures such as cut cables, loose connectors, or noise.

EtherNet/IP

EtherNet/IP (Ethernet Industrial Protocol) is an open communications protocol based on the Common Industrial Protocol (CIP) layer used in both DeviceNet and ControlNet. It allows users to link information seamlessly between devices running the EtherNet/IP protocol without custom hardware, as illustrated in Figure 14-43.

The following are some of the important features of EtherNet/IP:

- Sharing a common application layer between ControlNet, DeviceNet, and Ethernet/IP will make plug-and-play interoperability possible among complex devices from multiple vendors. *Plug and play* refers to the ability of a computer system to automatically configure devices. This allows you to plug in a device and play (operate) it without worrying about setting DIP switches, jumpers, and other configuration elements.

- EtherNet/IP provides standardized full-duplex operation which gives a single node, in a peer-to-peer connection, full attention and therefore maximum possible bandwidth. *Bandwidth* refers to the data rate supported by a network, commonly expressed in terms of bits per second. The greater the bandwidth the greater the overall performance.

- EtherNet/IP allows interoperability of industrial automation devices and control equipment on the same network used for business applications and browsing the Internet.

Modbus

Modbus is a serial communication protocol originally developed by Modicon for use with its PLCs. Basically, it is a method used for transmitting information over serial lines between electronic devices. The device requesting the information is called the Modbus Master and the devices supplying information are Modbus Slaves. Modbus is an open protocol, meaning that it's free for manufacturers to

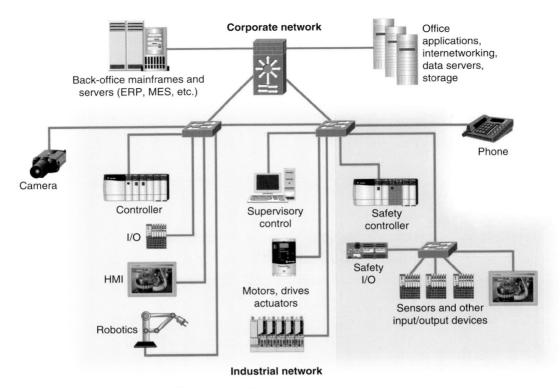

Figure 14-43 EtherNet/IP information links.
Source: Image Used with Permission of Rockwell Automation, Inc.

build into their equipment without having to pay royalties. It has become a standard communications protocol in industry, and is one of the most commonly available means of connecting industrial electronic devices. Figure 14-44 shows an Omron PLC with Modbus-RTU network communication capabilities via RS-232C and RS-422/485 serial ports.

Figure 14-44 Omron PLC with Modbus-RTU network communication capabilities.
Source: Photo courtesy Omron Industrial Automation, **www.ia.omron.com**.

Fieldbus

Fieldbus is an open, serial, two-way communications system that interconnects measurement and control equipment such as sensors, actuators, and controllers. At the base level in the hierarchy of plant networks, it serves as a network for field devices used in process control applications.

There are several possible topologies for fieldbus networks. Figure 14-45 illustrates the *daisy-chain* topology. With this topology, the fieldbus cable is routed from device to device. Installations using this topology require

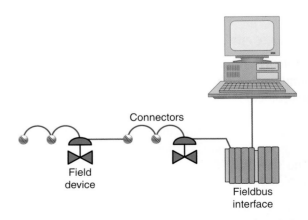

Figure 14-45 Fieldbus implemented using daisy-chain topology.

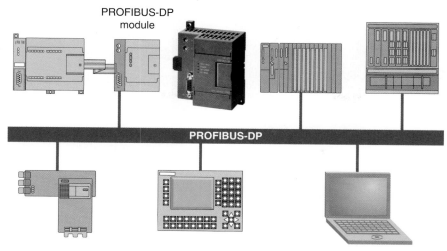

Figure 14-46 Micro PLC system connection to a PROFIBUS-DP network.
Source: Courtesy Siemens.

connectors or wiring practices such that disconnection of a single device is possible without disrupting the continuity of the whole segment.

PROFIBUS-DP

PROFIBUS-DP (where DP stands for Decentralized Periphery) is an open, international fieldbus communication standard that supports both analog and discrete signals. It is functionally comparable to DeviceNet. The physical media are defined via the RS-485 or fiber optic transmission technologies. PROFIBUS-DP communicates at speeds up to 12 Mbps over distances up to 1200 meters. Figure 14-46 illustrates a Siemens S7-200 Micro PLC system connection to a PROFIBUS-DP network.

14.7 Supervisory Control and Data Acquisition (SCADA)

In some applications, in addition to its normal control functions, the PLC is responsible for collecting data, performing the necessary processing, and structuring the data for generating reports. As an example, you could have a PLC count parts and automatically send the data to a spreadsheet on your desktop computer.

Data collection is simplified by using a SCADA (supervisory control and data acquisition) system, shown in Figure 14-47. Exchanging data from the plant floor to a supervisory computer allows data logging, data display, trending, downloading of recipes, setting of selected parameters, and availability of general production data. The additional supervisory control output capabilities allow you to tweak your processes accurately for maximum efficiency. In general, unlike distributive control systems, a SCADA system usually refers to a system that coordinates but does not control processes in real time.

In a typical SCADA system, independent PLCs perform I/O control functions on field devices while being supervised by a SCADA/HMI software package running

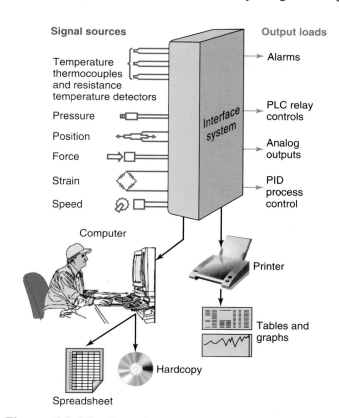

Figure 14-47 Supervisory control and data acquisition (SCADA).

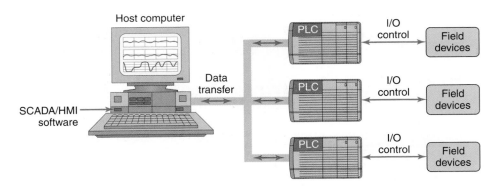

Figure 14-48 Typical SCADA system.

on a host computer, as illustrated in Figure 14-48. Process control operators monitor PLC operation on the host computer and send control commands to the PLCs if required. The great advantage of a SCADA system is that data are stored automatically in a form that can be retrieved for later analysis without error or additional work. Measurements are made under processor control and then displayed onscreen and stored to a hardcopy. Accurate measurements are easy to obtain, and there are no mechanical limitations to measurement speed.

1. Compare continuous and batch processes.
2. Compare centralized and distributive control systems.
3. State the basic function of each of the following as part of a process control system:
   a. Sensors
   b. Human-machine interface
   c. Signal conditioning
   d. Actuators
   e. Controller
4. State the purpose of each of the following types of screens associated with HMIs:
   a. Trend values
   b. Operational summary
   c. Alarm summary
5. What is the main characteristic of a closed-loop control system?
6. State the function of each of the following parts of a closed-loop control system:
   a. Set-point
   b. Process variable
   c. Error amplifier
   d. Controller
   e. Output actuator
7. Explain how on/off control works.
8. How does the proportional controller eliminate the cycling associated with on/off control?
9. Explain how a motor-driven control valve action can provide analog control.
10. How does time proportioning provide analog control?
11. What process error or deviation is produced by a proportional controller?
12. What term of a PID control is designed to eliminate offset?
13. What does the derivative action of a controller respond to?
14. List the three gain adjustments used in tuning the response of a PID control loop.
15. Compare manual, autotune, and intelligent tuning of a PID controller.
16. How many input and output values are normally referenced in a PLC PID instruction?
17. What information is contained in the process variable and control variable elements of a PID instruction?
18. State the function of each of the following elements of a PLC motion control system:
    a. Programmable controller
    b. Motion module
    c. Servo drive
    d. Servo motor
19. What does each axis of a robot arm function as?
20. List four types of communication tasks provided by local area networks.
21. Name three common types of transmission media.
22. What are the three general levels of functionality of industrial networks?
23. Define the term *node* as it applies to a network.
24. Explain the physical layout of devices on a network for each of the following network topologies:
    a. Star
    b. Bus
25. Compare device and process bus networks.
26. Define the term *protocol* as it applies to a network.
27. What is the function of a network gateway?
28. Define the term *access method* as it applies to a network.
29. Summarize the token passing network access method.
30. Summarize the collision detection network access method.
31. Summarize the polling network access method.
32. Compare parallel and serial data transmission.
33. Compare half-duplex and full-duplex data transmission.
34. Explain how networking schemes minimize the amount of wiring required.
35. What type of access control is used with DH+?
36. Compare the transmitting distances of RS-232 and RS-422/485 serial types.
37. What is DeviceNet used for?
38. List three pieces of information obtained from DeviceNet devices by the network scanner.
39. What is ControlNet used for?

40. Explain how redundant media works.

41. Define the term *bandwidth* as it applies to a network.

42. What is Ethernet/IP used for?

43. What type of protocol does Modbus use?

44. What is Fieldbus used for?

45. Summarize the two main functions of a SCADA system.

46. In what way does distributive control differ from the supervisory control of a SCADA system?

CHAPTER 14 PROBLEMS

1. Distributive control systems have to be network based. Why?

2. Assume an alarm is sounded in a control system with an electronic HMI interface. How would you proceed to identify and solve the problem?

3. How would an on/off controller respond if the deadband were too narrow?

4. In a home heating system with on/off control, what will be the effect of widening the deadband?

5. **a.** Calculate the proportional band of a temperature controller with a 5% bandwidth and a set-point of 500°F.

   b. Calculate the upper and lower limits beyond which the controller functions as an on/off unit.

6. Explain the advantage of using a 4- to 20-mA current loop as an input signal compared to a 0- to 5-V input signal.

7. What does the term *deterministic* mean, and why is it important in industrial communications?

8. How might a SCADA system be applied to determine the production rate of a bottled product over a two-week period?

15
ControlLogix Controllers

Programmable logic controllers continue to evolve as new technologies are added to their capabilities. The PLC started out as a replacement for banks of relays used to turn outputs on and off as well as for timing and counting functions. Gradually, various math and logic manipulation functions were added. In order to serve today's expanding industrial control system needs, leading automation companies have created a new class of industrial controllers called *programmable automation controllers* or *PACs* (Figure 15-1). They look like PLCs in their physical appearance but incorporate advanced control of communication, data logging, and signal processing, motion, process control, and machine vision in a single programming environment.

The Allen-Bradley programmable automation controller family includes the ControlLogix system, CompactLogix system, FlexLogix system, SoftLogix 5800 controller, and DriveLogix system. *Software* is the essential difference between PACs and PLCs. Basically, the ladder logic configuration does not change but the addressing of the instructions changes. Application of the software that pertains to the Logix control platform of controllers will be covered in the various sections of this chapter. Knowledge of basic ladder logic instructions and functions (bit, timer, counter, etc.) covered in previous chapters of the text is assumed and is thus not repeated in this chapter.

Figure 15-1 Programmable automation controllers (PACs).
Source: Image Used with Permission of Rockwell Automation, Inc.

Memory Layout

ControlLogix processors provide a flexible memory structure. There are no fixed areas of memory allocated for specific types of data or for I/O. The internal memory organization of a ControlLogix controller is configured by the user when creating a project with RSLogix 5000 software (Figure 15-2). This feature allows the program data to be constructed to meet the needs of your applications rather than requiring your application to fit a particular memory structure. A ControlLogix (CLX) system can consist of anything from a stand-alone controller and I/O modules in a single chassis, to a highly distributed system consisting of multiple chassis and networks working together.

Configuration

Configuration of a modular CLX system involves establishing a communications link between the controller and the process. The programming software needs to know what CLX hardware is being used in order to be able to send or receive data. Configuration information includes information about the type of processor and I/O modules used.

Part Objectives

After completing this part, you will be able to:

- Outline project organization
- Define tasks, programs, and routines
- Identify data file types
- Organize and apply the various data file types

RSLogix 5000 programming software is used to set up or *configure* the memory organization of an Allen-Bradley ControlLogix controller. *RSLinx* communication software is used to set up a communications link between RSLogix 5000 programming software and the ControlLogix hardware as illustrated in Figure 15-3. To establish communications with a controller, a driver must be created in RSLinx software. This driver functions as the software interface to a hardware device. The *RSWho* is the network browse interface that provides a single window to view all configured network drivers.

Figure 15-4 shows an example of the ControlLogix's *controllers properties and modules properties* dialog boxes used as part of the configuration process. The parameters shown are typical of what general information is required. After first configuring the controller,

Figure 15-2 RSLogix 5000 screen.

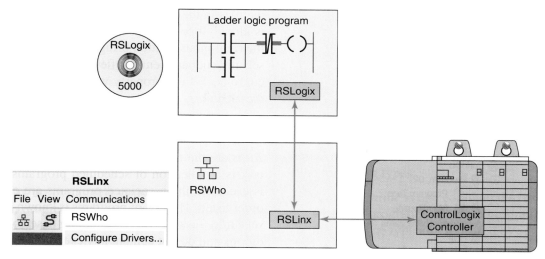

Figure 15-3 RSLinx and RSLogix software.

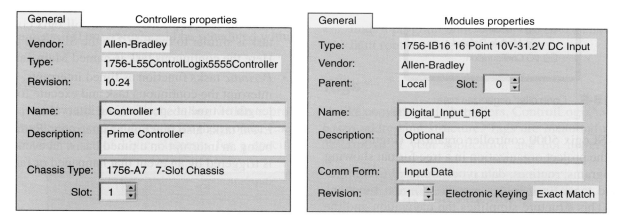

Figure 15-4 Controllers properties and modules properties dialog boxes.

the I/O modules are configured using RSLogix 5000 software. Modules will not work unless they have been properly configured. The software contains all the hardware information needed to configure any ControlLogix module.

Project

RSLogix software stores a controller's programming and configuration information in a file called a *project*. The block diagram of the processor's project file is shown in Figure 15-5. A project file contains all information relating to the project. The main components of the project file are tasks, programs, and routines. A controller can hold and execute only one project at a time.

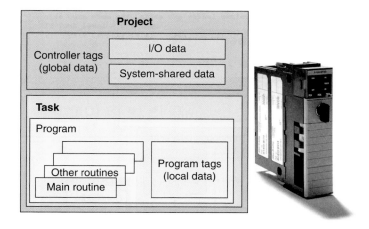

Figure 15-5 ControlLogix processor program file.
Source: Image Used with Permission of Rockwell Automation, Inc.

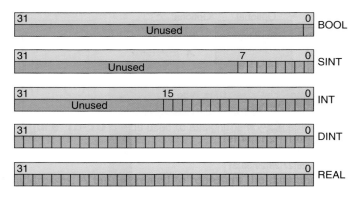

Figure 15-16 Types of base tag data.

- A **SINT** or Single Integer base tag uses 8 bits of memory and stores the data in bits 0 to 7. These bits are sometimes called the low byte. The other 3 bytes, bits 8 to 31, are unused. SINTs have a range of -128 (negative values) to 127 (positive values).

- An **INT** or Integer base tag is 16 bits, bits 0 to 15, sometimes called the lower bytes. Bits 16 to 31 are unused. INTs have a range between $-31,768$ and 32,767.

- A **DINT** or Double Integer base tag uses 32 bits, or all 4 bytes, and has the following range: -2^{31} to $2^{31}-1$ ($-2,147,483,648$ to $2,147,483,647$).

- A **REAL** base tag also uses 32 bits of a memory location and has a range of values based on the IEEE Standard for Floating-Point Arithmetic.

Structures

There is another class of data types called structures. A structure-type tag is a grouping of different data types that function as a single unit and serve a specific purpose. An example of an RSLogix structure is shown in Figure 15-17. Each element of a structure is referred to as a member and each member of a structure can be a different data type.

Name	Data Types	Style	Description
PRE	DINT	Decimal	
ACC	DINT	Decimal	
EN	BOOL	Decimal	
TT	BOOL		
DN	BOOL		
FS	BOOL	Decimal	
LS	BOOL	Decimal	
OV	BOOL	Decimal	
ER	BOOL		

Figure 15-17 Structure-type tag.

Data type : COUNTER

Name	Counter		
Description			

Members — Data type size : 12 byte(s)

Name	Data Type	Style	Description
PRE	DINT	Decimal	
ACC	DINT	Decimal	
CU	BOOL	Decimal	
CD	BOOL	Decimal	
DN	BOOL	Decimal	
OV	BOOL	Decimal	
UN	BOOL	Decimal	

Figure 15-18 Predefined structure.

There are three different types of structures in a Control-Logix controller: *predefined, module-defined,* and *user-defined.* The controller creates *predefined* structures for you that include timers, counters, messages and PID types. An example of a predefined counter instruction structure is shown in Figure 15-18. It is made up of the preset value, the accumulated value, and the instruction's status bits.

Module-defined structures are automatically created when the I/O modules are configured for the system. When you add input or output modules a number of defined tags are automatically added to the controller tags. Figure 15-19 shows the two tags (Local:1:C and Local:1:I) created after a digital input module has been

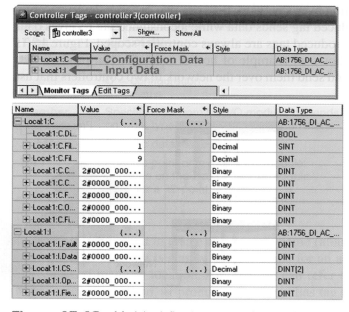

Figure 15-19 Module-defined structure for a digital input module.

of an array must be of the *same data type* (e.g., BOOL, SINT, or INT). An array occupies a contiguous block of controller memory. Arrays are similar to tables of values. The use of arrayed data types offers the fastest data throughput (output) from a ControlLogix processor. Because arrays are numerically sequenced tags of the same data that occupy a contiguous memory location, large amounts of data can be retrieved efficiently. Arrays can be built using 1, 2 or 3 dimensions, as illustrated in Figure 15-24, to represent the data they are intended to contain.

A single tag within the array is one element. The element may be a basic data type or a structure. The elements start with 0 and extend to the number of elements minus 1. Figure 15-25 is an example of the memory

Array - Temp
Data Type - INT[5]

Temp[0]	297
Temp[1]	200
Temp[2]	180
Temp[3]	120
Temp[4]	100

Figure 15-25 Memory layout for a 1-dimensional array.

layout for a 1-dimensional (one column of values) array created to hold five temperatures. The tag name is Temp and the array consists of 5 elements numbered 0 through 4.

1. Compare the memory configuration of a Logix 5000 controller with that of an SLC 500 controller.

2. What does a project contain?

3. List four programming functions that can be carried out using the program organizer.

4. Explain the function of tasks within the project.

5. State the three main types of tasks.

6. What type of tasks function as timed interrupts?

7. Explain the function of programs within the project.

8. Explain the function of routines within the project.

9. Which routine is configured to execute first?

10. Name the four types of programming languages that can be used to program Logix 5000 controllers.

11. What are tags used for?

12. Compare the accessibility of program scope and controller scope tags.

13. Name the tag type used for each of the following:
    a. Create an alternate name for a tag.
    b. Share information over a network.
    c. Store various types of data.

14. What is the difference between a produced tag and a consumed tag?

15. List the five types of base tag data.

16. State the data type used for each of the following:
    a. 32-bit memory storage
    b. On/Off toggle switch
    c. 16-bit memory storage
    d. 8-bit memory storage

17. Describe the make-up of a predefined structure.

18. Describe the make-up of a module-defined structure.

19. Describe the make-up of a user-defined structure.

20. Explain two ways of creating tags.

21. When defining tags what limitations are placed on the entering of a tag name?

22. What is meant by the tag display style?

23. Write an example of an array tag used to hold 4 speeds.

Part 2 Bit-Level Programming

Part Objectives

After completing this part, you will be able to:

- Know what happens during the program scan
- Demonstrate an understanding of input, output, and internal relay addressing format for a tag-based Logix controller
- Develop ladder logic programs with input instructions and output coil combinations
- Develop ladder logic programs with latched outputs

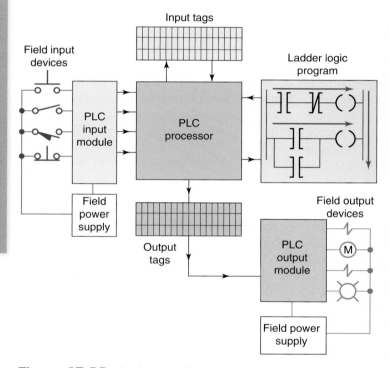

Figure 15-26　Logix controller operating cycle.

Program Scan

When a CLX controller executes a program, it must know—in real time—when external devices controlling a process are changing. During each operating cycle, the processor reads all the inputs, takes these values, and energizes or de-energizes the outputs according to the user program. This process is known as the *program scan*.

Figure 15-26 illustrates the signal flow into and out of a Logix controller during a controller's operating cycle when ladder logic is executing. During the program scan, the controller reads rungs and branches from left to right and top to bottom as follows:

- Only one rung at a time is scanned.
- As the program is scanned, the status of inputs are checked for True (1 or ON) or False (0 of OFF) conditions.

- The status signals from the inputs are sent to the input tags where they are stored.
- As the program is scanned by the processor, inputs are checked for True or False conditions and the ladder logic is evaluated based on these values.
- The resulting ON or OFF action, as a result of evaluating each rung, is then sent to the output tags for storage.
- During the output update portion of the scan, corresponding output values are sent to the process or machine by way of the output module.

329

- I/O updates occur asynchronously to the scan of the logic. With a ControlLogix processor two separate 32-bit unsynchronized processes gone on simultaneously—that is, asynchronously. This means that the module can update the input tag from the field and write the output tag to the field at any point (or at several points) during the processor's execution of the ladder rungs. The result is more efficiency and control over when the input field device data are updated in the input tag and when the output data resulting from the solved logic are sent to the output modules and their respective field devices.

Creating Ladder Logic

Although other programming languages are available, ladder logic is the most common programming language for PLCs. The instructions in ladder logic programming can be divided into two broad categories: input and output instructions. The most common input instruction is equivalent to a relay contact and the most common output instruction is the equivalent of a relay coil (Figure 15-27). When creating ladder I/O bit instructions, the following rules apply:

- All input instructions must be to the left of an output instruction.
- A rung cannot begin with an output instruction if it also contains an input instruction. This is because the controller tests all inputs for true or false before deciding what value the output instruction should be.
- A rung does not need to contain any input instructions, but it must contain at least one output instruction.

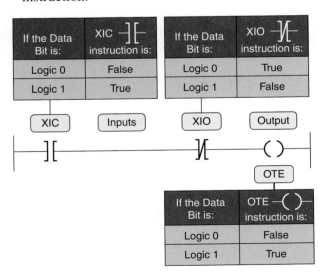

Figure 15-27 Contacts and coil instructions.

- When a rung has only one output instruction it will always be true.
- The last instruction on a rung must always be an output instruction.
- The XIC, or Examine If Closed contact instruction, checks to see if the input has a value of one. If the input is one, the XIC instruction returns a true value.
- The XIO, or Examine If Open contact instruction, checks to see if the input has a value of zero. If the input is zero, the XIO instruction returns a true value.
- The OTE or Output Energize coil instruction sets the tag associated with it to true or one when the rung has logic continuity. When true it can be used to energize an output device or simply set a value in memory to one.

ControlLogix PLCs support multiple outputs on one rung. CLX controllers allow the use of serial logic that does not conform to traditional electrical hardwired circuits or ladder logic. For example, both of the rungs shown in Figure 15-28 are valid in RSLogix 5000. However the series connection of outputs would not work if wired that way in an equivalent electrical circuit or programmed that way in RSLogix 500. In both instances in RSLogix 5000, instructions tagA and tagB must be true to energize output tag1 and tag2.

In ControlLogix output instructions can be placed between input instructions as illustrated in Figure 15-29. In this example instructions tagA and tagB must be true to energize output tag1. Instructions tagA and tagB and tagC must all be true before output tag2 is set to energize.

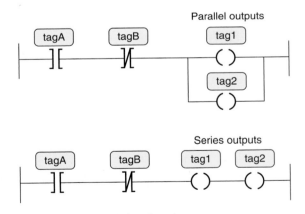

Figure 15-28 Parallel and series outputs.

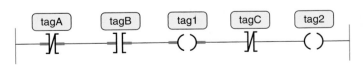

Figure 15-29 Output instruction placed between input instructions.

Tag-Based Addressing

Logix 5000 controllers use a tag-based addressing structure. A tag is a text-based name for an area of the controller where data is stored. An example of how a tag-based address is implemented using a ControlLogix controller is shown in Figure 15-30. Tag names use a meaningful description of the variable. In this application when the normally closed high limit switch is activated the program will switch the high limit output light on. The addressing format can be summarized as follows:

- The physical address for the tag Limit_switch is Local:1:I.Data.2(C). Local indicates that the module is in the same rack as the processor, 1 indicates that the module is in slot 1 in the rack, I indicates that the module is an input type, Data indicates that it

is a digital input, 2 indicates that the limit switch is connected to terminal 2 on the module, and C indicates that it is a controller tag with global access.

- The physical address for the tag High_limit_light is Local:2:O.Data.4(C). Local indicates that the module is in the same rack as the processor, 2 indicates that the module is in slot 2 in the rack, O indicates that the module is an output type, Data indicates that it is a digital input, 4 indicates that the high limit light is connected to terminal 4 on the module, and C indicates that it is a controller tag with global access.

One advantage of the use of tag-based addressing is that the allocation of variable names for program values is not tied to specific memory locations in the memory structure, as is the case with rack/slot and rack/group type

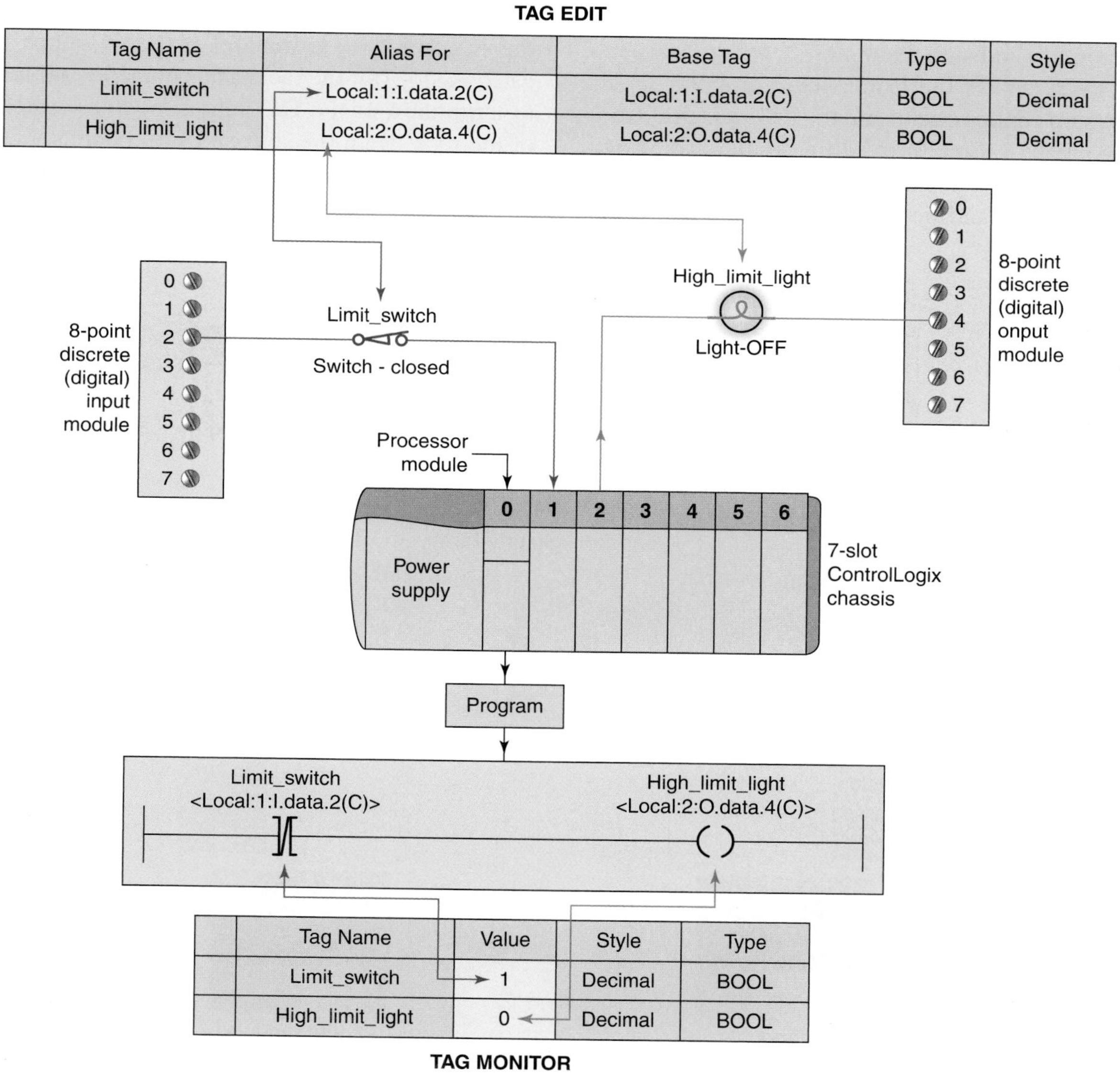

Figure 15-30 Tag-based address implementation.

systems. Initially, all program development can proceed with just the tag names and data types assigned. Using tag aliases, programmers can write code independent of electrical connection assignments. At a later date, input and output field devices are easily matched to the pin numbers on the respective module they are connected to.

Adding Ladder Logic to the Main Routine

Figure 15-31 shows the diagram for a hardwired contactor operated motor start/stop control circuit. The normally open start button is momentarily closed to energize the contactor coil and close its main contacts to start the motor. The seal-in auxiliary contact of the contactor is connected in parallel with the start button to keep the starter coil energized when the start button is released. The normally closed start button is momentarily opened to de-energize the contactor coil and stop the motor.

Figure 15-32 shows the ladder logic program for the motor start/stop control circuit and the RSLogix 5000 toolbar used to create it. Free form editing found in RSLogix 5000 helps speed development in that you do not have to place an instruction and tie an address to the instruction

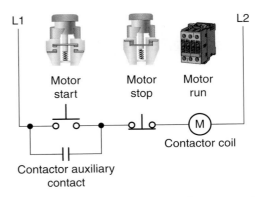

Figure 15-31 Hardwired motor start/stop control circuit.

before adding the next instruction. In this example we have chosen to use question marks [?] in place of tag names and assign the tags later. Field device wiring for the two pushbutton inputs and the single contactor coil output are as illustrated. The stop button is connected to terminal 3 and the start button to terminal 4 of the DC input module located in slot 1 of the rack. The contactor coil is connected to terminal 4 of the DC output module located in slot 2 of the rack. Both the start and stop buttons are examined for a closed condition (XIC) because both buttons must be closed to cause the motor starter to operate.

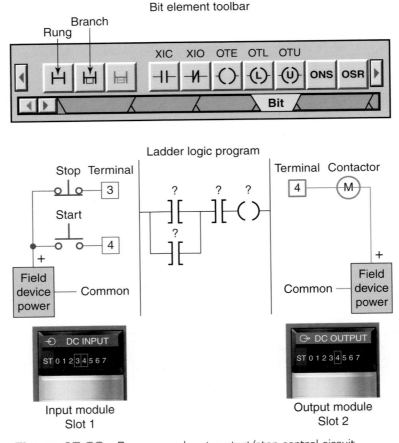

Figure 15-32 Programmed motor start/stop control circuit.

With text-based Logix systems you can use the name of the tag to document your ladder code and organize your data to mirror your application. For the programmed motor start/stop control circuit three tags Motor_Start, Motor_Stop, and Motor_Run are created. Figure 15-33 illustrates how the Motor_Start tag is created in the New Tag window. This window can be accessed by right clicking the ? mark above the XIC instruction in the ladder logic program. Since this tag represents a value from an input field device a link through the module to the field device must be created. When Local:1:I.Data is selected a dialog box for all of the terminal numbers on the input module appears. The tag name (Motor_Start) used in the program is then linked to input terminal number 3 where the field device represented by the tag name is connected.

Figure 15-34 shows what the ladder logic program would look like after all three tags have been created. Users have the ability to reference data via multiple names using Aliases. This allows the flexibility to name data differently depending on their use. The tag description provides for a more meaningful description of the tag name. Tag names are downloaded and stored in the controller

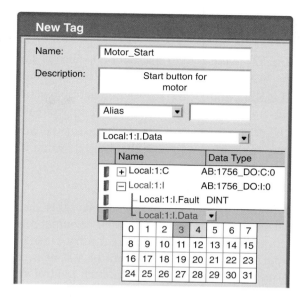

Figure 15-33   Creating the Motor_Start tag.

but the description is not as it is part of the documentation of the project.

Figure 15-35 shows the state of the tags created for the motor start/stop program as seen in the program and

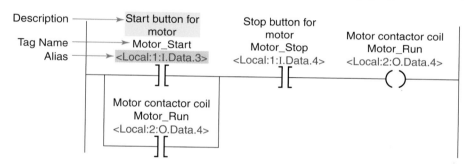

Figure 15-34   Ladder logic program after all tags have been created.

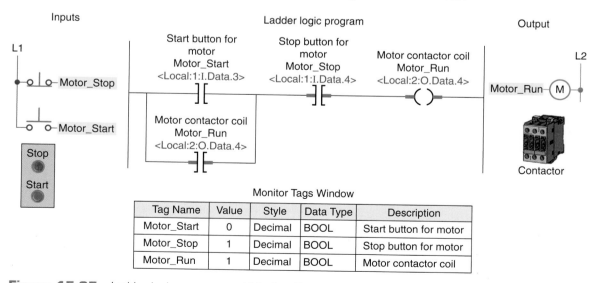

Tag Name	Value	Style	Data Type	Description
Motor_Start	0	Decimal	BOOL	Start button for motor
Motor_Stop	1	Decimal	BOOL	Stop button for motor
Motor_Run	1	Decimal	BOOL	Motor contactor coil

Figure 15-35   Ladder logic program and Monitor Tags window with motor operating.

Monitor Tags window, when the motor is operating. When the motor is operating:

- The XIC Motor_Start instruction is false because the NO start button is open; therefore its value is 0.
- The XIC Motor_Stop instruction is true because the NC stop button is closed; therefore its value is 1.
- The OTE Motor_Run instruction is true because the rung has logic continuity; therefore its value is 1.

Internal Relay Instructions

Internal relay instructions are used when other than real-world field devices are needed as input or output reference instructions. For example, an internal relay bit is used as an output when the logical resultant of a rung is used to control other internal logic. An internal control relay is programmed in the ControlLogix system by creating a tag (either program or controller type) and assigning a Boolean type to the tag.

Figure 15-36 shows a ControlLogix program that uses an internal relay to implement on/off control of a room light from three different entrances or positions. Three single pole switches are used for inputs in place of the two 3-way and one 4-way switches normally required for an equivalent hardwired control circuit. The operation of the program can be summarized as follows:

- An internal relay is used to execute the logic of the circuit without having to use a real-world output.
- The status value stored in memory for all tags, when all input switches are open, is 0 and so the room light will be off.
- Closing Position_1_Switch changes the status of its XIC instruction from false to true thereby establishing logic continuity for Rung 1.
- As a result, the status of the internal relay coil and its XIC contact change from false to true.
- This establishes logic continuity for Rung 2 and switches the room light on.
- A change in the state of any of input switches will change the current state of the light.

Latch and Unlatch Instructions

The *output latch (OTL)* instruction is a retentive output instruction that is used to maintain, or latch, an output. If this output is turned on, it will stay on even if the status

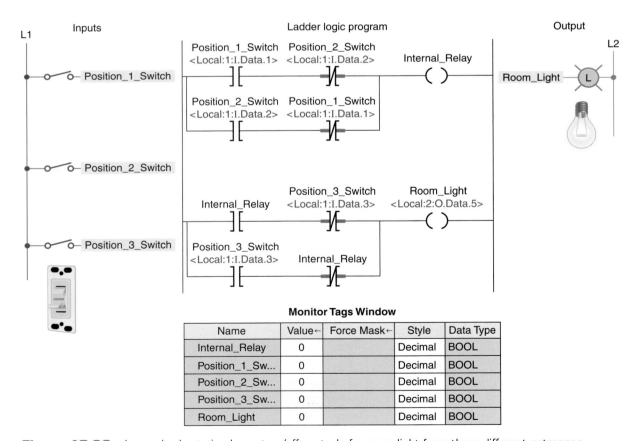

Name	Value←	Force Mask←	Style	Data Type
Internal_Relay	0		Decimal	BOOL
Position_1_Sw...	0		Decimal	BOOL
Position_2_Sw...	0		Decimal	BOOL
Position_3_Sw...	0		Decimal	BOOL
Room_Light	0		Decimal	BOOL

Figure 15-36 Internal relay to implement on/off control of a room light from three different entrances.

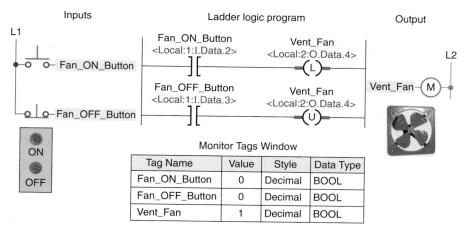

Figure 15-37 Output latch and unlatch instructions used to control a vent fan motor.

of the input logic that caused the output to energize becomes false. The OTL instruction will remain in a latched on condition until an unlatch instruction (OTU) with the same referenced tag is energized. The OTL instruction is often used in programs where the value of a variable must be maintained in instances where there is a shutdown due to a power failure or system fault. Retentive memory permits the system to be restarted with memory locations holding the values that were present when the program execution was halted.

Figure 15-37 shows a ControlLogix program that uses an output latch and unlatch instruction pair to implement the control of a vent fan motor. The operation of the program can be summarized as follows:

- The OTL instruction will write a 1 to its address when true.
- When the OTL goes false, the output address will remain a 1.
- This is true even if the processor powers down and then back up.
- The output address will remain a 1 until reset to 0 by the unlatch instruction.

- If the output address is off, both the latch and unlatch instructions are not intensified, but once the bit is turned on, you will see both the latch and unlatch intensified even though both inputs are shut off.

One-Shot Instruction

The CLX *One-Shot (ONS)* instruction is an input instruction used to turn an output on for one program scan only. The program of Figure 15-38 uses the ONS instruction with a math instruction to perform a calculation once per scan. This program is used to execute the ADD math function only once per actuation of the limit switch, no matter how long the limit switch is held closed. The operation of the program can be summarized as follows:

- On any scan for which *limit_switch_1* is cleared or *storage_1* is set, this rung has no effect.
- On any scan for which *limit_switch_1* is set and *storage_1* is cleared, the ONS instruction sets *storage_1* and the ADD instruction increments *sum* by 1.
- As long as *limit_switch_1* stays set, *sum* stays the same value. The *limit_switch_1* must go from cleared to set again for *sum* to be incremented again.

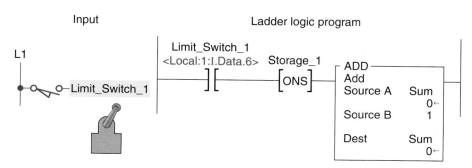

Figure 15-38 ONS instruction used to perform a calculation once per scan.

1. What operations are performed by the processor during the program scan?

2. With a ControlLogix processor I/O updates occur asynchronously. Explain what this means.

3. In ladder logic programming into what two broad categories can instruction types be classified?

4. A field input switch is examined using an XIC instruction.
   a. What is the value (0 or 1) stored in its memory bit when the switch is opened and closed?
   b. What is the state of the instruction (true or false) when the switch is opened and closed?

5. A field input switch is examined using an XIO instruction.
   a. What is the value (0 or 1) stored in its memory bit when the switch is opened and closed?
   b. What is the state of the instruction (true or false) when the switch is opened and closed?

6. The value of an OTE instruction as it appears in the Monitor Tags window is 1. Explain what this means as far as the status of a real-world field output and programmed XIC and XIO instructions associated with this tag are concerned.

7. Define a tag in the ControlLogix system.

8. What advantage do tag-based addressing systems have over rack/slot and rack/group types?

9. How is an internal relay programmed in the ControlLogix system?

10. The output latch instruction is a retentive output instruction. Explain what retentive means.

11. The ControlLogix ONS instruction is a one-shot instruction. Explain what this means.

1. Modify the original ControlLogix start/stop motor control program with a second start and stop button added to the program. The additional start button is to be connected to pin 1 and the stop button to pin 2 of the digital input module.

2. Extend control of the original ControlLogix internal relay program used to control a room light from 3 entrances to 4. The additional single-pole switch is to be connected to pin 4 of the digital input module.

3. Implement the hardwired latching relay alarm circuit of Figure 15-39 in Logix format. The alarm will be latched on anytime:
   • The normally open temperature switch closes.
   • Both normally open float switches 1 and 2 close.
   • Either normally open sensor switch 1 or 2 closes while the normally closed pressure switch is closed.

4. Implement the hardwired tank filling and emptying operation shown in Figure 15-40 in Logix format.

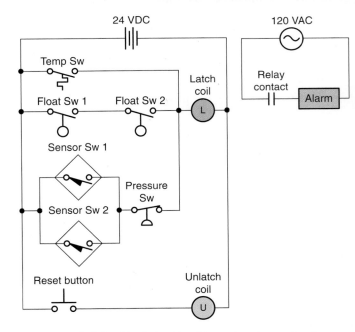

Figure 15-39 Hardwired latching relay alarm circuit for Problem 3.

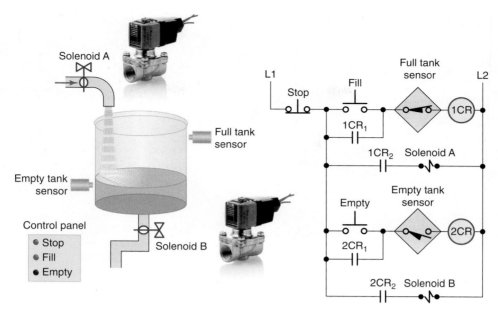

Figure 15-40 Hardwired tank filling and emptying operation for Problem 4.
Source: Photo courtesy ASCO Valve Inc., **www.ascovalve.com**.

The operation of the control circuit can be summarized as follows:

- Assuming the liquid level of the tank is at or below the empty level mark, momentarily pressing the FILL pushbutton will energize control relay 1CR.
- Contacts $1CR_1$ and $1CR_2$ will both close to seal in the 1CR coil and energize normally closed solenoid valve A to start filling the tank.
- As the tank fills, the normally open empty-level sensor switch closes.
- When the liquid reaches the full level, the normally closed full-level sensor switch opens to open the circuit to the 1CR relay coil and switch solenoid valve A to its de-energized closed state.

- Anytime the liquid level of the tank is above the empty-level mark, momentarily pressing the EMPTY pushbutton will energize control relay 2CR.
- Contacts $2CR_1$ and $2CR_2$ will both close to seal in the 2CR coil and energize normally closed solenoid valve B to start emptying the tank.
- When the liquid reaches the empty level, the normally open empty-level sensor switch opens to open the circuit to the 2CR relay coil and switch solenoid valve B to its de-energized closed state.
- The stop button may be pressed at any time to halt the process.

Part 3 Programming Timers

Timer Predefined Structure

Timers are used to turn outputs on and off after a time delay, turn outputs on or off for a set amount of time, and keep track of the time an output is on or off. The timer address in the SLC 500 controller is a data table address or symbol, whereas the timer address in the ControlLogix controller is a predefined structure of the TIMER data type. The TIMER structure is shown in Figure 15-41. Timer parameters and status bits include:

- **Tag Name**—User-friendly tag name for the timer (e.g., Pump_Timer). If you want to use a timer, you must create a tag of type timer.
- **Preset (PRE)**—The number of time increments that the timer must accumulate to reach the desired time delay. Specifies the value (in milliseconds) which the timer must reach before the done bit (DN) changes state. The preset value is stored as a binary

Part Objectives

After completing this part, you will be able to:

- Understand ControlLogix timer tags and their members
- Utilize status bits from timers in logic
- Develop ladder logic programs using ControlLogix timers

number (DINT). The time base is always 1 msec. For example, for a 3 second timer, enter 3000 for the PRE value.

- **Accumulator (ACC)**—The accumulator value is the number of milliseconds the instruction has been enabled. The accumulator value stops changing when ACC value = PRE value.
- **Enable Bit (EN)**—The enable bit indicates the TON instruction is enabled. The EN bit is true when the rung input logic is true, and false when the rung input logic is false.
- **Timer Timing Bit (TT)**—The timing bit indicates that a timing operation is in process. The TT bit is true only when the accumulator is incrementing. TT remains true until the accumulator reaches the preset value.
- **Done Bit (DN)**—The done bit indicates that accumulated value (ACC) is equal to the preset (PRE)

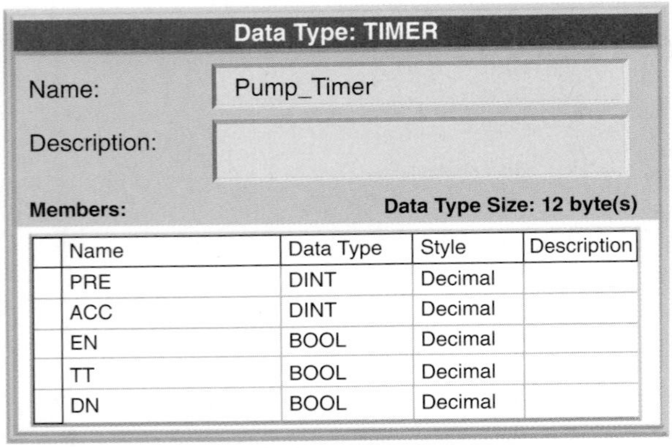

Data Type: TIMER			
Name:	Pump_Timer		
Description:			
Members:		**Data Type Size: 12 byte(s)**	
Name	Data Type	Style	Description
PRE	DINT	Decimal	
ACC	DINT	Decimal	
EN	BOOL	Decimal	
TT	BOOL	Decimal	
DN	BOOL	Decimal	

Figure 15-41 TIMER predefined structure.

value. The DN bit signals the end of the timing process by changing states from false-to-true or from true-to-false depending on the type of time contact instruction used. The DN bit is the most commonly used timer status bit.

On-Delay Timer (TON)

The *on-delay timer (TON)* is a nonretentive output instruction used when the application requires an action to occur at some time after the rung conditions for the timer become true. The ControlLogix TON on-delay instruction and timer selection toolbar are shown in Figure 15-42. When you want to use a timer, you must create a tag of type TIMER (it is a predefined data type) and enter the preset and the accumulated value. The tag must be defined before the preset and accumulated values can be entered. A value can be entered for the accumulator while programming. When the program is downloaded this value will be in the timer for the first scan. If the TON timer is not enabled the value will be set back to zero. Normally zero will be entered for the accumulator value.

The timer tag name is declared using the new tag properties dialog box shown in Figure 15-43. Tag name, description (optional), tag type, data type, and scope are selected or typed to complete the validation. A descriptive tag name, such as Solenoid_Delay, makes it easier to know what function the timer serves in the control system.

The program of Figure 15-44 is an example of a 10000 ms (10 s) TON timer. Timers generate both word level (DINT) and bit level (BOOL) data and status. The operation of the program can be summarized with reference to the Monitor Tags window.

- The status of all instruction is shown after the timer input switch has been switched from off to on (1) and accumulated 5000 ms (5 s) of time.
- At this halfway point the EN bit is 1 since the rung is true, the TT bit is 1 since the accumulated value is

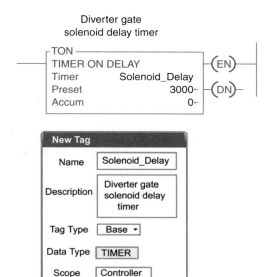

Figure 15-43 Timer tag validation.

changing, and the DN bit is 0 since the accumulated value does not yet equal the preset value.

- When the ACC equals PRE, the accumulated value stops incrementing, EN stays on for as long as the rung remains true, TT equals 0 since the accumulated value is not changing, and DN equals 1 since ACC = PRE.
- This will result in the DN pilot light switching on at the same time as the TT pilot light switches off.
- The EN pilot light remains on as long as the input switch is closed.
- Opening the input switch at any time causes the TON instruction to go false resetting the counter ACC value to 0 and EN, TT, and DN bits to 0. This in turn switches off all output pilot lights.
- The TON instruction is a self-resetting timer. When the rung goes false, the timer is automatically reset. A reset instruction can be used, but usually is not.

Figure 15-45 shows a TON timer used to delay the operation of a diverter gate solenoid for 3 seconds after a target has been sensed by the solenoid energize sensor. The operation of the program can be summarized as follows:

- Detection of the target causes closure of the SOL_Energize_Sensor contacts making the timer rung true and start timing.
- With passage of the target the SOL_Energize_Sensor contacts open but the rung remains true through the EN bit of the TON timer.
- After 3000 ms (3 s) delay time has elapsed, delay timer DN bit is set to 1 to energize the SOL_Gate.

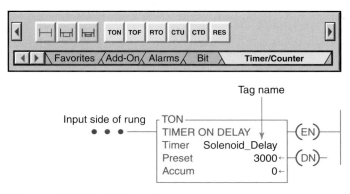

Figure 15-42 TON on-delay instruction.

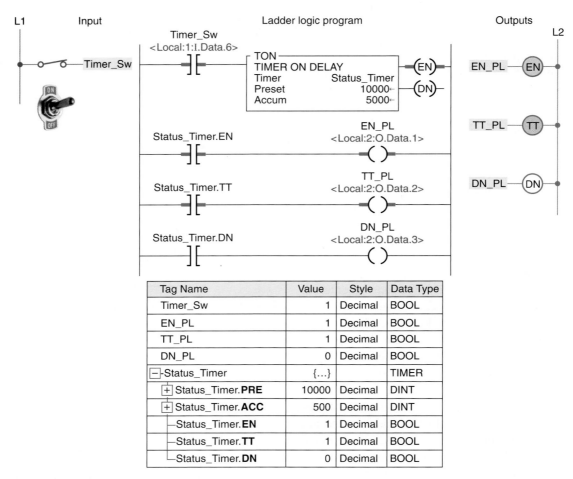

Figure 15-44 Ten-second TON timer program.

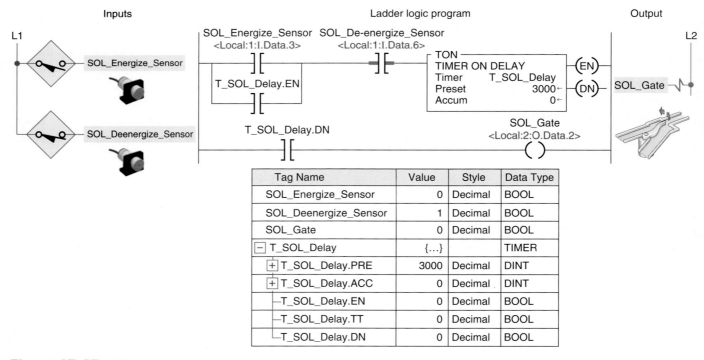

Figure 15-45 TON timer used to delay the operation of a diverter gate solenoid.

Source: Photos courtesy Omron Industrial Automation, **www.ia.omron.com**.

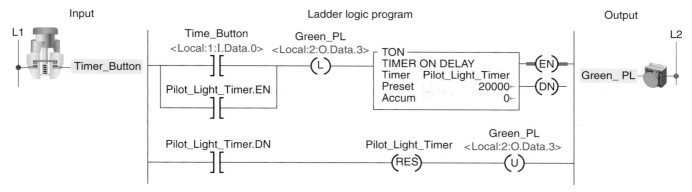

Figure 15-46 Pilot light TON timer.

- Momentary detection of the target by the SOL_Deenergize_Sensor causes the opening of its contacts and resets the program to its original state.

Figure 15-46 shows a program that uses a TON timer to illuminate a green pilot light for 20 seconds each time a momentary button is pressed. In addition to the TON timer this program uses multiple outputs on one rung, output latch and unlatch instructions, as well as a timer reset instruction. The operation of the program can be summarized as follows:

- Initially closing the Timer_Button sets (latches) the Green_PL on and enables the Pilot_Light_Timer.

- When the button is then opened the timer rung remains true through the logic path created by the Pilot_Light_Timer.EN bit.
- After 20000 ms (20 s) have elapsed the timer DN bit is set to reset the timer to its original state and unlatch the Green_PL and switch it off.

The ControlLogix program of Figure 15-47 shows three TON timers cascaded (connected together) for traffic light control. The ladder logic used is the same as that used to program the traffic lights using the SLC 500 controller. The different tags created to fit the program are

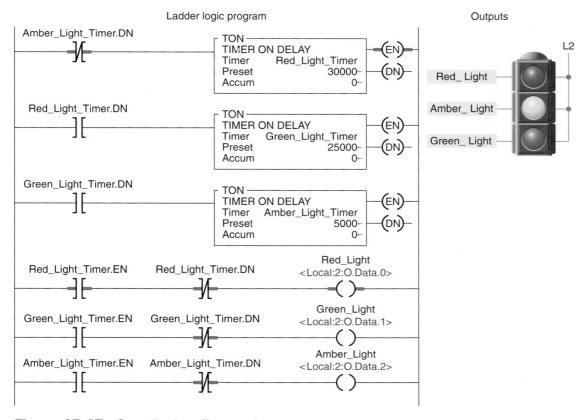

Figure 15-47 ControlLogix traffic control program.

Tag Name	Value	Style	Data Type
+-Amber_Light_Timer	{...}		TIMER
+-Green_Light_Timer	{...}		TIMER
−-Red_Light_Timer	{...}		TIMER
+-Red_Light_Timer.PRE	30000	Decimal	DINT
+-Red_Light_Timer.ACC	0	Decimal	DINT
−Red_Light_Timer.EN	1	Decimal	BOOL
−Red_Light_Timer.TT	1	Decimal	BOOL
−Red_Light_Timer.DN	0	Decimal	BOOL
Red_Light	1	Decimal	BOOL
Green_Light	0	Decimal	BOOL
Amber_Light	0	Decimal	BOOL

Figure 15-48 Tags created for traffic light program.

shown in Figure 15-48. Operation of the program can be summarized as follows:

- Transition from red light to green light to amber light is accomplished by the interconnection of the EN and DN bits of the three TON timer instructions.
- The input to the Red_Light_Timer is controlled by the Amber_Light_Timer.DN bit.
- The input to the Green_Light_Timer is controlled by the Red_Light_Timer.DN bit.
- The input to the Amber_Light_Timer is controlled by the Green_Light_Timer.DN bit.
- The timed sequence of the lights is:
  - Red—30 s on
  - Green—25 s on
  - Amber—5 s on
- The sequence then repeats itself.

Off-Delay Timer (TOF)

The *off-delay timer (TOF)* operates in a fashion opposite to the TON on-delay timer. An off-delay timer will turn on immediately when the rung of ladder logic is true, but it will delay before turning off after the rung goes false. The ControlLogix TOF off-delay timer instruction is shown in Figure 15-49. The description of the function block fields and tag references are the same as for that of a TON timer.

Figure 15-50 shows a program that uses a TOF timer to illuminate a green pilot light for 20 seconds each time a momentary button is pressed. The program code is simpler than that used to accomplish the same task using a TON timer. The operation of the program can be summarized as follows:

- When the Timer_Button is initially closed the timer rung and instruction and DN bit all become true.
- The DN bit switches on the Green_PL and the program remains in this state as long as the button is held closed.
- When the button is released the Timer_Button instruction goes false and starts the timing cycle.
- The light remains on and the timer begins accumulating time.
- When the accumulator reaches 20000 ms (20 s) the timer DN bit becomes false and the light is switched off.

The program of Figure 15-51 uses both on-delay and off-delay timers for control of a heating oven process. The different tags created to fit the program are shown

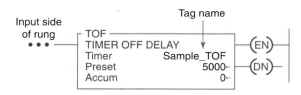

Figure 15-49 ControlLogix TOF off-delay timer instruction.

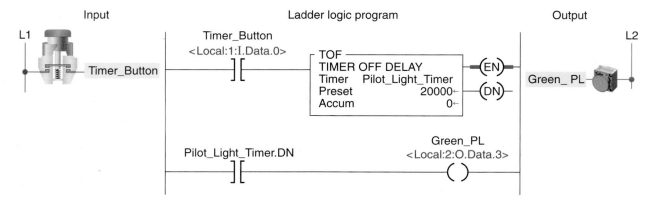

Figure 15-50 Pilot light TOF timer.

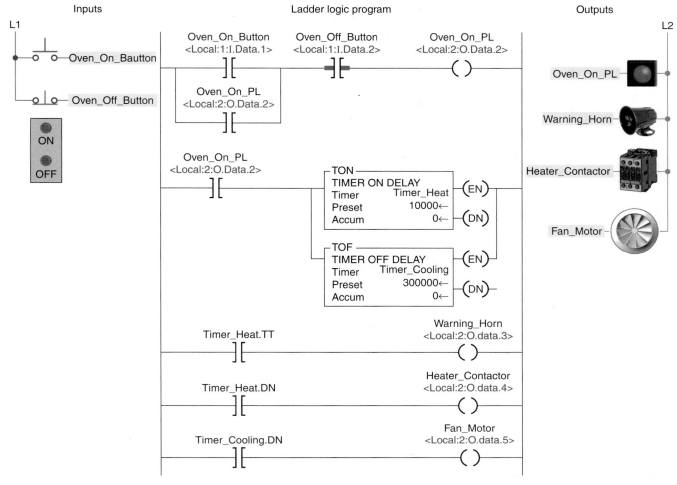

Figure 15-51 Timer control of a heating oven process.

Tag Name	Alias For	Base Tag	Data Type	Style
Warning_Horn	Local:2:O.Data.3	Local:2:O.Data.3	BOOL	Decimal
Heater_Contactor	Local:2:O.Data.4	Local:2:O.Data.4	BOOL	Decimal
Fan_Motor	Local:2:O.Data.5	Local:2:O.Data.5	BOOL	Decimal
Oven_On_PL	Local:2:O.Data.2	Local:2:O.Data.2	BOOL	Decimal
Oven_On_Button	Local:1:I.Data.1	Local:1:I.Data.1	BOOL	Decimal
Oven_Off_Button	Local:1:I.Data.2	Local:1:I.Data.2	BOOL	Decimal
+-Timer_Heat			TIMER	
+-Timer_Cooling			TIMER	

Figure 15-52 Tags created for heating oven process.

in Figure 15-52. Operation of the program can be summarized as follows:

- Pressing the Oven_On_Button energizes the Oven_On_PL output which seals itself in and enables the TON and TOF timer instructions.
- The Timer_Heat.TT bit of the TON timer becomes true which sounds the Warning_Horn to warn that the oven is about to come on.

- The Timer_Cooling.DN bit of the TOF timer becomes true which energizes the Fan_Motor.
- After 10 s (10000 ms) have elapsed the Timer_Heat.TT bit becomes false to turn off the Warning_Horn and the Timer_Heat.DN bit becomes true to energize the Heater_Contactor and turn on the heating coils.
- When the Oven_Off_Button is momentarily actuated the Oven_On_PL output goes false which turns the pilot light off and opens the continuity of its seal-in logic path.
- The Timer_Heat timer instruction and its DN bit instruction become false which de-energizes the Heater_Contactor and turns off the heating coils.
- The Timer_Cooling timer begins accumulating time and the fan continues to operate for the 5 minute (300000 ms) delay period after which the Timer_Cooling.DN bit becomes false to turn the fan off.

Retentive Timer On (RTO)

A *retentive on-delay timer (RTO)* operates the same as a TON timer, except that the retentive timer retains (remembers) its ACC value even if:

- The rung goes false.
- The processor is placed in the program mode.
- The processor faults.
- Power to the processor is temporarily interrupted and the processor battery is functioning properly.

The ControlLogix RTO retentive on-delay timer instruction is shown in Figure 15-53. The description of the function block fields and tag references are the same as for that of a TON timer; however, a RES reset instruction must be used to reset the accumulated value of a retentive timer. The RES instruction must have the same tag name as the timer you want to reset.

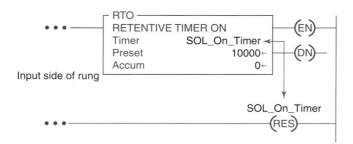

Figure 15-53 RTO retentive on-delay timer instruction.

An example application of a limit switch 2 minute (120000 ms) RTO timer program is shown in Figure 15-54. The different tags created to fit the program are shown in Figure 15-55. The operation of the program can be summarized as follows:

- The status and value of all instructions, with the timer initially reset, are as shown in the monitor tags window.
- When the Limit_Switch has been closed for 1 minute, the status and value of the instructions would be:
  - PRE – 120000
  - ACC – 60000
  - LS_Timer.EN – 1
  - LS_Timer.TT – 1
  - LS_Timer.DN – 0
  - LS_EN_PL – 1
  - LS_TT_PL – 1
  - LS_Alarm – 0
- When the Limit_Switch is opened after 1.5 minutes, the status and value of the instructions would be:
  - PRE – 120000
  - ACC – 90000
  - LS_Timer.EN – 0
  - LS_Timer.TT – 0
  - LS_Timer.DN – 0
  - LS_EN_PL – 0
  - LS_TT_PL – 0
  - LS_Alarm – 0

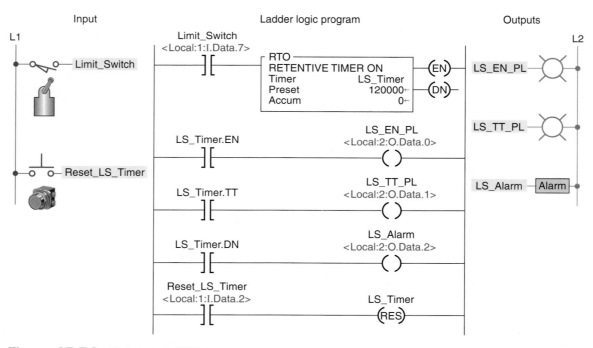

Figure 15-54 Limit switch RTO timer program.

Tag Name	Value	Style	Data Type
⊟-LS_Timer	{...}		TIMER
⊞-LS_Timer.PRE	120000	Decimal	DINT
⊞-LS_Timer.ACC	0	Decimal	DINT
⌐LS_Timer.EN	0	Decimal	BOOL
⌐LS_Timer.TT	0	Decimal	BOOL
∟LS_Timer.DN	0	Decimal	BOOL
Limit_Switch	0	Decimal	BOOL
LS_EN_PL	0	Decimal	BOOL
LS_TT_PL	0	Decimal	BOOL
LS_Alarm	0	Decimal	BOOL

Figure 15-55 Tags created for the RTO retentive on-delay timer program.

- When the Limit_Switch is closed and stays closed until the timer times out, the status and value of the instructions would be:
  - PRE – 120000
  - ACC –120000

- LS_Timer.EN – 1
- LS_Timer.TT – 0
- LS_Timer.DN – 1
- LS_EN_PL – 1
- LS_TT_PL – 0
- LS_Alarm – 1

- When the Limit_Switch is opened after the timer times out, the status and value of the instructions would be:
  - PRE – 120000
  - ACC –120000
  - LS_Timer.EN – 0
  - LS_Timer.TT – 0
  - LS_Timer.DN – 1
  - LS_EN_PL – 0
  - LS_TT_PL – 0
  - LS_Alarm – 1

- When the Reset_LS_Timer is closed, the status and value of the instructions are reset to their original values.

1. Compare the methods used to address timers in an SLC 500 and a ControlLogix controller.

2. List the five different members of a TIMER structure.

3. What type of timing application may require you to use a TON on-delay timer?

4. What PRE value is used for a timer?

5. To what value is the accumulated value of a timer normally set?

6. What timer status bit is set to 1 when the TON timer times out?

7. The TON instruction is self-resetting. Explain what this means.

8. What number would be entered into the PRE value of a ControlLogix timer for a timing period of 4.5 minutes?

9. Compare the operation a TOF and a TON timer.

10. When does the rung of a TOF timer begin accumulating time?

11. The RTO timer is a retentive timer. Explain what this means.

12. How are the retentive timer and reset instruction related?

1. Modify the original CLX ten-second TON timer program with an additional rung added to the program that will energize a solenoid whenever the timer is enabled and timing. The solenoid is to be connected to pin 6 of the digital output module.

2. With reference to the ladder logic of the CLX diverter gate program, assume the solenoid gate fails to energize as programmed. You suspect the problem is due to an open in the solenoid coil or wiring to it. How might observation of the solenoid output status light help confirm this?

3. You are required to extend the Green light-on time of the CLX traffic control program to 40 seconds. What changes would have to be made to the program?

4. With reference to the CLX heating oven process program, assume the oven-on pilot light burns out. In what way would the operation of the program be affected?

5. With reference to the CLX limit switch RTO program, in addition to the alarm you are required to install a warning pilot light to indicate that the timer has timed out. How would you proceed?

6. Implement the hardwired TON alarm circuit of Figure 15-56 in Logix format.

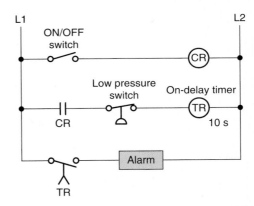

Figure 15-56   Hardwired TON alarm circuit for Problem 6.

Part 4 Programming Counters

Part Objectives

After completing this part, you will be able to:

- Understand ControlLogix counter tags and their members
- Utilize status bits from counters in logic
- Develop ladder logic programs using ControlLogix counters

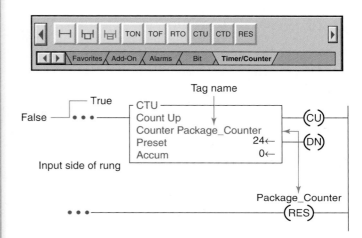

Figure 15-57 CTU count-up counter instruction.

Counters

Counters are similar to timers, except that a counter accumulates (counts) the changes in state of an external trigger signal whereas timers increment using an internal clock. PLC counters are generally triggered by a change in an input field device that causes a false-to-true transition of the counter ladder rung. It does not matter how long the rung stays true or false—it is only the transition that counts.

There are two basic counter types: count-up (CTU) and count-down (CTD). The ControlLogix CTU instruction and counter selection toolbar are shown in Figure 15-57. When you want to use a timer, you must create a tag of type COUNTER (it is a predefined data type) and enter the preset and the accumulated value. When entering the instruction, this tag must be defined before the preset and accumulated values can be entered. A RES reset instruction that has the same tag name as the

counter must be used to reset the accumulated value of the counter to zero.

All counters are retentive in that the accumulated value of any counter is retained, even during a power failure, until reset. The on/off status of the counter done, overflow, and underflow bits are retentive as well. ControlLogix counter parameters and status bits are shown in the edit tags window of Figure 15-58 and can be summarized as follows:

- **Preset (PRE) Value**—Specifies the value the counter must reach before the done (DN) bit turns on (1).
- **Accumulated (ACC) Value**—Is the number of false-to-true transitions of the counter run. ACC is reset to zero when a reset (RES) instruction (of the same counter address) is executed.
- **CU (Count-Up Enable Bit)**—The count-up enable bit indicates the CTU instruction is enabled.

Tag Name	Data Type	Style
−Part_Counter	COUNTER	Decimal
+−Part_Counter.PRE	DINT	Decimal
+−Part_Counter.ACC	DINT	Decimal
− Part_Counter.CU	BOOL	Decimal
− Part_Counter.CD	BOOL	Decimal
− Part_Counter.DN	BOOL	Decimal
− Part_Counter.OV	BOOL	Decimal
− Part_Counter.UN	BOOL	Decimal

Figure 15-58 ControlLogix counter parameters and status bits.

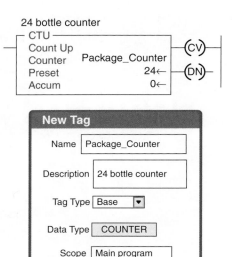

Figure 15-59 Counter tag validation.

- **CD (Count-Down Enable Bit)**—The count-down enable bit indicates the CTD instruction is enabled.
- **DN (Count-Up Done Bit)**—Is set (1) when ACC value is equal to or greater than the PRE value. Is reset by the RES instruction.
- **OV (Overflow Bit)**—The overflow bit indicates the counter exceeded the upper limit. Is set when the ACC value is greater than +2,147,483,647 and reset when the reset instruction is executed. Note that the accumulated value keeps incrementing even after the ACC value equals the PRE value.
- **UN (Underflow Bit)**—Indicates that the counter exceeded the lower limit of −2,147,483,648.

The counter tag name is declared using the new tag properties dialog box shown in Figure 15-59. Tag name, description (optional), tag type, data type (base type is used most often), and scope are selected or typed to complete the validation.

Count-Up (CTU) Counter

Count-up (CTU) counters will cause the accumulated count to increase by 1 every time there is a false-to-true transition of the counter ladder rung. An example application of a count-up counter program used to count packets of bottles is shown in Figure 15-60. The operation of the program can be summarized as follows:

- Each open-to-close transition of the Bottle_Sensor proximity switch causes the counter to increment by 1.

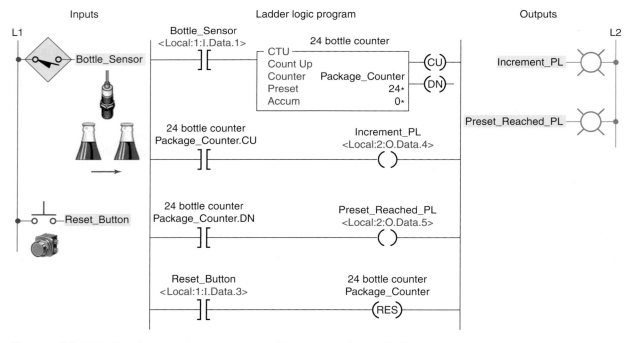

Figure 15-60 Count-up counter program used to count packets of bottles.

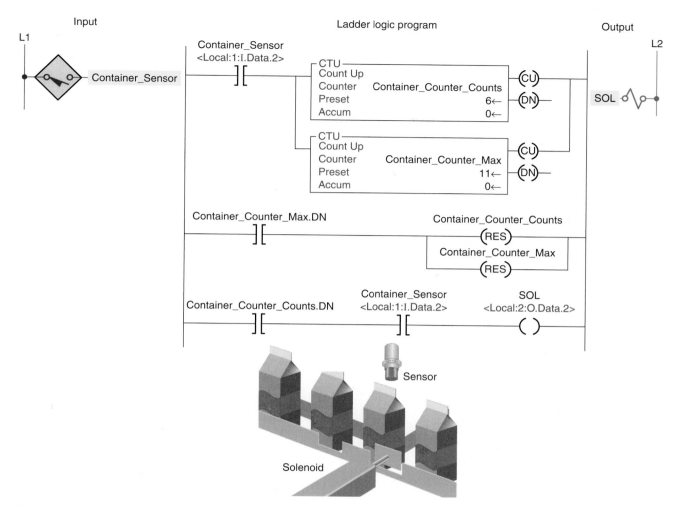

Figure 15-61 CTU program used to remove containers from a conveyor line.

- The Increment_PL controlled by the Package_Counter.CU status bit turns on and off as each bottle passes to show that the counter is incrementing.
- When the accumulated value of the counter is 24 the DN bit of the counter is set and switches on the Preset_Reached_PL.
- The counter is reset by momentarily closing the Reset_Button.

The program shown in Figure 15-61 uses two CTU instructions as part of a program to remove 5 out of every 10 containers from a conveyor line using an electric solenoid. The different tags created to fit the program are shown in Figure 15-62. The operation of the program can be summarized as follows:

- The preset for the Container_Counter_Counts is set for 6 and that for the Container_Counter_Max is set to 11.

Tag Name	Value	Style	Data Type
⊟ Container_Counter_Counts	{...}		COUNTER
⊞ Container_Counter_Counts .PRE	6	Decimal	DINT
⊞ Container_Counter_Counts .ACC	0	Decimal	DINT
Container_Counter_Counts .CU	0	Decimal	BOOL
Container_Counter_Counts .CD	0	Decimal	BOOL
Container_Counter_Counts .DN	0	Decimal	BOOL
Container_Counter_Counts .OV	0	Decimal	BOOL
Container_Counter_Counts .UN	0	Decimal	BOOL
⊟ Container_Counter_Max	{...}		COUNTER
⊞ Container_Counter_Max .PRE	11	Decimal	DINT
⊞ Container_Counter_Max .ACC	0	Decimal	DINT
Container_Counter_Max .CU	0	Decimal	BOOL
Container_Counter_Max .CD	0	Decimal	BOOL
Container_Counter_Max .DN	0	Decimal	BOOL
Container_Counter_Max .OV	0	Decimal	BOOL
Container_Counter_Max .UN	0	Decimal	BOOL
Container_Sensor	0	Decimal	BOOL
SOL	0	Decimal	BOOL

Figure 15-62 Tags created for the CTU program used to remove containers from a conveyor line.

- When the container is detected both counters will increase their accumulated values by 1.
- When the sixth part arrives the Container_Counter_Counts counter will then be done, thereby allowing the solenoid to actuate for any container after the fifth.
- The Container_Counter_Max counter will continue until the eleventh part is detected and then both of the counters will be reset.

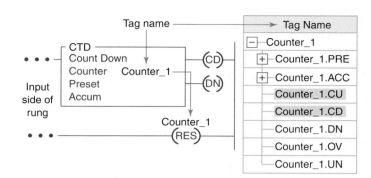

Figure 15-63   Count-down CTD counter instruction.

Count-Down (CTD) Counter

The *count-down (CTD) counter* operates in a fashion opposite to the count-up CTU counter. CTD counters will cause the accumulated count to decrease instead of increase by one every time there is a false-to-true transition of the counter ladder rung. The ControlLogix CTD down-counter instruction is shown in Figure 15-63. The descriptions of the function block fields and the tag references are the same as those associated with the CTU function block. The CTD instruction is typically used with a CTU instruction that references the same counter structure.

The application program shown in Figure 15-64 is used to limit the number of parts that can be stored in the buffer zone to a maximum of 50. A CTU counter and a CTD counter are used together *with the same*

address to form an Up/Down counter. This is the most common type of application of the CTD counter. The different tags created to fit the program are shown in Figure 15-65. The operation of the program can be summarized as follows:

- The Restart_Button is momentarily actuated at any time to reset the accumulated value of the counter to zero.
- Conveyor brings parts into a buffer zone.
- Each time a part enters the buffer zone, the Enter_Limit_Sw is actuated and Counter_1 increments by 1.

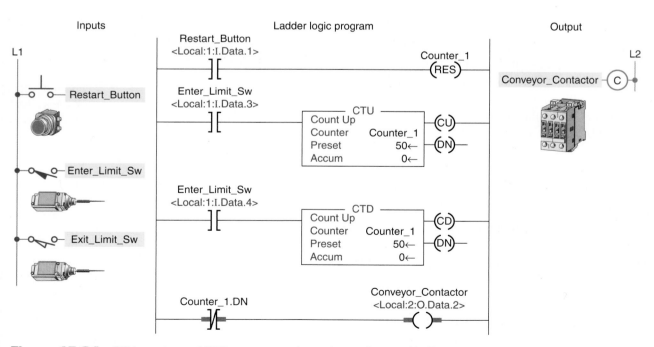

Figure 15-64   CTU counter and CTD counter used together to form an Up/Down counter.

Tag Name	Value	Style	Data Type
⊟—Counter_1	{ . . . }		COUNTER
⊞—Counter_1.PRE	50	Decimal	DINT
⊞—Counter_1.ACC	0	Decimal	DINT
—Counter_1.CU	0	Decimal	BOOL
—Counter_1.CD	0	Decimal	BOOL
—Counter_1.DN	0	Decimal	BOOL
—Counter_1.OV	0	Decimal	BOOL
—Counter_1.UN	0	Decimal	BOOL
Restart_Button	0	Decimal	BOOL
Enter_Limit_Sw	0	Decimal	BOOL
Exit_Limit_Sw	0	Decimal	BOOL
Conveyor_Contactor	1	Decimal	BOOL

Figure 15-65 Tags created for the Up/Down counter program.

- Each time a part leaves the buffer zone, the Exit_Limit_Sw is actuated and Counter_1 decrements by 1.
- When the number of parts in the buffer zone, at any one time, reaches 50, the Counter_1.DN bit is set.
- As a result the Conveyor_Contactor rung goes false to de-energize the conveyor contactor, automatically stopping the conveyor from bringing in any more parts until the accumulated count drops below 50.

PART 4 REVIEW QUESTIONS

1. In what way are timers and counters similar?
2. Outline the procedure followed to create a tag when you want to use a counter.
3. All counters are retentive. In what way does this affect their operation?
4. What is specified by the preset value of a counter?
5. When is each of the following counter bits set?
   a. CU
   b. DN
   c. CD

6. Compare the operations of a CTU and a CTD counter.
7. What is an Up/Down counter?
8. Explain how you go about creating tags for an Up/Down counter that uses a CTU and CTD instruction.

PART 4 PROBLEMS

1. With reference to the CTU packets of bottles program, what changes to the program would be required to count 6 bottle packets?
2. With reference to the CTU program used to remove containers from a conveyor line, assume the output solenoid coil failed open. In what way would the operation of the program be affected?
3. Modify the original Up/Down counter program to include:
   a. A red pilot light to indicate entry of a part into the buffer zone. Light to be connected to pin 4 of the digital output module.
   b. A green pilot light to indicate exit of a part from the buffer zone. Light to be connected to pin 3 of the digital output module.

4. Write a ControlLogix program, complete with tags, for an Up/Down counter used to keep track of cars entering and exiting a parking lot. The program requirements for this application can be summarized as follows:
   - The parking lot holds 30 vehicles.
   - There is an entrance vehicle sensor and an exit vehicle sensor.
   - When the parking lot is full a Lot Full sign is illuminated.
   - Whenever a car exits the lot, a Caution Buzzer/Light is activated to warn pedestrians.

Part 5 Math, Comparison, and Move Instructions

Part Objectives

After completing this part, you will be able to:

- Utilize ControlLogix math instructions in programs
- Utilize ControlLogix comparison instructions in programs
- Utilize ControlLogix move instructions in programs
- Develop and follow the operation of programs that use math, comparison, and move instructions

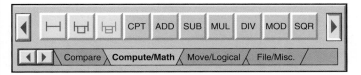

Figure 15-66 Compute/Math toolbar for the ControlLogix controller.

Math Instructions

ControlLogix basic math instructions include addition, subtraction, multiplication, division, square root, and clear. Figure 15-66 shows the Compute/Math toolbar for the ControlLogix controller.

The *ADD* instruction is used to add two numbers. This instruction adds these values from Source A and Source B. The source can be a constant value or a tag. The result of the ADD instruction is put in the destination (Dest) tag.

Figure 15-67 shows an example of an ADD instruction rung along with its Monitor Tags window. The operation of the rung can be summarized as follows:

- When the ADD_Sw is closed the rung will be true.
- The ADD instruction will execute to add the number from Source A (Value_A) and the value from Source B (Value_B).

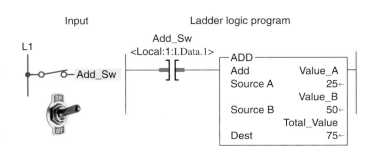

Tag Name	Value	Style	Data Type
+–Total_Value	75	Decimal	DINT
+–Value_A	25	Decimal	DINT
+–Value_B	50	Decimal	DINT
ADD_Sw	1	Decimal	BOOL

Figure 15-67 ADD instruction rung and its Monitor Tags window.

- The result will be stored in the Dest tag (Total_Value).
- In this example, the 25 was added to 50 and the result (75) was stored in Total_Value.

The *SUB* instruction is used to subtract two numbers. Figure 15-68 shows an example of a SUB instruction rung along with its Monitor Tags window. The operation of the rung can be summarized as follows:

- When the SUB_Sw or Calculate tag is true the SUB instruction is executed.
- Source B (Shipped_Parts) is subtracted from Source A (Parts_Stock) and the result is stored in the Dest tag named Current_Inventory.

- In this example, the 200 was subtracted from 900 and the result (700) was stored in Current_Inventory.
- Source A and Source B can be constants (numbers) or tags.

The *MUL* instruction is used to multiply two numbers. Figure 15-69 shows an example of a MUL instruction rung along with its Monitor Tags window. When multiple bottles are packed in cases, the number of bottles per case, the number of cases, and the multiply instruction will give you the total number of bottles. The operation of the rung can be summarized as follows:

- When the Sw_1 and Sw_2 are both true the MUL instruction is executed.

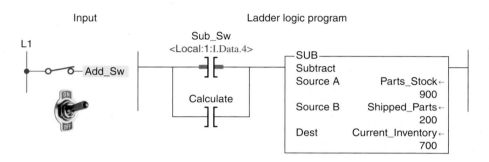

Tag Name	Value	Style	Data Type
+–Parts_Stock	900	Decimal	DINT
+–Shipped_Parts	200	Decimal	DINT
+–Current_Inventory	700	Decimal	DINT
Sub_Sw	1	Decimal	BOOL
Calculate	0	Decimal	BOOL

Figure 15-68 SUB instruction rung and its Monitor Tags window.

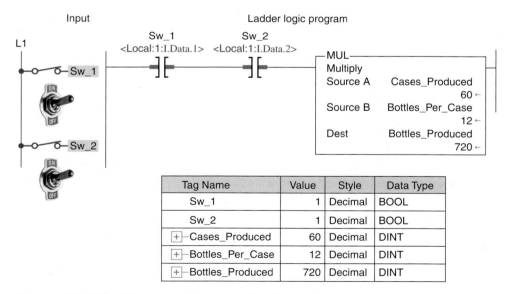

Tag Name	Value	Style	Data Type
Sw_1	1	Decimal	BOOL
Sw_2	1	Decimal	BOOL
+–Cases_Produced	60	Decimal	DINT
+–Bottles_Per_Case	12	Decimal	DINT
+–Bottles_Produced	720	Decimal	DINT

Figure 15-69 MUL instruction rung and its Monitor Tags window.

Ladder logic program

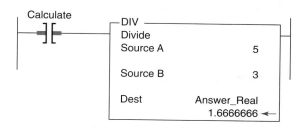

Tag Name	Value	Style	Data Type
Calculate	1	Decimal	BOOL
Answer_Real	1.6666666	Float	REAL

Figure 15-70 DIV instruction rung and its Monitor Tags window.

- Source A (the value in tag Cases_Produced) is multiplied by Source B (the value in tag Bottles_Per_Case) and the result is stored in the Dest tag Bottles_Produced.
- Source A and Source B can be constants (numbers) or tags.

The *DIV* instruction is used to divide two numbers. Figure 15-70 shows an example of a DIV instruction rung along with its Monitor Tags window. The operation of the rung can be summarized as follows:

- A constant (5) is used for Source A and a constant (3) for Source B. Note that tags could have been used for Source A or Source B.
- When the Calculate tag is true the DIV instruction is executed.
- Source A (5) is divided by Source B (3) and the result (1.6666666) is stored in the Dest tag Answer_Real. Note that in this example a Real-type tag has been used for its destination.

The program of Figure 15-71 is used as part of a parts tracking system with three conveyors. The number of parts in conveyor 1 and the number of parts in conveyor 2 are added to get the number of parts on conveyor 3. The operation of the program can be summarized as follows:

- Each time Conveyor_1_Sensor is actuated the accumulated value of Counter_1_Parts is incremented by 1.
- Each time Conveyor_2_Sensor is actuated the accumulated value of Counter_2_Parts is incremented by 1.

- The addition in the ADD instruction places the sum of the accumulated values of the two counters in the Conveyor_3_Parts tag.
- When the accumulated value for either counter is equal to 150 the reset (RES) instructions for both counters are enabled to automatically reset both counter ACC values to zero.
- Both counters can also be reset manually at any time by actuation of the Manual_Conveyor_Reset button.

Comparison Instructions

Compare instructions are used to compare two values. They can be used to see if two values are equal, if one value is greater or less than the other, and so on. In ControlLogix controllers compare instructions are input instructions that do comparisons by either using an expression or doing the comparison indicated by the specific instruction. Figure 15-72 shows the Compare toolbar for the ControlLogix controller.

The *equal (EQU)* instruction is used to test if two values are equal. Values compared can be actual values or tags that contain values. Figure 15-73 shows an example of an EQU instruction rung along with its Monitor Tags window. The operation of the rung can be summarized as follows:

- The value stored at Source A is compared to the value stored at Source B.
- If the values are equal, the instruction is logically true.
- If the values are unequal, the instruction is logically false.
- In this example Source A (25) is equal to Source B (25) so the instruction is true and output Equal_PL is on.
- Source A and Source B may be SINT, INT, DINT, or REAL data types.

The *not equal (NEQ)* instruction is used to test two values for inequality. Figure 15-74 shows an example of an NEQ instruction rung. When Source A is not equal to Source B, the instruction is logically true; otherwise, it is logically false. In this example the two values are not equal so the Not_Equal_PL is energized.

The *less than (LES)* instruction is used to check if a value from one source is less than the value from a second source. Figure 15-75 shows an example of an LES instruction rung. When Source A is less than Source B, the instruction is logically true; otherwise, it is logically false. In this example Value_1 (100) is less than Value_2 (300) so the Less_Than_PL is energized.

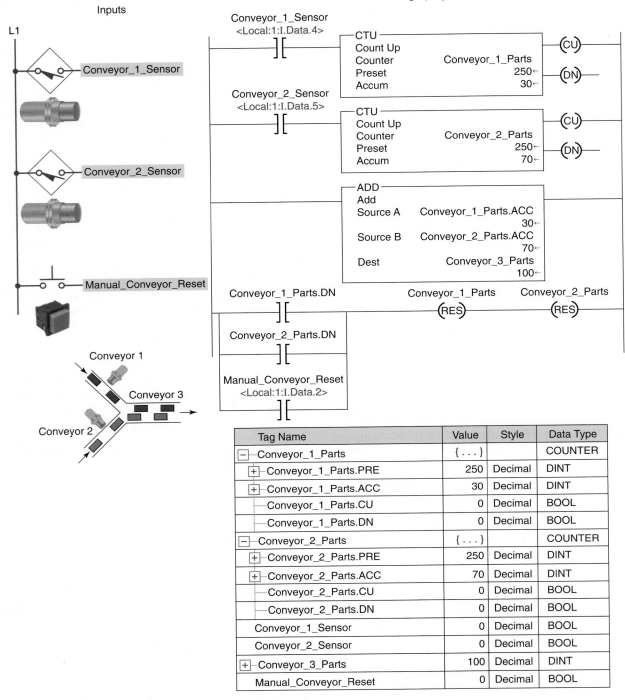

Figure 15-71 Program used as part of a parts tracking system.

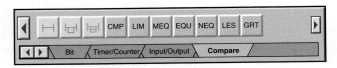

Figure 15-72 Compare toolbar for the ControlLogix controller.

The *greater than (GRT)* instruction is used to check if a value from one source is greater than the value from a second source. Figure 15-76 shows an example of a GRT instruction rung. When Source A is greater than Source B, the instruction is logically true; otherwise, it is logically false. In this example Value_1 (1420) is

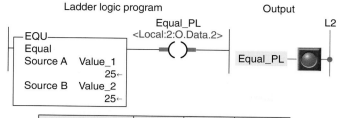

Figure 15-73 EQU instruction rung and its Monitor Tags window.

Tag Name	Value	Style	Data Type
Equal_PL	1	Decimal	BOOL
+-Value_1	25	Decimal	DINT
+-Value_2	25	Decimal	DINT

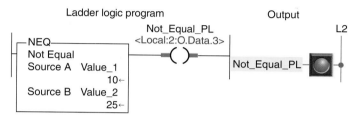

Figure 15-74 NEQ instruction rung.

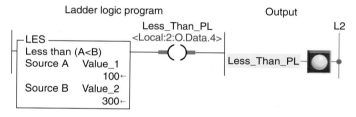

Figure 15-75 LES instruction rung.

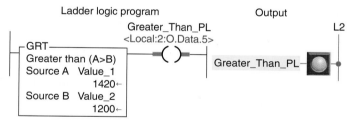

Figure 15-76 GRT instruction rung.

greater than Value_2 (1200) so the Greater_Than_PL is energized.

The *compare (CMP)* instruction performs a comparison on the arithmetic operations specified by the expression. The expression may contain arithmetic operators, comparison operators, and tags. The execution of a CMP instruction is slightly slower and uses more memory than the execution of the other comparison instructions. The advantage of the CMP instruction is that it allows you to enter complex expressions in one instruction. Figure 15-77 shows an example of a CMP instruction rung. In this example the comparison operator found in the expression is the equivalent of an EQU instruction. The comparison instruction is true because Value_1 (300) is equal to Value_2 (300).

The program of Figure 15-78 is an example of the use of comparison instructions used to test the accumulated value of a counter. The operation of the program can be summarized as follows:

- When the accumulated count is between 5 and 10 the GRT and LES instructions will both be logically true so the PL_1 pilot light will be on.
- When the accumulated count is equal to 15, the EQU instruction will be logically true so the PL_2 pilot light will be on.
- The PL_3 pilot light will be on at all times except when the accumulated count is 20 at which time the NEQ instruction is logically false.
- The counter is reset automatically when the accumulated count reaches 25 or manually anytime the Reset_PB is actuated.

Move Instructions

The *move (MOV)* instruction is an output instruction that can move a constant or the contents of one memory location to another location. Figure 15-79 shows the Move toolbar and instruction for the ControlLogix

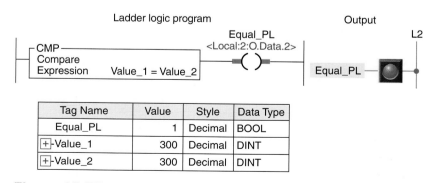

Tag Name	Value	Style	Data Type
Equal_PL	1	Decimal	BOOL
+-Value_1	300	Decimal	DINT
+-Value_2	300	Decimal	DINT

Figure 15-77 CMP instruction rung.

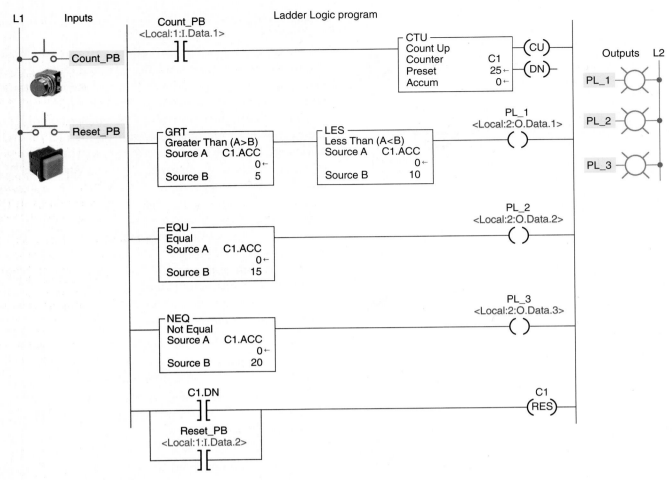

Figure 15-78 Comparison instructions used to test the accumulated value of a counter.

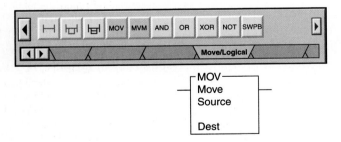

Figure 15-79 Move toolbar for the ControlLogix controller.

controller. The MOV instruction is used to copy data from a source to a destination. Both the source and the destination data type of a MOV instruction may be INT, DINT, SINT, or REAL.

The program of Figure 15-80 is an example of how the MOV instruction can be used to create a variable preset timer. The operation of the program can be summarized as follows:

- Actuating the PB_10s button executes its MOV instruction to transfer 10000 to the timer preset value setting the delay period for 10 seconds.
- Actuating the PB_15s button executes its MOV instruction to transfer 15000 to the timer preset value setting the delay period for 15 seconds.
- Closing the Timer_Start switch starts the timer timing.
- While the timer is timing, the pilot light PL_1 is on for the duration of the timer preset period.
- When the timer times out, PL_1 turns off and PL_2 turns on.

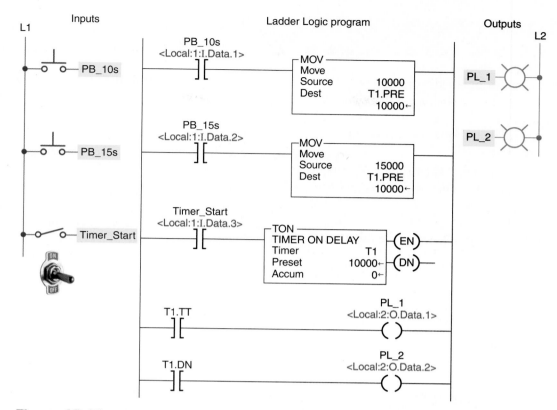

Figure 15-80 MOV instruction used to create a variable preset timer.

1. Construct a ControlLogix ladder rung with a math instruction that executes when a toggle switch is closed to add the tag named Pressure_A (value 680) to the constant of 50 and store the answer in the tag named Result.

2. Construct a ControlLogix ladder rung with a math instruction that executes when two normally open limit switches are closed to subtract the tag named Count_1 (value 60) from the tag named Count_2 (value 460) and store the answer in the tag named Count_Total.

3. Construct a ControlLogix ladder rung with a math instruction that executes when either one of two normally open pushbuttons is closed to multiply the tag named Cases (value 10) by the constant 24 and store the answer in the tag named Cans.

4. Construct a ControlLogix ladder rung with a compare instruction that will energize a pilot light output anytime the value stored at Data_3 is 60.

5. Construct a ControlLogix ladder rung with a compare instruction that will energize a pilot light output anytime the value stored at Data_2 is not the same as that stored at Data_6.

6. Construct a ControlLogix ladder rung with compare instructions that will energize a pilot light output anytime the pressure of a system goes above 300 psi or below 100 psi.

1. While checking the operation of the parts tracking system with the Monitor Tags window, you note that the value of Conveyor_Sensor_1 remains at 1 with parts passing by. What can you surmise from this? Why?

2. Three conveyors are delivering the same parts in different packages. A package can hold 12, 24, or 18 parts. Proximity switches installed on each of the conveyor lines are used to advance the accumulated value of the three counters. Write a ControlLogix program that uses multiply and add instructions to calculate the sum of the parts.

3. A single pole switch is used in place of the two pushbuttons for the variable preset timer program. When this switch is closed the timer is to be set for 10 seconds and when open to 15 seconds. Make the necessary changes to the program.

Part 6 Function Block Programming

Part Objectives

After completing this part, you will be able to:

* Describe the difference between ladder logic and function block diagram programming
* Recognize the basic elements of a function block diagram
* Write and read a function block diagram

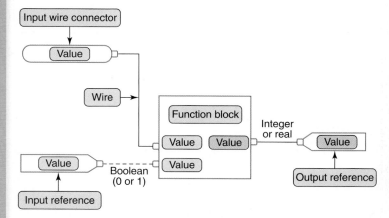

Figure 15-81   Structure of function block or routine.

Function Block Diagram (FBD)

A *function block diagram (FBD)* is a graphical depiction of process flow using simple and complex interconnecting blocks. It is similar to a ladder logic diagram, except that function blocks replace the interconnection of contacts and the coils. In addition, there are no power rails.

A function block circuit is analogous to an electrical circuit where links and wires depict signal paths between components. The workplace is known as a sheet and consists of function blocks joined together with lines called wires. The structure of a function block program, or routine, is shown in Figure 15-81. A function block diagram consists of four basic elements: function block, references, wire connectors, and wires. Data flow on a wire from wire connectors or input references, move through the function block, and then pass on to an output reference. The line type of the link between function blocks

indicates what type of data are present. A dash line indicates a Boolean signal path (e.g., 0 or 1) and a solid line indicates an integer or real value.

Function blocks are graphical representations of executable code. A function block can take one or more inputs and make decisions or calculations and then generate one or more outputs. There are many different types of function blocks included in the programming software to perform various common tasks. In addition, customized *Add-On* instructions can be created by the programmer for sets of commonly used logic. Once an Add-On instruction is defined in a project, it appears on the instruction toolbar and behaves like the standard instructions.

Figure 15-82 shows an example of a BAND (Boolean AND) function block. The information associated with a function block can be summarized as follows:

* Inputs are shown entering from the left and outputs exiting on the right.

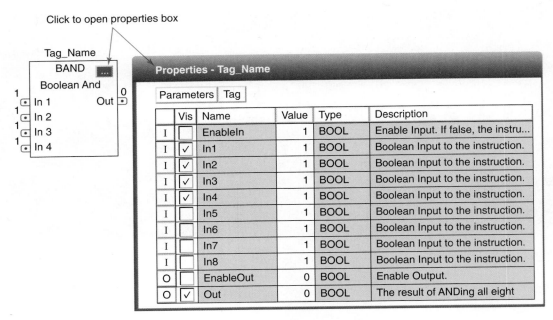

Figure 15-82 Example of a BAND (Boolean AND) function block.

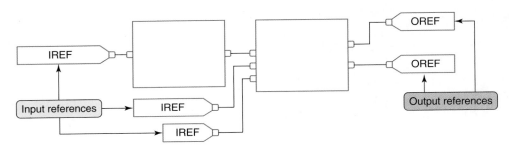

Figure 15-83 Input and output references.

- The function block type is shown within the block.
- A tag name for the block is placed above it.
- The names of the inputs and outputs are shown within the block.
- The default view of the block has some but not all of the input and output parameters visible when the box is placed into the program.
- The properties box, used to set the option of input and output parameters, is displayed by clicking the selection button located at the upper right hand corner of the block.
- The 1 and 0 next to the inputs and outputs identifies the logical state of the input and output pins for the instruction.
- The dots on the input and output pins indicate BOOL type data is required.

References represent tags that are linked to values stored in a controller's memory. The two types of references, input and output, are illustrated in Figure 15-83. An input reference, or IREF, is used to receive a value from an input device or tag. An output reference, or OREF, is used to send a value to an output device or tag. When you use an IREF or an OREF you must create a tag or assign an existing tag to the element. You may use any of the data types for an IREF or OREF.

Function blocks can be connected to other function blocks by connecting their outputs to the input of another function block using *wires and pins* (Figure 15-84). Wires map a signal's path and show the flow of controller execution. Each element in a function block diagram contains pins. Elements are connected by moving wires from input pins to output pins or vice versa. The pins on the left of a function block are input pins, and those on the

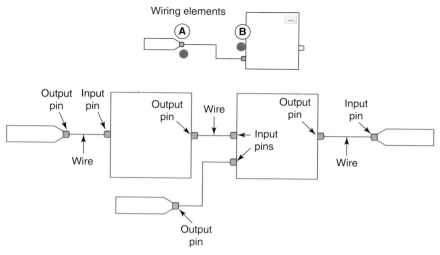

Figure 15-84 Function block diagram wire and pins.

right are output pins. To wire two elements together, click the output pin of the first element (A) and then click the input pin of the other element (B). A green dot shows a valid connection point.

Wire connectors are used to create a path without using a wire. When there are many function blocks on a sheet, or the function blocks are far apart, wire connectors used in place of wires can make the logic harder to read. Wire connectors are also used to connect function blocks that are on a different sheet of the same function block routine, as illustrated in Figure 15-85. The use of wire connectors can be summarized as follows:

- An output wire connector, or **OCON**, sends a value or signal to an input wire connector, or **ICON**.
- Each output wire connector must have at least one corresponding input wire connector.
- Each output wire connector requires a unique tag name and the corresponding input connector must have the same name.

- Multiple input wire connectors can reference the same output wire connector. This lets you share data at several points in your function block diagram.

Figure 15-86 illustrates the signal flow and execution of an FBD program. The operation can be summarized as follows:

- Each program scan sets all the FBD blocks starting on the left side of the signal flow and continues to evaluate all blocks according to the signal flow until the final output is determined.
- The location of a block does not affect the order in which the blocks execute.
- The inputs of a block require data to be available before the controller can execute that block.
- If function blocks are not wired together, it does not matter which block executes first as there is no data flow between the blocks.
- The interconnected line between the blocks indicates what type of signal is present.

Data latching refers to how the controller verifies that the data present at the input to a function block are valid. If you use an IREF to specify input data for a function block instruction, as illustrated in Figure 15-87, the data in that IREF are latched (won't change) for the scan of the function block routine. The IREF latches data from program-scoped and controller-scoped tags. The controller updates all IREF data at the beginning of each scan. A function block routine executes in the following order:

- The controller latches all data values in IREFs.

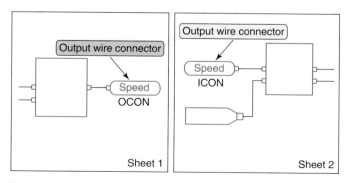

Figure 15-85 OCON and ICON wire connectors.

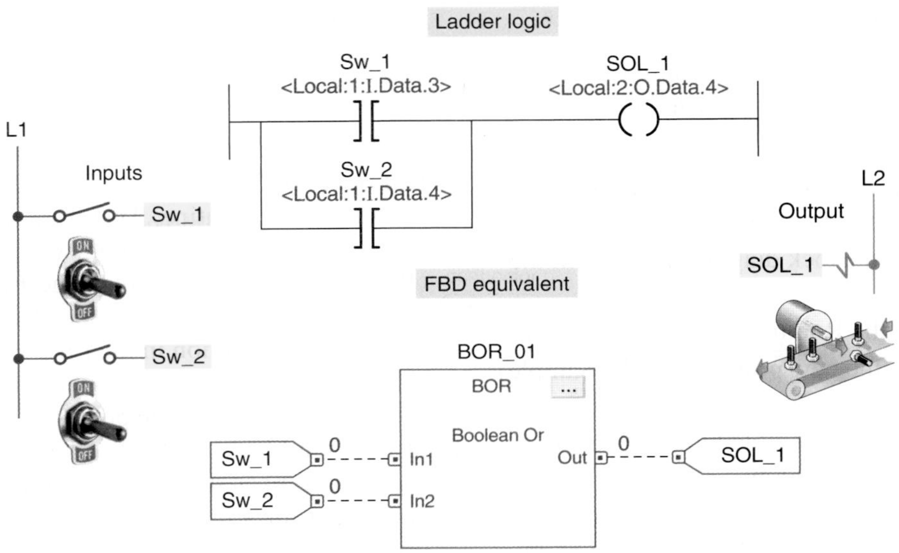

Figure 15-92 Comparison between ladder logic and the FBD equivalent for a two-input OR ladder logic rung.

- When the inputs represented by Sensor_1, Sensor_2, and Sensor_3 are true (value 1) the BAND (Boolean AND) function block will be true.
- The BAND block executes to set output Caution_PL true and switch the pilot light on.
- The 0 to the right of the input reference and out pin indicates its logic state. A 0 indicates the state of the tag is false, while a 1 signifies it is true.
- The same field input sensors and output pilot light devices and tags can be used with either program.
- The XIC and OTE contact and coil instructions have been replaced by the BAND function block.

Figure 15-92 shows a comparison between ladder logic and the FBD equivalent for a two-input OR ladder logic rung. As with ladder OR logic, if any of the two inputs is true the BOR function block will be true. In this example, with the BOR function block true, the output reference tag SOL_1 will be true energizing the solenoid.

Figure 15-93 shows a comparison between ladder logic and the FBD equivalent for a combination of multiple inputs. The operation of the FBD can be summarized as follows:

- The alarm will be energized if either input In1 or In2 to the BOR block is true.

- Input In2 of the BOR block will be true only when all three of the sensor switches are closed.
- Input In1 of the BOR block will be true only when the Temp_Sw is closed at the same time as the Press_Sw is open.
- The BNOT function block executes similarly to an XIO ladder logic contact instruction. When In is 0, Out is 1 and vice versa.

Figure 15-94 shows a comparison between ladder logic and the FBD equivalent for the motor start/stop control circuit. The logic sequence for starting and stopping the motor can be summarized as follows:

- When Motor_Start button is closed the BOR output will become true making the BAND output true.
- Motor_Run output energizes the contactor coil, the contacts of which close to start the motor operating.
- When the Motor_Start button is then opened the output of the BOR block remains true due to the 1 status of the feedback signal from the Motor_Run tag.
- When the Motor_Stop button is opened the output of the BAND block turns false to de-energize the contactor coil and stop the motor.

Figure 15-95 shows a comparison between ladder logic and the FBD equivalent for the 10 second TON (on-delay

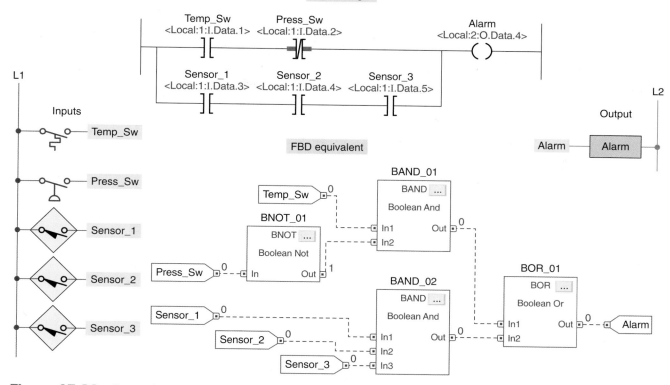

Figure 15-93 Comparison between ladder logic and the FBD equivalent for a combination of multiple inputs.

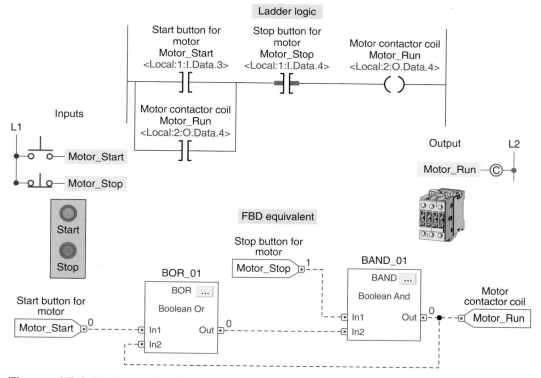

Figure 15-94 Comparison between ladder logic and the FBD equivalent for a motor start/stop control circuit.

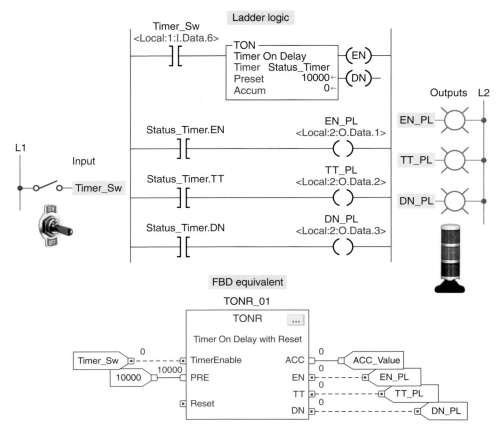

Figure 15-95 Comparison between ladder logic and the FBD equivalent for a 10 second TON and TONR timer.

timer) and TONR (on-delay with reset). The operation of the FBD can be summarized as follows:

- When the Timer_Sw is closed, the TONR function block timer turns true and starts accumulating time.
- The accumulated time is monitored by the output reference tag named ACC.
- The EN (enable bit) output changes to 1 to turn on the EN_PL.
- The TT (timer timing bit) output changes to 1 to turn on the TT_PL.
- The timer times out after 10 seconds to set the DN (done bit) to 1 and turn on the DN_PL and reset the TT bit to zero and turn off the TT_PL.
- The EN bit and EN_PL remain on as long as the Timer_Sw stays toggled closed.
- Opening the Timer_Sw resets all outputs as well as the accumulated value to zero.
- The timer can also be reset by way of the Reset input.

Figure 15-96 shows a comparison between ladder logic and the FBD equivalent for the Up/Down counter used to limit the number of parts stored in a buffer zone to 50. The operation of the FBD can be summarized as follows:

- The CTUD up/down counter function block accumulated value is initially reset by momentary actuation of the Restart_Button.
- The accumulated count is monitored by the output reference tag named ACC.
- Each time a part enters the buffer zone, the Enter_Limit_Sw is actuated and the CUEnable input turns true to increment the count by 1.
- Each time a part exits the buffer zone, the Exit_Limit_Sw is actuated and the CDEnable input turns true to decrement the count by 1.
- Whenever the number of parts in the buffer zone reaches 50 the DN bit is set to 1 and the output of the BNOT block is reset to zero. This de-energizes the Conveyor_Contactor to stop the conveyor motor from delivering more parts to the buffer zone.

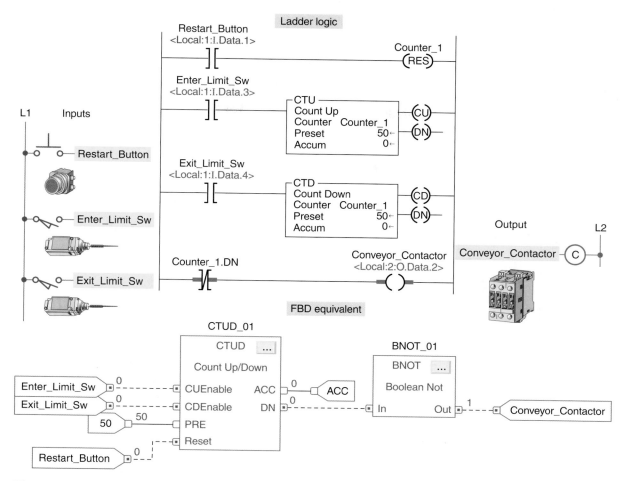

Figure 15-96 Comparison between ladder logic and the FBD equivalent for an Up/Down counter application.

Figure 15-97 shows a comparison between ladder logic and the FBD equivalent for the program used to test the accumulated value of a counter. The operation of the FBD can be summarized as follows:

- The function block routine is broken into four sheets.
- The order of the sheets does not affect the order in which the function blocks execute.
- When a function block routine executes, all sheets execute.

- Using one sheet for each device that is to be programmed helps organize your program and make it easier to understand.
- The use of the OCON and ICON named ACC enables the function blocks to be on different sheets of the same function block routine.
- The numbers and letters under the ACC output indicate the sheet number and location on the sheet where the output is used.

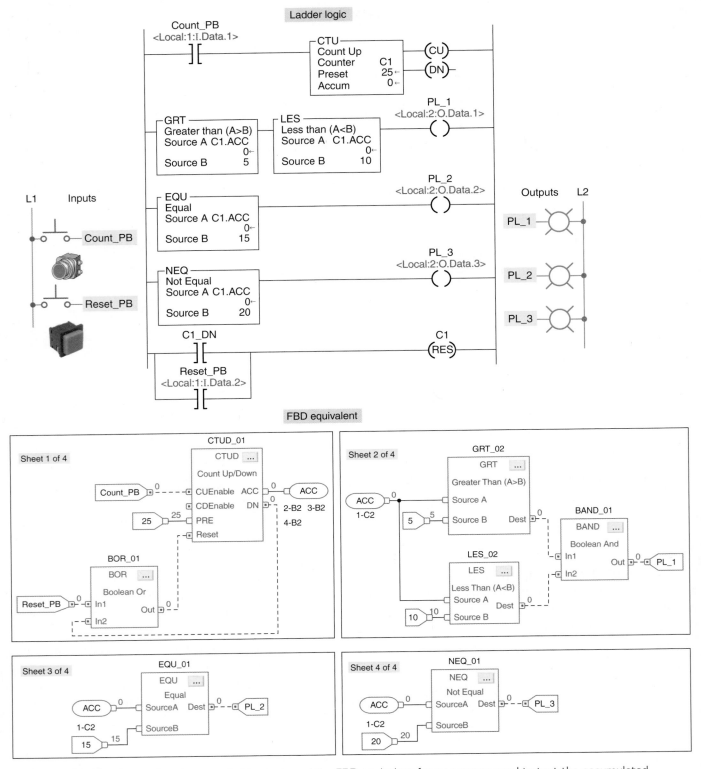

Figure 15-97 Comparison between ladder logic and the FBD equivalent for a program used to test the accumulated value of a counter.

PART 6 REVIEW QUESTIONS

1. Compare the graphical representation of a function block diagram to that of a logic ladder diagram.

2. Name the four basic elements of an FBD.

3. What do the solid and dashed interconnecting lines between FBD function blocks indicate?

4. What is an Add-On instruction?

5. How are the input and output parameter options for a function block set?

6. What does the dot on an input or output pin of a function block indicate?

7. Compare the functions of input and output reference tags.

8. Which pins of a function block are inputs and which are outputs?

9. Explain the role of input and output wire connectors.

10. How does the program scan function for an FBD program?

11. Explain data latching as it applies to function block inputs.

12. How is a function block feedback loop created?

13. What is the Assume Data Available indicator used for?

14. Outline how an FBD program is initiated.

PART 6 PROBLEMS

1. Write an FBD program that will cause the output, solenoid SOL_1, to be energized when pushbutton PB_1 is open and PB_2 is closed, and either limit switch LS_1 is open or limit switch LS_2 is closed. Assume all pushbuttons and limit switches are of the normally open type.

2. Modify the motor start/stop FBD program to include a second start/stop pushbutton station.

3. You are required to change the on-delay time of the 10 second timer program to 1 minute. What changes would have to be made to the FBD program?

4. Modify the Up/Down counter FBD program to include the following pilot lights:
   • PL_1 to come on when a part enters
   • PL_2 to come on when a part exits
   • PL_3 to come on when the buffer zone is full

5. Modify the test accumulated value of a counter FBD program as follows:
   • PL_1 to be on for an accumulated count between 0 and 5
   • PL_2 to be on for an accumulated count of 12
   • PL_3 to be on at all times except for when the accumulated count is 15

Glossary

1's complement   The system used to represent negative numbers in a personal computer and a programmable logic controller.

2's complement   A numbering system used to express positive and negative binary numbers.

A

Access   To locate data stored in a programmable logic controller system or in computer-related equipment.

Accumulated value   The number of elapsed timed intervals or counted events.

Actuator   An output device normally connected to an output module. Examples are an air valve and cylinder.

Address   A code that indicates the location of data to be used by a program, or the location of additional program instructions.

Algorithm   Mathematical procedure for problem solving.

Alias tag   References a memory location that has been defined by another tag.

Alphanumeric   Term describing character strings consisting of any combination of letters, numerals, and/or special characters (e.g., A15$) used for representing text, commands, numbers, and/or code groups.

Alternating current (AC) input module   An input module that converts various alternating current signals originating at user devices to the appropriate logic level signal for use within the processor.

Alternating current (AC) output module   An output module that converts the logic level signal of the processor to a usable output signal to control a user alternating current device.

Ambient temperature   The temperature of the air surrounding a module or system.

American National Standard Code for Information Interchange (ASCII)   An 8-bit (7 bits plus parity) code that represents all characters of a standard typewriter keyboard, both uppercase and lowercase, as well as a group of special characters used for control purposes.

American National Standards Institute (ANSI)   A clearinghouse and coordinating agency for voluntary standards in the United States.

American wire gauge (AWG)   A standard system used for designating the size of electrical conductors. Gauge numbers have an inverse relationship to size; larger numbers have a smaller diameter.

Analog device   Apparatus that measures continuous information (e.g., voltage or current). The measured analog signal has an infinite number of possible values. The only limitation on resolution is the accuracy of the measuring device.

Analog input module   An input circuit that employs an analog-to-digital converter to convert an analog value, measured by an analog measuring device, to a digital value that can be used by the processor.

Analog output module   An output circuit that employs a digital-to-analog converter to convert a digital value, sent from the processor, to an analog value that will control a connected analog device.

Analog signal   Signal having the characteristic of being continuous and changing smoothly over a given range rather than switching suddenly between certain levels, as with discrete signals.

Analog-to-digital (A/D) converter   A circuit for converting a varying analog signal to a corresponding representative binary number.

AND (logic)   A Boolean operation that yields a logic 1 output if all inputs are 1, and a logic 0 if any input is 0.

Arithmetic capability   The ability to do addition, subtraction, multiplication, division, and other advanced math functions with the processor.

Array   A combination of panels, such as LEDs, coordinated in structure and function.

ASCII-input module   Converts ASCII-code input information from an external peripheral into alphanumeric information a PLC can understand.

ASCII-output module   Converts alphanumeric information from the PLC into ASCII code to be sent to an external peripheral.

Asynchronous   Recurrent or repeated operations that occur in unrelated patterns over time.

Automatic control   A process in which the output is kept at a desired level by using feedback from the output to control the input.

Auxiliary power supply   A power supply not associated with the processor. Auxiliary power supplies are usually required to supply logic power to input/output racks and to other processor support hardware and are often referred to as *remote power supplies.*

B

Backplane   A printed circuit board, located in the back of a chassis, that contains a data bus, power bus, and mating connectors for modules to be inserted in the chassis.

Base tag   A definition of the memory location at which a data element is stored.

BASIC   A computer language that uses brief English-language statements to instruct a computer or microprocessor.

Battery indicator   A diagnostic aid that provides a visual indication to the user and/or an internal processor software indication that the memory power-fail support battery is in need of replacement.

Baud   A unit of signaling speed equal to the number of discrete conditions or signal events per second; often defined as the number of binary digits transmitted per second.

BCD-input module   Allows the processor to accept 4-bit BCD digital codes.

BCD-output module   Enables a PLC to operate devices that require BCD-coded signals to operate.

Binary   A number system using 2 as a base. The binary number system requires only two digits, zero (0) and one (1), to express any alphanumeric quantity desired by the user.

Binary-coded decimal (BCD) A system of numbering that expresses each individual decimal digit (0 through 9) of a number as a series of 4-bit binary notations. The binary-coded decimal system is often referred to as *8421 code.*

Binary word A related group of 1s and 0s that has meaning assigned by position or by numerical value in the binary system of numbers.

Bit An abbreviated term for the words *binary digit.* The bit is the smallest unit of information in the binary numbering system. It represents a decision between one of two possible and equally likely values or states. It is often used to represent an off or on state as well as a true or false condition.

Bit manipulation instructions A family of programmable logic controller instructions that exchange, alter, move, or otherwise modify the individual bits or groups of processor data memory words.

Bit storage A user-defined data table area in which bits can be set or reset without directly affecting or controlling output devices. However, any storage bit can be monitored as necessary in the user program.

Block diagram A method of representing the major functional subdivisions, conditions, or operations of an overall system, function, or operation.

Block format Format that uses a box shape to display instructions.

Block transfer An instruction that copies the contents of one or more contiguous data memory words to a second contiguous data memory location; an instruction that transfers data between an intelligent input/output module or card and specified processor data memory locations.

BOOL A data type that stores the state of a single bit, where 0 equals off and 1 equals on.

Boolean algebra A mathematical shorthand notation that expresses logic functions, such as AND, OR, EXCLUSIVE OR, NAND, NOR, and NOT.

Branch A parallel logic path within a rung.

Buffer In software terms, a register or group of registers used for temporary storage of data; a buffer is used to compensate for transmission rate differences between the transmitter and receiving device. In hardware terms, a buffer is an isolating circuit used to avoid the reaction of one circuit with another.

Bug A system defect or error that causes a malfunction; can be caused by either software or hardware.

Burn The process by which information is entered into programmable read-only memory.

Bus A group of lines used for data transmission or control; power distribution conductors.

Bus topology A network configuration in which all stations are connected in parallel with the communication medium and all stations can receive information from any other station on the network.

Byte A group of adjacent bits usually operated on as a unit, such as when moving data to and from memory. There are 8 bits per byte.

C

Cascading In the programming of timers and counters, a technique used to extend the timing or counting range beyond what would normally be available. This technique involves the driving of one timer or counter instruction from the output of another, similar instruction.

Cell controller A specialized computer used to control a work cell through multiple paths to the various cell devices.

Central processing unit (CPU) The electronic circuitry that controls all the data activity of the PLC, performs calculations, and makes decisions with its operation controlled by a sequence of instructions. The central processing unit is also referred to as the *processor* or the *CPU.*

Channel A designated path for a signal.

Character A symbol that is one of a larger group of similar symbols and that is used to represent information on a display device. The letters of the alphabet and the decimal numbers are examples of characters used to convey information.

Chassis A rack that serves as an electrical backplane for a PLC processor and I/O modules.

Chip A tiny piece of semiconductor material on which electronic components are formed. Chips are normally made of silicon and are typically less than 1/4 in. square and 1/100 in. thick.

Clear An instruction or a sequence of instructions that removes all current information from a programmable logic controller's memory.

Clock A circuit that generates timed pulses to synchronize the timing of computer operations.

Clock rate The speed at which the microprocessor system operates.

Closed loop A control system that uses feedback from the process to maintain outputs at a desired level.

Coaxial cable A transmission line constructed such that an outer conductor forms a cylinder around a central conductor. An insulating dielectric separates the inner and outer conductors, and the complete assembly is enclosed in a protective outer sheath. Coaxial cables are not susceptible to external electric and magnetic fields and generate no electric or magnetic fields of their own.

Code A system of communications that uses arbitrary groups of symbols to represent information or instructions. A set of programmed instructions.

Coil Represents the output of a programmable logic controller. In the output devices, it is the electrical coil that, when energized, changes the status of its corresponding contacts.

Coil format Format that uses coils to display instructions.

Comment Text that is included with each PLC ladder rung and is used to help individuals understand how the program operates or how the rung interacts with the rest of the program.

Communication module Allows the user to connect the PLC to high-speed local networks that may differ from the network communication provided with the PLC.

Compare An instruction that compares the contents of two designated data memory locations of a programmable logic controller for equality or inequality.

Compatibility The ability of various specified units to replace one another with little or no reduction in capability; the ability of units to be interconnected and used without modification.

Complement A logical operation that inverts a signal or bit.

Complementary metal-oxide semiconductor (CMOS) A logic base that offers lower power consumption and high-speed operation.

Computer Any electronic device that can accept information, manipulate it according to a set of preprogrammed instructions, and supply the results of the manipulation.

Computer integrated manufacturing (CIM) A manufacturing system controlled by an easily reprogrammable computer for flexibility and speed of changeover.

Computer interface A device designed for data communication between a programmable logic controller and a computer.

Consumed tag References data that come from another controller.

Contact The current-carrying part of an electric relay or switch. The contact engages to permit power flow and disengages to interrupt power flow to a load device.

Contact bounce The uncontrollable making and breaking of a contact during the initial engaging or disengaging of the contact.

Contact histogram An instruction sequence that monitors a designated memory bit or a designated input or output point for a change of state. A listing is generated by the instruction sequence that displays how quickly the monitored point is changing state.

Contactor A special-purpose relay designed to establish and interrupt the power flow of high-current electric circuits.

Contact symbology A set of symbols used to express the control program with conventional relay symbols.

Continuous current per module The maximum current for each module. The sum of the output current for each point should not exceed this value.

Continuous current per point The maximum current each output is designed to supply continuously to a load.

Controlled variable The output variable that the automatic control adjusts to keep the process value at a set-point.

Control logic The control plan for a given system; the program.

Control loop The method of adjusting the control variable in a process control system by analyzing the process variable data and then comparing it to the set-point to determine the amount of error in the system.

ControlNet An open, high-speed, deterministic network that transfers on the same network time-critical I/O updates, controller-to-controller interlocking data, and non-time-critical data such as data monitoring and program uploads and downloads.

Control relay A relay used to control the operation of an event or a sequence of events.

Counter An electromechanical device in relay-based control systems that counts numbers of events for the purpose of controlling other devices based on the current number of counts recorded; a programmable logic controller instruction that performs the functions of its electromechanical counterpart.

Cross reference In ladder diagrams, letters or numbers to the right of coils or functions. The letters or numbers indicate where on other ladder lines contacts of the coil or function are located.

Crosstalk Undesired energy appearing in one signal path as a result of coupling from other signal paths or use of a common return line.

Current The rate of electrical electron movement, measured in amperes.

Current-carrying capacity The maximum amount of current a conductor can carry without heating beyond a predetermined safe limit.

Current sinking Refers to an output device (typically an NPN transistor) that allows current flow from the load through the output to ground.

Current sourcing Output device (typically a PNP transistor) that allows current flow from the output through the load and then to ground.

Cycle A sequence of operations repeated regularly; the time it takes for one such sequence to occur.

D

Data Information encoded in a digital form, which is stored in an assigned address of data memory for later use by the processor.

Data address A location in memory where data can be stored.

Data file A group of data memory words acted on as a group rather than singly.

Data highway A communications network that allows devices such as PLCs to communicate. They are normally proprietary, which means that only like devices of the same brand can communicate over the highway.

Data latching A technique used to read the value of the input data that will be operated on by the instructions with a function block.

Data link The equipment that makes up a data communications network.

Data manipulation The process of exchanging, altering, or moving data within a programmable logic controller or between programmable logic controllers.

Data manipulation instructions A classification of processor instructions that alter, exchange, move, or otherwise modify data memory words.

Data table The part of processor memory that contains input and output values as well as files where data are monitored, manipulated, and changed for control purposes.

Data transfer The process of moving information from one location to another, in other words, from register to register, from device to device, and so forth.

Data transmission line A medium for transferring signals over a distance.

Debouncing The act of removing intermediate noise states from a mechanical switch.

Debug The process of locating and removing mistakes from a software program or from hardware interconnections.

Decimal number system A number system that uses ten numeral digits (decimal digits): 0, 1, 2, 3, 4, 5, 6, 7, 8, 9. Each digit position has a place value of 1, 10, 100, 1000, and so on, beginning with the least significant (rightmost) digit; base 10.

Decrement The act of reducing the contents of a storage location or value in varying increments.

Determinism The ability to reliably predict when data will be delivered.

DeviceNet An open communication network designed to connect factory-floor devices together without interfacing through an I/O system. Up to 64 intelligent nodes can be connected to one DeviceNet network.

Diagnostic program A user program designed to help isolate hardware malfunctions in the programmable logic controller and the application equipment.

Diagnostics The detection and isolation of an error or malfunction.

Digital device One that processes discrete electric signals.

Digital gate A device that analyzes the digital states of its inputs and outputs and an appropriate output state.

Digital signals A system of discrete states: high or low, on or off, 1 or 0.

Digital-to-analog converter An electrical circuit that converts binary bits to a representative, continuous, analog signal.

DINT A data type that stores a 32-bit (4-byte) signed integer value.

DIP switch A group of small, in-line on-off switches. From *d*ual *i*n-line *p*ackage.

Direct addressing An addressing mode in which the memory address of the data is supplied with the instruction.

Discrete I/O A group of input and/or output modules that operate with on/off signals, as contrasted to analog modules that operate with continuously variable signals.

Disk drive The device that writes or reads data from a magnetic disk.

Diskette The flat, flexible disk on which a disk drive writes and reads.

Display The image that appears on a cathode-ray tube screen or on other image projection systems.

Display menu The list of displays from which the user selects specific information for viewing.

Distributed control A method of dividing process control into several subsystems. A PLC oversees the entire operation.

Divide A programmable logic controller instruction that performs a numerical division of one number by another.

Documentation An orderly collection of recorded hardware and software data such as tables, listings, and diagrams to provide reference information for programmable logic controller application operation and maintenance.

Done bit (DN) Bit that is set to 1 when the instruction has completed its task, such as reaching its preset value.

Double precision The system of using two addresses or registers to display a number too large for one address or register; allows the display of more significant figures because twice as many bits are used.

Down-counter A counter that starts from a specified number and increments down to zero.

Download Loading data from a master listing to a readout or another position in a computer system.

Dry-contact-output module Enables a PLC's processor to control output devices by providing a contact isolated electrically from any power source.

E

Edit The act of modifying a programmable logic controller program to eliminate mistakes and/or simplify or change system operation.

Electrically erasable programmable read-only memory (EEPROM) A type of programmable read-only memory that is programmed and erased by electrical pulses.

Electrical optical isolator A device that couples input to output using a semiconductor light source and detector in the same package.

Electromagnetic interference (EMI) A phenomenon responsible for noise in electric circuits.

Element A single instruction of a relay ladder diagram program.

Emergency stop relay A relay used to inhibit all electric power to a control system in an emergency or other event requiring that the controlled hardware be brought to an immediate halt.

Enable To permit a particular function or operation to occur under natural or preprogrammed conditions.

Enclosure A steel box with a removable cover or hinged door used to house electric equipment.

Encoder A rotary device that transmits position information; a device that transmits a fixed number of pulses for each revolution.

Energize The physical application of power to a circuit or device to activate it; the act of setting the on, true, or 1 state of a programmable logic controller's relay ladder diagram output device or instruction.

Erasable programmable read-only memory (EPROM) A programmable read-only memory that can be erased with ultraviolet light, then reprogrammed with electrical pulses.

Error signal A signal proportional to the difference between the actual output and the desired output.

Ethernet An network type or protocol that uses the Carrier Sense Multiple Access Collision Detection (CSMA/CD) network access method.

Ethernet/IP An open industrial networking standard that takes advantage of commercial off-the-shelf Ethernet communication devices and physical media; IP refers to industrial protocol.

Even parity When the sum of the number of 1s in a binary word is always even.

Examine if closed (XIC) Refers to a normally open contact instruction in a logic ladder program. An examine if closed instruction is true if its addressed bit is on (1). It is false if the bit is off (0).

Examine if open (XIO) Refers to a normally closed contact instruction in a logic ladder program. An examine if open instruction is true if its addressed bit is off (0). It is false if the bit is on (1).

Exclusive-OR gate A logic device requiring one or the other, but not both, of its inputs to be satisfied before activating its output.

Execution The performance of a specific operation accomplished through processing one instruction, a series of instructions, or a complete program.

Execution time The total time required for the execution of one specific operation.

F

False As related to programmable logic controller instructions, a disabling logic state.

Fault Any malfunction that interferes with normal operation.

Fault indicator A diagnostic aid that provides a visual indication and/or an internal processor software indication that a fault is present in the system.

Fault-routine file A special subroutine that, if assigned, executes when the processor has a major fault.

Feedback In analog systems, a correcting signal received from the output or an output monitor. The correcting signal is fed to the controller for process correction.

Fiber optic cable Transmits information via light pulses down optical fibers.

Fieldbus An open, all-digital, serial, two-way communications system that interconnects measurement and control equipment such as sensors, actuators, and controllers.

File A formatted block of data treated as a unit.

Fixed I/O Input/output terminals on a programmable logic controller that are built into the unit and are not changeable. A fixed I/O PLC has no removable modules.

Floating-point data file   Used to store integers and other numerical values that cannot be stored in an integer file.

Flowchart   A graphical representation for the definition, analysis, or solution of a problem. Symbols are used to represent a process or sequence of decisions and events.

Force function   A mode of operation or instruction that allows an operator to override the processor to control the state of a device.

Force off function   A feature that allows the user to reset an input image table file bit or de-energize an output independently of the programmable logic controller program.

Force on function   A feature that allows the user to set an image table file bit or energize an output independently of the programmable logic controller program.

Full duplex   A mode of data communications in which data may be transmitted and received simultaneously.

Function block   Rectangular block with inputs entering from the left and outputs exiting on the right.

Function block diagram (FBD)   Graphical language where the basic programming elements appear as blocks.

Function keys   Keys on a personal computer, electronic operator device, or hand-held programmer keyboard that are labeled F1, F2, and so on. The operation of each of these keys is defined on many electronic operator interface devices.

G

Gate   A circuit having two or more input terminals and one output terminal, where an output is present when and only when the prescribed inputs are present.

Gateway   A device or pair of devices that connects two or more communication networks. This device may act as a host to each network and may transfer messages between the networks by translating their protocols.

Glitch   A voltage or current spike of short duration that adversely affects the operation of a PLC.

Gray code   A binary coding scheme that allows only 1 bit in the data word to change state at each increment of the code sequence.

Gray-encoder module   Converts the Gray-code signal from an input device into straight binary.

Ground   A conducting connection between an electric circuit or equipment chassis and the earth ground.

Ground loop   A condition in which two or more electrical paths exist within a ground line.

Ground potential   Zero voltage potential with respect to the ground.

H

Half-duplex   A mode of data transmission that communicates in two directions but in only one direction at a time.

Handshaking   The method by which two digital machines establish communication.

Hard contacts   Any type of physical switch contacts.

Hard copy   Any form of a printed document such as a ladder diagram program listing, paper tape, or punched cards.

Hard drive   An inflexible recording disk used as a computer disk drive.

Hardware   The mechanical, electric, and electronic devices that make up a programmable logic controller and its application.

Hardwired   The physical interconnection of electric and electronic components with wire.

Hexadecimal   A number system having a base of 16. This numbering system requires 16 elements for representation, and thus uses the decimal digits zero (0) through nine (9) and the first six letters of the alphabet, A through F.

High-speed counter encoder module   A module that enables you to count and encode faster than you could with a regular control program written on a PLC in which the control program's execution is too slow.

Histogram   A graphic representation of the frequency at which an event occurs.

Host computer   A main computer that controls other computers, PLCs, or computer peripherals.

Human machine interface (HMI)   Graphical display hardware in which machine status, alarms, messages, diagnostics, and data entry are available to the operator in graphical display format.

I

IEC 1131 programming standard   The international standard for programmable controller programming languages.

Image table   An area in programmable logic controller memory dedicated to input/output data. Ones and zeros (1s and 0s) represent on and off conditions, respectively. During every input/output scan, each input controls a bit in the input image table file; each output is controlled by a bit in the output image table file.

Immediate input instruction   A programmable logic controller instruction that temporarily halts the user program scan so that the processor can update the input image table file with the current status of one or more user-specified input points.

Immediate output instruction   A programmable logic controller instruction that temporarily halts the user program scan so that the current status of one or more user-specified output points can be updated to current output image table file status by the processor.

Impedance   The total resistive and inductive opposition that an electric circuit or device offers to a varying current at a specified frequency. Impedance is measured in ohms (V) and is denoted by the symbol Z.

Increment   The act of increasing the contents of a storage location or value in varying amounts.

Index   A reference used to specify an element within an array.

Indirect addressing   An addressing mode in which the address of the instruction serves as a reference point instead of the actual address.

Inductance   A circuit property that opposes any current change. Inductance is measured in henrys and is represented by the letter H.

Industrial terminal   The device used to enter and monitor the program in a PLC.

Input   Information transmitted from a peripheral device to the input module and then to the data table.

Input devices   Devices such as limit switches, pressure switches, pushbuttons, and analog and/or digital devices that supply data to a programmable logic controller.

Input/output (I/O) address   A unique number assigned to each input and output. The address number is used when programming, monitoring, or modifying a specific input or output.

Input/output (I/O) module   A plug-in assembly that contains more than one input or output circuit. A module usually contains two or more identical circuits. Normally, it contains 2, 4, 8, 16 32, or 64 circuits.

Input/output (I/O) scan time   The time required for the processor to monitor inputs and control outputs.

Input/output (I/O) update   The continuous process of revising each and every bit in the input and output tables, based on the latest results from reading the inputs and processing the outputs according to the control program.

Input scan   One of three parts of the PLC scan. During the input scan, input terminals are read and the input table is updated accordingly.

Instruction   A command that causes a programmable logic controller to perform one specific operation. The user enters a combination of instructions into the programmable logic controller's memory to form a unique application program.

Instruction set   The set of general-purpose instructions available with a given controller. In general, different machines have different instruction sets.

INT   Two-byte integer.

Integer   A positive or negative whole number.

Integrated circuit (IC)   A circuit in which all components are integrated on a single tiny silicon chip.

Intelligent field devices   Microprocessor-based devices used to provide process-variable, performance, and diagnostic information to the PLC processor. These devices are able to execute their assigned control functions with little interaction, except communications, with their host processor.

Intelligent input/output module   A microprocessor-based module that performs processing or sophisticated closed-loop application functions.

Interface   A circuit that permits communication between the central processing unit and a field input or output device. Different devices require different interfaces.

Interlock   A system for preventing one element or device from turning on while another device is on.

Internal coil instruction   A relay coil instruction used for internal storage or buffering of an on/off logic state. An internal coil instruction differs from an output coil instruction because the on/off status of the internal coil is not passed to the input/output hardware for control of a field device.

Inversion   Conversion of a high level to a low level, or vice versa.

Inverter   The digital circuit that performs inversion.

I/O module   A plug-in assembly, containing two or more identical input or output circuits, that contain the connections between a processor and connected devices.

Interrupt   The act of redirecting a program's execution to perform a more urgent task.

IP address   A specified Internet protocol address for every Ethernet device that is unique and is assigned by the manufacturer.

Isolated input module   A module that receives dry contacts as inputs, which the processor can recognize and change into two-state digital signals.

Isolated input/output (I/O) circuits   Input and output circuits that are electrically isolated from any and all other circuits of a module. Isolated input/output circuits are designed to allow field devices that are powered from different sources to be connected to one module.

J

Jumper   A short length of conduit used to make a connection between terminals around a break in a circuit.

Jump instruction   An instruction that permits the bypassing of selected portions of the user program. Jump instructions are conditional whenever their operation is determined by a set of preconditions and unconditional whenever they are executed to occur every time they are programmed.

K

K $2^{10} = 1K = 1024$; used to denote size of memory and can be expressed in bits, bytes, or words; example: $2K = 2048$.

k   Kilo; a prefix used with units of measurement to designate quantities 1000 times as great.

Keyboard   The alphanumeric keypad on which the user types instructions to the PLC.

Keying   Bands installed on backplane connectors to ensure that only one type of module can be inserted into a keyed connector.

L

Label instruction   A programmable logic controller instruction that assigns an alphanumeric designation to a particular location in a program. This location is used as the target of a jump, skip, or jump to subroutine instruction.

Ladder diagram   An industry standard for representing relay logic control systems. The diagram resembles a ladder because the vertical supports of the ladder appear as power feed and return buses and the horizontal rungs of the ladder appear as series and/or parallel circuits connected across the power lines.

Ladder diagram programming   A method of writing a user program in a format similar to a relay ladder diagram.

Ladder matrix   A rectangular array of programmed contacts that defines the number of contacts that can be programmed across a row and the number of parallel branches allowed in a single ladder rung.

Language   A set of symbols and rules for representing and communicating information among people or between people and machines; the method used to instruct a programmable device to perform various operations.

Language module   Enables the user to write programs in a high-level language. BASIC is the most popular language module. Other language modules available include C, Forth, and PASCAL.

Latching relay   A relay that maintains a given position by mechanical or electrical means until released mechanically or electrically.

Latch instruction   One-half of an instruction pair (the second instruction of the pair being the unlatch instruction) that emulates the latching action of a latching relay. The latch instruction for a programmable logic controller energizes a specified output point or internal coil until it is de-energized by a corresponding unlatch instruction.

Leakage   The small amount of current that flows in a semiconductor device when it is in the off state.

Least significant bit (LSB)   The bit that represents the smallest value in a byte or word.

Least significant digit (LSD)   The digit that represents the smallest value in a byte or word.

Light-emitting diode (LED)   A semiconductor junction that emits light when biased in the forward direction.

Light-emitting diode (LED) display   A display device incorporating light-emitting diodes to form the segments of the displayed characters and numbers.

Limit switch   An electric switch actuated by some part and/or motion of a machine or equipment.

Limit test   A test that determines if a value is inside or outside a specified range.

Line   A component part of a system used to link various subsystems located remotely from the processor; the source of power for operation; example: 120 V alternating current line.

Line-powered sensor   Normally, three-wire sensors, although four-wire sensors also exist. The line-powered sensor is powered from the power supply. A separate wire (the third) is used for the output line.

Liquid-crystal display (LCD)   A display device using reflected light from liquid crystals to form the segments of the displayed characters and numbers.

Load   The power used by a machine or apparatus; to place data into an internal register under program control; to place a program from an external storage device into central memory under operator control.

Load-powered sensor   A two-wire sensor. A small leakage current flows through the sensor even when the output is off. The current is required to operate the sensor electronics.

Local area network (LAN)   A system of hardware and software designed to allow a group of intelligent devices to communicate within a fairly close proximity.

Local input/output (I/O)   A programmable logic controller whose input/output distance is physically limited. The PLC must be located near the process; however, the PLC may still be mounted in a separate enclosure.

Local power supply   The power supply used to provide power to the processor and a limited number of local input/output modules.

Location   In reference to memory, a storage position or register identified by a unique address.

Logic   A process of solving complex problems through the repeated use of simple functions that can be either true or false. The three basic logic functions are AND, OR, and NOT.

Logic diagram   A diagram that represents the logic elements and their interconnections.

Logic level   The voltage magnitude associated with signal pulses representing 1s and 0s in binary computation.

Loop control   A control of a process or machine that uses feedback. An output status indicator modifies the input signal effect on the process control.

M

Machine language   A programmable language using the binary form.

Major fault   A fault condition that is severe enough for the controller to shut down, unless the condition is cleared.

Manufacturing automation protocol (MAP)   Standard developed to make industrial devices communicate more easily.

Masking   A means of selectively screening out data. Masking allows unused bits in a specific instruction to be used independently.

Mass storage   A means of storing large amounts of data on magnetic tape, floppy disks, and so on.

Master control relay (MCR)   A mandatory hardwired relay that can be de-energized by any series-connected emergency stop switch. Whenever the master control relay is de-energized, its contacts open to de-energize all application input and output devices.

Master control relay (MCR) zones   User program areas in which all nonretentive outputs can be turned off simultaneously. Each master control relay zone must be delimited and controlled by master control relay fence codes (master control relay instructions).

Matrix   A logic network that is an intersection of input and output connection points.

Memory   That part of the programmable logic controller in which data and instructions are stored either temporarily or semi-permanently. The control program is stored in memory.

Memory map   A diagram showing a system's memory addresses and what programs and data are assigned to each section of memory.

Menu   A list of programming selections displayed on a programming terminal.

Metal-oxide semiconductor (MOS)   A semiconductor device in which an electric field controls the conductance of a channel under a metal electrode called a *gate*.

Metal oxide varistor (MOV)   Used for suppressing electrical power surges.

Microprocessor   A central processing unit manufactured on a single integrated-circuit chip (or several chips) by utilizing large-scale integration technology.

Microsecond   One millionth of a second $= 1 \times 10^{-6}$ second $= 0.000001$ second.

Millisecond   One thousandth of a second $= 1 \times 10^{-3}$ second $= 0.001$ second.

Mnemonic   A term, usually an abbreviation, that is easy to remember and pronounce.

Mnemonic code   A code in which information is represented by symbols or characters.

Modbus   A network that uses a master/slave communication technique.

Mode   A term used to refer to the selected operating method, such as automatic, manual, TEST, PROGRAM, or diagnostic.

Module   An interchangeable, plug-in item containing electronic components.

Module addressing   A method of identifying the input/output modules installed in a chassis.

Most significant bit (MSB)   The bit representing the greatest value of a byte or word.

Most significant digit (MSD)   The digit representing the greatest value of a byte or word.

Motor controller or starter   A device or group of devices that serve to govern, in a predetermined manner, the electric power delivered to a motor.

Motor starter   A special relay designed to provide power to motors; it has both a contactor relay and an overload relay connected in series and prewired so that, if the overload operates, the contactor is de-energized.

Move instruction   A programmable logic controller instruction that moves data from one location to another. Although a move instruction typically places the data in a new location, the original data still reside in their original location.

Multiplexing The time-shared scanning of a number of data lines into a single channel, and only one data line is enabled at any time; the incorporation of two or more signals into a single wave from which the individual signals can be recovered.

Multiply instruction A programmable logic controller instruction that provides for the mathematical multiplication of two numbers.

Multiprocessing A method of applying more than one microprocessor to a specific function to speed up operation time and reduce the possibility of system failure.

N

National Electrical Code (NEC) A set of regulations developed by the National Fire Protection Association that governs the construction and installation of electric wiring and electric devices. The National Electrical Code is recognized by many governmental bodies, and compliance is mandatory in much of the United States.

National Electrical Manufacturers Association (NEMA) An organization of electric device and product manufacturers. The National Electrical Manufacturers Association issues standards relating to the design and construction of electric devices and products.

NEMA Type 12 enclosure A category of industrial enclosures intended for indoor use and designed to provide a degree of protection against dust, falling dirt, and dripping noncorrosive liquids. They do not provide protection against conditions such as internal condensation.

Nested branches A branch that begins or ends within another branch.

Network A series of stations or devices connected by some type of communications medium.

Network access control The method of accessing the network media (cable) to ensure that data are transmitted in an organized manner in order to reduce the possibilities of data corruption.

Node In hardware, a connection point on the network; in programming, the smallest possible increment in a ladder diagram.

Noise Random, unwanted electric signals, normally caused by radio waves or electric or magnetic fields generated by one conductor and picked up by another.

Noise filter or **noise suppressor** An electronic filter network used to reduce and/or eliminate any noise that may be present on the leads to an electric or electronic device.

Noise immunity A measure of insensitivity of an electronic system to noise.

Noise spike A short burst of electric noise with more magnitude than the background noise level.

Nonretentive output An output controlled continuously by a program rung. Whenever the rung changes state (true or false), the output turns on or off; contrasted with a retentive output, which remains in its last state (on or off) depending on which of its two rungs, latch or unlatch, was last true.

Nonvolatile memory A memory designed to retain its data while its power supply is turned off.

NOR The logic gate that results in zero unless both inputs are zero.

Normally closed contact (NC) A contact that is conductive when its operating coil is not energized.

Normally open contact (NO) A contact that is nonconductive when its operating coil is not energized.

NOT A logical operation that yields a logic 1 at the output if a logic 0 is entered at the input, and a logic 0 at the output if a logic 1 is entered at the input. The NOT, also called the *inverter,* is normally used in conjunction with the AND and OR functions.

O

Octal number system A base eight numbering system that uses numbers 0–7, 10–17, 20–27, and so on. There are no 8s or 9s in the octal number system.

Odd parity Condition when the sum of the number of 1s in a binary word is always odd.

Off-delay timer An electromechanical relay with contacts that change state a predetermined time period after power is removed from its coil; on re-energization of the coil, the contacts return to their shelf state immediately; also, a programmable logic controller instruction that emulates the operation of the electromechanical off-delay relay.

Offline programming and/or **offline editing** A method of programmable logic controller programming and/or editing in which the operation of the processor is stopped and all output devices are switched off. Offline programming is the safest way to develop or edit a programmable logic controller program since the entry of instructions does not affect operating hardware until the program can be verified for accuracy of entry.

On-delay timer An electromechanical relay with contacts that change state a predetermined time period after the coil is energized; the contacts return to their shelf state immediately on de-energization of the coil; also, a programmable logic controller instruction that emulates the operation of the electromechanical on-delay timer.

One-shot A programmed technique that sets a storage bit or output for only one program scan.

Online data change Allows the user to change various data table values using a peripheral device while the application is operating normally.

Online programming and/or **online editing** The ability of a processor and programming terminal to make joint user-directed additions, deletions, or changes to a user program while the processor is actively solving and executing the commands of the existing user program. Extreme care should be exercised when performing online programming to ensure that erroneous system operation does not result.

Open loop A system that has no feedback or auto correction.

Operand A number used in an arithmetic operation as an input.

Operational amplifier (op-amp) A high-gain DC amplifier used to increase signal strength for devices such as analog input modules.

Optical coupler A device that couples signals from one circuit to another by means of electromagnetic radiation, usually infrared or visible. A typical optical coupler uses a light-emitting diode to convert the electric signal of the primary circuit into light and uses a phototransistor in the secondary circuit to reconvert the light back into an electric signal; sometimes referred to as *optical isolation.*

Optical isolation Electrical separation of two circuits with the use of an optical coupler.

OR A logical operation that yields a logic 1 output if one of any number of inputs is 1, and a logic 0 if all inputs are 0.

Output Information sent from the processor to a connected device via some interface. The information could be in the form of control data that will signal some device such as a motor to switch on or off or to vary the speed of a drive.

Output device Any connected equipment that will receive information or instructions from the central processing unit, such as control devices (e.g., motors, solenoids, alarms) or peripheral devices (e.g., line printers, disk drives, displays). Each type of output device has a unique interface to the processor.

Output image table file A portion of a processor's data memory reserved for the storage of output device statuses. A 1, on, or true state in an output image table file storage location is used to switch on the corresponding output point.

Output instruction The term applied to any programmable logic controller instruction capable of controlling the discrete or analog status of an output device connected to the programmable logic controller.

Output register or **output word** A particular word in a processor's output image table file in which numerical data are placed for transmission to a field output device.

Output scan One of three parts of the PLC scan. During the output scan, data associated with the output status table are transferred to the output terminals.

Overflow Exceeding the numerical capacity of a device such as a timer or counter. The overflow can be either a positive or negative value.

Overload A load greater than the one that a component or system is designed to handle.

Overload relay A special-purpose relay designed so that its contacts transfer whenever its current exceeds a predetermined value. Overload relays are used with electric motors to prevent motor burnout due to mechanical overload.

P

Parallel circuit A circuit in which two or more of the connected components or contact symbols in a ladder program are connected to the same pair of terminals so that current may flow through all the branches; contrasted with a series connection, in which the parts are connected end to end so that current flow has only one path.

Parallel instruction A programmable logic controller instruction used to begin and/or end a parallel branch of instructions programmed on a programming terminal.

Parallel operation A type of information transfer in which all bits, bytes, or words are handled simultaneously.

Parallel transmission A computer operation in which two or more bits of information are transmitted simultaneously.

Parity The use of a self-checking code that employs binary digits in which the total number of 1s is always even or odd.

PC Personal computer.

Peer-to-peer network A network in which nodes are given an equal chance of initiating and controlling communications.

Peripheral equipment Units that communicate with the programmable logic controller but are not part of the programmable logic controller; example: a programming device or computer.

PID Proportional-integral-derivative closed-loop control that lets the user hold a process variable at a desired set-point.

Pilot-type device Used in a circuit as a control apparatus to carry electric signals for directing performance. This device does not carry primary current.

PLC processor A computer designed specifically for programmable controllers. It supervises the action of the modules attached to it.

Polarity The directional indication of electrical flow in a circuit; the indication of charge as either positive or negative, or the indication of a magnetic pole as either north or south.

Polling A network access method where a master controller manages the communication process by interrogating each slave controller under it to determine whether the slave has any information to send.

Port A connector or terminal strip used to access a system or circuit. Generally, ports are used for the connection of peripheral equipment.

Power supply The unit that supplies the necessary voltage and current to a system's circuitry.

Preset value (PRE) The number of time intervals or events to be counted.

Pressure switch A switch activated at a specified pressure.

Process A continuous manufacturing operation.

Program A sequence of instructions to be executed by the processor to control a machine or process.

Program files The area of processor memory in which the ladder logic programming is stored.

Programmable controller A computer that has been hardened to work in an industrial environment and is equipped with special I/O and a control programming language.

Programmable read-only memory (PROM) A retentive memory used to store data. This type of memory device can be programmed only once and cannot be altered afterward.

Programming terminal A combination of keyboard and monitor used to insert, modify, and observe programs stored in a PLC.

Program scan One of three parts of the PLC scan. During the program scan, the CPU scans each rung of the user program.

Project file Contains all data associated with the PLC project. A project comprises five major pieces: help folder, controller folder, ladder folder, data folder, and data base folder.

Proportional-integral-derivative (PID) A mathematical formula that provides a closed-loop control of a process. Inputs and outputs are continuously variable and typically will be analog signals.

Protocol A formal definition of criteria for receiving and transmitting data through communications channels.

Proximity switch An input device that senses the presence or absence of a target without physical contact.

Pulse A short change in the value of a voltage or current level. A pulse has a definite rise and fall time and a finite duration.

R

Rack A housing or framework used to hold assemblies; a plastic and/or metal assembly that supports input/output modules and provides a means of supplying power and signals to each input/output module or card.

Random-access memory (RAM) A memory system that permits the random accessing of any storage location for the purpose of either storing (writing) or retrieving (reading) information. Random-access memory systems allow the data to be retrieved and stored at speeds independent of the storage locations being accessed.

Read The accessing of information from a memory system or data storage device; the gathering of information from an input device or devices or a peripheral device.

Read-only memory (ROM)   A permanent memory structure in which data are placed at time of fabrication or by the user at a speed much slower than it will be read. Information entered in a read-only memory is usually not changed once it is entered.

Read/write memory   Memory in which data can be stored (write mode) or accessed (read mode). The write mode replaces previously stored data with current data; the read mode does not alter stored data.

Real numbers   Numbers that have both integer and fractional parts.

Real-time clock (RTC)   A device that continually measures time in a system without respect to what tasks the system is performing.

Rectifier   A solid-state device that converts alternating current to pulsed direct current.

Register   A memory word or area for the temporary storage of data used within mathematical, logical, or transfer functions.

Relay   An electrically operated device that mechanically switches electric circuits.

Relay contacts   The contacts of a relay that are either opened or closed according to the condition of the relay coil. Relay contacts are designated as either normally open or normally closed in design.

Relay logic   A representation of the program or other logic in a form normally used for relays.

Remote input/output (I/O) system   Any input/output system that permits communication between the processor and input/output hardware over a coaxial or twin axial cable. Remote input/output systems permit the placement of input/output hardware at any distance from the processor.

Resolution   The smallest distinguishable increment into which a quantity is divided.

Response time   The amount of time required for a device to react to a change in its input signal or to a request.

Retentive instruction   Any programmable logic controller instruction that does not need to be continuously controlled for operation. Loss of power to the instruction does not halt execution or operation of the instruction.

Retentive timer   An electromechanical relay that accumulates time whenever the device receives power and maintains the current time should power be removed from the device. Loss of power to the device after reaching its preset value does not affect the state of the contacts.

Retentive timer instruction   A programmable logic controller instruction that emulates the timing operation of the electromechanical retentive timer.

Retentive timer reset instruction   A programmable logic controller instruction that emulates the reset operation of the electromechanical retentive timer.

Ring topology   A network topology that that forms a data path in a ring.

Routine   A series of instructions that perform a specific function or task.

RS-232   An Electronic Industries Association (EIA) standard for data transfer and communication for serial binary communication circuits.

Run   The single, continuous execution of a program by a programmable logic controller.

Rung   A group of programmable logic controller instructions that controls an output or storage bit, or performs other control functions such as file moves, arithmetic, and/or sequencer instructions. A rung is represented as one section of a ladder logic diagram.

S

SCADA   An acronym for supervisory control and data acquisition.

Scan time   The time required to read all inputs, execute the control program, and update local and remote input and output statuses. Scan time is, in effect, the time required to activate an output controlled by programmed logic.

Schematic   A diagram of graphic symbols representing the electrical scheme of a circuit.

Search function   Allows the user to display quickly any instruction in the programmable logic controller program.

Self-diagnostic   The hardware and firmware within a controller that monitors its own operation and indicates any fault it can detect.

Sensor   A device used to gather information by the conversion of a physical occurrence to an electric signal.

Sequencer   A mechanical, electric, or electronic device that can be programmed so that a predetermined set of events occurs repeatedly.

Sequence table   A table or chart indicating the sequence of operation of output devices.

Sequential control   A process that dictates the correct order of events and allows one event to occur only after the completion of another.

Sequential Function Chart (SFC)   Graphical language whose basic language elements are steps or states with associated actions and transitions with associated conditions used to move from the current state to the next.

Serial communication   A type of information transfer in which the bits are handled sequentially; contrasted with parallel communication.

Series circuit   A circuit in which the components or contact symbols are connected end to end, and all must be closed to permit current flow.

Servo module   The device whose feedback is used to accomplish closed-loop control. Though programmed through a PLC, once programmed it can control a device independently without interfering with the PLC's normal operation.

Set-point   The value that the process value is to be held to by the automatic control function.

Shield   A barrier, usually conductive, that substantially reduces the effect of electric and/or magnetic fields.

Shift   To move binary data within a shift register or other storage device.

Shift register   A PLC function capable of storing and shifting binary data.

Short circuit   An undesirable path of very low resistance in a circuit between two points.

Short-circuit protection   Any fuse, circuit breaker, or electronic hardware used to protect a circuit or device from severe overcurrent conditions or short circuits.

Signal   The event or electrical quantity that conveys information from one point to another.

Significant digit   A digit that contributes to the precision of a number. The number of significant digits is counted beginning with the digit contributing the most value, called the *most significant digit* (leftmost), and ending with the digit contributing the least value, called the *least significant digit* (rightmost).

Silicon-controlled rectifier (SCR)   A semiconductor device that functions as an electronic switch.

Single-scan function   A supervisory instruction that causes the control program to be executed for one scan, including input/output update. This troubleshooting function allows step-by-step inspection of occurrences while the machine is stopped.

Sink mode output   A mode of operation of solid-state devices in which the device controls the current from the load. For example, when the output is energized, it connects the load to the negative polarity of the supply.

SINT   A data type that stores an 8-bit (1-byte) signed integer value.

Snubber   A circuit generally used to suppress inductive loads; it consists of a resistor in series with a capacitor (RC snubber) and/or a MOV placed across the alternating current load.

Software   Programs that control the processing of data in a system, as contrasted to the physical equipment itself (hardware).

Solid-state switch   Any electronic device incorporating a transistor, silicon-controlled rectifier, or triac semiconductor switch to control the on/off flow of electric power.

Source mode output   A mode of operation of solid-state output devices in which the device controls the current to the load. For example, when the output is energized, it connects the load to the positive polarity of the supply.

Star topology   A network architecture in which all network nodes are connected to a central device that routes the nodes' messages.

State   The logic 0 or 1 condition in programmable logic controller memory or at a circuit input or output.

Station   Any programmable logic controller, computer, or data terminal connected to, and communicating by means of, a data highway.

Status indicators   LEDs that indicate the on-off status of an input or output point and are visible on the outside of the PLC.

Stepper-motor module   Provides pulse trains to a stepper-motor translator that enables control of a stepper motor.

STI   An acronym for selectable time interrupt, a subroutine that executes on a time basis rather than an event basis.

Storage bit   A bit in a data table word that can be set or reset but that is not associated with a physical input or output terminal point.

Structure Text (ST)   High-level, text-based language with commands that support a highly structured program development and the ability to evaluate complex mathematical expressions.

Subroutines   Program files that are scanned only when called on by logic and can be used to break the program into smaller segments.

Subtract   A programmable logic controller instruction that performs the mathematical subtraction of one number from another.

Suppression device   A unit that attenuates the magnitude of electrical noise.

Surge   A transient wave of current or power.

Synchronous shift register   A shift register in which only one change of state occurs per control pulse.

Synchronous transmission   A type of serial transmission that maintains a constant time interval between successive events.

T

Tag   A text-based name for an area of the controller's memory where data are stored.

Tap   A device that provides mechanical and electrical connections to a trunk cable. A tap allows the signals on the trunk to be passed to a station and the signals transmitted by the stations to be passed to the trunk.

Task   It holds the information necessary to schedule the program's execution and sets the execution priority for one or more programs.

Terminal address   The alphanumeric address assigned to a particular input or output point. It is also related directly to a specific image table bit address.

Thermocouple   A temperature-measuring device that utilizes two dissimilar metals for temperature measurement. As the junction of the two dissimilar metals is heated, a proportional voltage difference, which can be measured, is generated.

Thumbwheel switch   A rotating switch used to input numeric information into a controller.

Time base   A unit of time generated by a microprocessor's clock circuit and used by PLC timer instructions. Typical time bases are 0.01, 0.1, and 1.0 second.

Timed contact   A normally open and/or normally closed contact that is actuated at the end of a timer's time-delay period.

Timer   In relay-panel hardware, an electromechanical device that can be wired and preset to control the operating interval of other devices. In a programmable logic controller, a timer is internal to the *processor;* that is, it is controlled by a user-programmed instruction.

Toggle switch   A panel-mounted switch with an extended lever; normally used for on/off switching.

Token   The logical right to initiate communications in a communication network.

Token passing   A technique in which tokens are circulated among nodes in a communication network.

Topology   The structure of a communications network; examples are bus, ring, and star.

Transducer   A device used to convert physical parameters such as temperature, pressure, and weight into electric signals.

Transformer   An electric device that converts a circuit's electrical energy into a circuit or circuits with different voltages and current ratings.

Transistor   A three-terminal active semiconductor device composed of silicon or germanium that is capable of switching or amplifying an electric current.

Transistor-transistor logic (TTL)   A semiconductor logic family in which the basic logic element is a multiple-emitter transistor. This family of devices is characterized by high speed and medium power dissipation.

Transitional contact   A contact that, depending on how it is programmed, will be on for one program scan every 0 to 1 transition, or every 1 to 0 transition, of the referenced coil.

Transmission line   A system of one or more electric conductors used to transmit electric signals or power from one place to another.

Triac   A solid-state component capable of switching alternating current.

True   As related to programmable logic controller instructions, an on, enabled, or 1 state.

Truth table   A table listing that shows the state of a given output as a function of all possible input combinations.

TTL-input module   Enables devices that produce TTL-level signals to communicate with a PLC's processor.

TTL-output module   Enables a PLC to operate devices requiring TTL-level signals to operate.

Twisted pair cable   A pair of wires that can transmit data; the wires are twisted to provide protection against crosstalk.

U

Unlatch instruction   One-half of a programmable logic controller instruction pair that emulates the unlatching action of a latching relay. The unlatch instruction de-energizes a specified output point or internal coil until re-energized by a latch instruction. The output point or internal coil remains de-energized regardless of whether or not the unlatch instruction is energized.

Up-counter   An event that starts from 0 and increments up to the preset value.

V

Variable   A factor that can be altered, measured, or controlled.

Variable data   Numerical information that can be changed during application operation. It includes timer and counter accumulated values, thumbwheel settings, and arithmetic results.

Volatile memory   A memory structure that loses its information whenever power is removed. Volatile memories require a battery backup to ensure memory retention during power outages.

W

Watchdog timer   Monitors logic circuits controlling the processor. If the watchdog timer, which is reset every scan, ever times out, the processor is assumed to be faulty and is disconnected from the process.

Word   A grouping or a number of bits in a sequence treated as a unit.

Word length   The total number of bits that make up a word. Most programmable logic controllers use either 8, 16, or 32 bits to form a word.

Work cell   A group of machines that work together to manufacture a product; normally includes one or more robots. The machines are programmed to work together in appropriate sequences. Work cells are often controlled by one or more PLCs.

Write   Refers to the process of loading information into memory; can also refer to block transfer, that is, a transfer of data from the processor data table to an intelligent input/output module.

Z

Zone   The portion of a PLC ladder program that can be enabled or disabled by a control function.

Index

Commissioning (programs), 275
Common Industrial Protocol (CIP), 309–311
Common-mode rejection, 33
Communications
    configuring, 286–287
    data; *see* Data communications
Communications capability, 2–4
Communications modules, 19, 31
Communications protocols, 287
Commutative law, 63
CompactLogix system, 317
Comparators, 54
Compare (CMP), 357
Comparison instructions (CLX controllers), 355–358
Compute (CPT) command, 227
Conducted noise, 271
Configuration
    CLX controllers, 318–319
    communications, 286–287
    control systems, 292–294
    display screens, 295
Constant voltage (CV) transformers, 274
Consumed tags, 323
Contact histogram, 277
Contact symbology, 81; *see also* Ladder diagram language
Contactors, in relay schematics, 97–98
Continuous processes, 292
Continuous tasks, 320
Continuous test mode, 276
Continuous-scan test mode, 91
Control data file, 208
Control file (file 6), 73–75
Control level functionality, 304
Control management applications, 13
Control systems
    centralized, 293
    closed loop, 295–296
    configurations, 292–294
    distributive, 293–294
    individual, 292–293
    motion, 301–303
    on/off, 296–297
    PID, 297–301
    process, 292–296
    structure of, 294–296
    supervisory control and data acquisition, 35, 304, 313–314
Control variable (CV), 221
Control word (counters), 153
Controller tags, 322
Controllers
    in closed-loop systems, 295
    in process control systems, 294
Controllers Properties and Modules Properties dialog box (CLX controllers), 318–319
ControlLogix controllers; *see* Allen-Bradley ControlLogix (CLX) controllers
ControlLogix format, 20, 21, 45, 206
ControlNet, 310–311
Convert from BCD (FRD), 227, 234
Convert to BCD (TOD), 227, 234
Cooling, 269
COP (file copy), 208–209
Cost, 2, 3
Count-down (CTD) counters, 14
    CLX controllers, 347, 350–351
    SLC 500 counters, 154

Count-Down Enable Bit (CD)
    CLX controllers, 348
    SLC 500 counters, 153
Counter data file, 208
Counter file (file 5), 73, 74
Counter number (SLC 500 counters), 154
Counters, 149–170
    cascading, 162–165
    CLX controllers, 347–351
    combining timer functions and, 166–170
    down-counters, 159–162
    incremental encoder-counter applications, 165–166
    instructions, 150–151
    up-counters, 152–159
    using LES instruction, 215
Count-up (CTU) counters, 14
    CLX controllers, 347–350
    SLC 500 counters, 154
Count-Up Done Bit (DN), 348
Count-Up Enable Bit (CU)
    CLX controllers, 347
    SLC 500 counters, 153
CPT (compute) command, 227
CPU (central processing unit), 4–6, 33–35
Cross reference function, 276
CTD (count-down) counters, 14
    CLX controllers, 347, 350–351
    SLC 500 counters, 154
CTU (count-up) counters, 14
    CLX controllers, 347–350
    SLC 500 counters, 154
CU (Count-Up Enable Bit)
    CLX controllers, 347
    SLC 500 counters, 153
Current, grounding, 272–273
Current leakage, 32, 272
Current sensing analog I/O modules, 27
CV (control variable), 221
CV (constant voltage) transformers, 274

D

Daisy-chain topology, 312–313
Data
    reading and writing, 35
    recording and retrieving, 38
    writing over, 201
Data communications
    access control, 306–307
    ControlNet, 310–311
    Data Highway networks, 308
    DeviceNet, 308–310
    EtherNet/IP, 311
    Fieldbus, 312–313
    Modbus, 311–312
    network topologies, 304–305
    networks, 303–304
    PROFIBUS-DP, 313
    protocols, 305–306
    serial, 308
    transmitting PLC data, 307–308
Data compare instructions, 209–213
Data files, 72–76
Data Files Window (RSLogix 500), 90
Data Highway, 307, 308
Data Highway DH-485, 308
Data Highway networks, 307, 308
Data Highway Plus (DH+), 308